CRC Series in Naturally Occurring Pesticides

Series Editor-in-Chief

N. Bhushan Mandava

Handbook of Natural Pesticides: Methods

Volume I: Theory, Practice, and Detection
Volume II: Isolation and Identification

Editor

N. Bhushan Mandava

Handbook of Natural Pesticides

Volume III: Insect Growth Regulators
Volume IV: Pheromones

Editors

E. David Morgan
N. Bhushan Mandava

Future Volumes

Handbook of Natural Pesticides

Insect Attractants, Deterrents, and Defensive Secretions

Editors

E. David Morgan
N. Bhushan Mandava

Plant Growth Regulators

Editor

N. Bhushan Mandava

Microbial Insecticides

Editors

Carl M. Ignoffo

CRC
Handbook
of
Natural Pesticides

Volume III
Insect Growth
Regulators

Part B

Editors

E. David Morgan, D.Phil.
Reader
Department of Chemistry
University of Keele
Staffordshire, England

N. Bhushan Mandava, Ph.D.
Senior Associate
Mandava Associates
Washington, D.C.

CRC Press
Taylor & Francis Group
Boca Raton London New York

CRC Press is an imprint of the
Taylor & Francis Group, an **informa** business

First published 1987 by CRC Press
Taylor & Francis Group
6000 Broken Sound Parkway NW, Suite 300
Boca Raton, FL 33487-2742

Reissued 2018 by CRC Press

Publisher's Note
The publisher has gone to great lengths to ensure the quality of this reprint but points out that some imperfections in the original copies may be apparent.

Disclaimer
The publisher has made every effort to trace copyright holders and welcomes correspondence from those they have been unable to contact.

ISBN 13: 978-1-138-59696-2 (hbk)
ISBN 13: 978-0-429-48724-8 (ebk)

Visit the Taylor & Francis Web site at http://www.taylorandfrancis.com and the
CRC Press Web site at http://www.crcpress.com

INTRODUCTION

The United States has been blessed with high quality, dependable supplies of low cost food and fiber, but few people are aware of the never-ending battle that makes this possible. There are at present approximately 1,100,000 species of animals, many of them very simple forms, and 350,000 species of plants that currently inhabit the planet earth. In the U.S. there are an estimated 10,000 species of insects and related acarinids which at sometime or other cause significant agricultural damage. Of these, about 200 species are serious pests which require control or suppression every year. World-wide, the total number of insect pests is about ten times greater. The annual losses of crops, livestock, agricultural products, and forests caused by insect pests in the U.S. have been estimated to aggregate about 12% of the total crop production and to represent a value of about $4 billion (1984 dollars). On a world-wide basis, the insect pests annually damage or destroy about 15% of total potential crop production, with a value of more than $35 billion, enough food to feed more than the population of a country like India. Thus, both the losses caused by pests and the costs of their control are considerably high. Insect control is a complex problem for there are more than 200 insects that are or have been subsisting on our main crops, livestock, forests, and aquatic resources. Today, in the U.S., conventional insecticides are needed to control more than half of the insect problems affecting agriculture and public health. If the use of pesticides were to be completely banned, crop losses would soar and food prices would also increase dramatically.

About 1 billion pounds of pesticides are used annually in the U.S. for pest control. The benefits of pesticides have been estimated at about $4/$1 cost. In other words, chemical pest control in U.S. crop production costs an estimated $2.2 billion and yields a gross return of $8.7 billion annually.

Another contributing factor for increased crop production is the effective control of weeds, nematodes, and plant diseases. Crop losses due to unwanted weed species are very high. Of the total losses caused by pests, weeds alone count for about 10% of the agricultural production losses valued at more than $12 billion annually. Farmers spend more than $6.2 billion each year to control weeds. Today, nearly all major crops grown in the U.S. are treated with herbicides. As in insect pest and weed control programs, several chemicals are used in the disease programs. Chemical compounds (e.g., fungicides, bactericides, nematicides, and viracides) that are toxic to pathogens are used for controlling plant diseases. Several million dollars are spent annually by American farmers to control the diseases of major crops such as cotton and soybeans.

Another aspect for improved crop efficiency and production is the use of plant growth regulators. These chemicals that regulate the growth and development of plants are used by farmers in the U.S. on a modest scale. The annual sale of growth regulators is about $130 million. The plant growth regulator market is made up of two distinct entities — growth regulators and harvest aids. Growth regulators are used to increase crop yield or quality. Harvest aids are used at the end of the crop cycle. For instance, harvest aids defoliate cotton before picking or desiccate potatoes before digging.

The use of modern pesticides has accounted for astonishing gains in agricultural production as the pesticides have reduced the hidden toll exacted by the aggregate attack of insect pests, weeds, and diseases, and also improved the health of humans and livestock as they control parasites and other microorganisms. However, the same chemicals have allegedly posed some serious problems to health and environmental safety, because of their high toxicity and severe persistence, and have become a grave public concern in the last 2 decades. Since the general public is very much concerned about their hazards, the U.S. Environmental

Protection Agency enforced strong regulations for use, application, and handling of the pesticides. Moreover, such toxic pesticides as DDT 2,4,5-T and toxaphene were either completely banned or approved for limited use. They were, however, replaced with less dangerous chemicals for insect control. Newer approaches for pest control are continuously sought, and several of them look very promising.

According to a recent study by the National Academy of Sciences, pesticides of several kinds will be widely used in the foreseeable future. However, newer selective and biodegradable compounds must replace older highly toxic persistent chemicals. The pest control methods that are being tested or used on different insects and weeds include: (1) use of natural predators, parasites, and pathogens, (2) breeding of resistant varieties of species, (3) genetic sterilization techniques, (4) use of mating and feeding attractants, (5) use of traps, (6) development of hormones to interfere with life cycles, (7) improvement of cultural practices, and (8) development of better biodegradable insecticides and growth regulators that will effectively combat the target species without doing damage to beneficial insects, wildlife, or man. Many leads are now available, such as the hormone mimics of the insect juvenile and molting hormones. Synthetic pyretheroids are now replacing the conventional insecticides. These insecticides, which are a synthesized version of the extract of the pyrethrum flower, are much more attractive biologically than the traditional insecticides. Thus, the application rates are much lower in some cases, one tenth the rates of more traditional insecticides such as organophosphorus pesticides. The pyrethroids are found to be very specific for killing insects and apparently exhibit no negative effects on plants, livestock, or humans. Another apparent benefit is that there is no resistance to these compounds accumulated in the insects. The use of these compounds is now widely accepted for use on cotton, field corn, soybean, and vegetable crops.

For the long term, integrated pest management (IPM) will have tremendous impact on pest control for crop improvement and efficiency. Under this concept, all types of pest control — cultural, chemical, inbred, and biological — are integrated to control all types of pests and weeds. The chemical control includes all of the traditional pesticides. Cultural controls consist of cultivation, crop rotation, optimum planting dates, and sanitation. Inbred plant resistance involves the use of varieties and hybrids that are resistant to certain pests. Finally, the biological control involves encouraging natural predators, parasites, and microbials. Under this system, pest-detection scouts measure pest populations and determine the best time for applying pesticides. If properly practiced, IPM could reduce pesticide use up to 75% on some crops.

The naturally occurring pesticides appear to have a prominent role for the development of future commercial pesticides not only for agricultural crop productivity but also for the safety of the environment and public health. They are produced by plants, insects, and several microorganisms, which utilize them for survival and maintenance of defense mechanisms, as well as for growth and development. They are easily biodegradable, often times species-specific and also sometimes less toxic (or nontoxic) on other non-target organisms or species, an important consideration for alternate approaches of pest control. Several of the compounds, especially those produced by crop plants and other organisms, are consumed by humans and livestock, and yet appear to have no detrimental effects. They appear to be safe and will not contaminate the environment. Hence, they will be readily accepted for use in pest control by the public and the regulatory agencies. These natural compounds occur in nature only in trace amounts and require very low dosage for pesticide use. It is hoped that the knowledge gained by studying these compounds is helpful for the development of new pest control methods such as their use for interference with hormonal life cycles and trapping insects with pheromones, and also for the development of safe and biodegradable chemicals (e.g., pyrethroid insecticides). Undoubtedly, the costs are very high as compared to the presently used pesticides. But hopefully, these costs would be compensated for by the benefits derived through these natural pesticides from the lower volume of pesticide use and elimination of risks. Furthermore, the indirect or external costs resulting from pesticide poisoning, fatalities, livestock losses, and increased control expenses (due to the destruction of natural enemies and beneficial insects as well as the environmental contamination and pollution from chlorinated, organophosphorus, and carbamate pesticides) could be assessed against benefits vs. risks. The development and use of such naturally occurring chemicals could become an integral part of IPM strategies.

As long as they remain endogenously, several of the natural products presented in this handbook series serve as hormones, growth regulators, and sensory compounds for growth, development, and reproduction of insects, plants, and microorganisms. Others are useful for defense or attack against other species or organisms. Once these chemicals or their analogs and derivatives are applied by external means to the same (where produced) or different species, they come under the label "pesticides" because they contaminate the environment. Therefore, they are subject to regulatory requirements, in the same way the other pesticides are handled before they are used commercially. However, it is anticipated that the naturally occurring pesticides would easily meet the regulatory and environmental requirements for their safe and effective use in pest control programs.

A vast body of literature has been accumulated on naturally occurring pesticides during the last 2 or 3 decades; we plan to assemble this information in this handbook series. However, we realize that it is a single handbook series. Therefore, we have limited our attempts to chemical and a few biological aspects concerned with biochemistry and physiology. Wherever possible, we tried to focus our attention on the application of these compounds for pesticidal use. We hope that the first volume which deals with theory and practice will serve as an introductory volume and will be useful to everyone interested in learning about the current technology that is being adapted from compound identification to the field trials. The subsequent volumes deal with the chemical, biochemical, and physiological aspects of naturally occurring compounds, grouped under such titles as insect growth regulators, plant growth regulators, etc.

In a handbook series of this type with diversified subjects dealing with plant, insect, and microbial compounds, it is very difficult to achieve either uniformity or complete coverage while putting the subject matter together. This goal was achieved to a large extent with the understanding and full cooperation of chapter contributors who deserve my sincere appreciation.

The editors of the individual volumes relentlessly sought to meet the deadlines and, more importantly, to bring a balanced coverage of the subject matter, but, however, that seems to be an unattainable goal. Therefore, they bear full responsibility for any pitfalls and deficiencies. We invite comments and criticisms from readers and users as they will greatly help to update future editions. It is hoped that this handbook series will serve as a source book for chemists, biochemists, physiologists, and other biologists alike — those engaged in active research as well as those interested in different areas of natural products that affect the growth and development of plants, insects, and other organisms.

The editors wish to acknowledge their sincere thanks to the members of the Advisory Board for their helpful suggestions and comments. Their appreciation is extended to the publishing staff, especially Pamela Woodcock, Amy Skallerup, and Sandy Pearlman for their ready cooperation and unlimited support from the initiation to the completion of this project.

N. Bhushan Mandava
Editor-in-Chief

FOREWORD

Pests of crops and livestock annually account for multi-billion dollar losses in agricultural productivity and costs of control. Insects alone are responsible for more than 50% of these losses.

For the past 40 years the principal weapons used against these troublesome insects have been chemical insecticides. The majority of such materials used during this period have been synthetic organic chemicals discovered, synthesized, developed, and marketed by commercial industry. In recent years, environmental concerns, regulatory restraints, and problems of pest resistance to insecticides have combined to reduce the number of materials available for use in agriculture. Replacement materials reaching the marketplace have been relatively few due to increased costs of development and the general lack of knowledge about new classes of chemicals having selective insecticidal activity.

In response to these trends, it is gratifying to note that scientists in both the public and private sectors have given significant attention to the discovery and evaluation of natural products as fertile sources of new insecticidal agents. Not only are these materials directly useful as insect control agents, but they also serve as models for new classes of chemicals with novel modes of action to attack selective target sites in pest species. Such new control agents may also be less susceptible to the cross resistance difficulties encountered with most classes of currently used synthetic pesticide chemicals to which insects have developed immunity.

Natural products originating in plants, animals, and microorganisms are providing a vast source of bioactive substances. The rapid development and application of powerful analytical instrumentation, such as mass spectrometry, nuclear magnetic resonance, high performance liquid chromatography, reverse phase liquid chromatography, immunoassay, and radioimmunoassay, have greatly facilitated the identification of miniscule amounts of active biological chemicals isolated from natural sources. These new science approaches and tools are addressed and reviewed extensively in these volumes.

Some excellent examples of success in this research involve the discovery of insect growth regulators, especially the so-called juvenoids, which are responsible for control of insect metamorphosis, reproduction, and behavior. Pheromones which play essential roles in insect communication, feeding, and sexual behavior represent another important class of natural products holding great promise for new pest insect control technology. All of these are discussed in detail in Volumes 1, 2, and 3.

It is hoped that the science described in these volumes will serve researchers in industry, government, and academia, and stimulate them to continue to seek even more useful natural materials that produce effective, safe, and environmentally acceptable materials for use against insect pests affecting agriculture and mankind.

Orville G. Bentley
Assistant Secretary
Science and Education
U.S. Department of Agriculture

PREFACE

Naturally occurring pesticides are those with which man began. Forty years ago science improved on these with the introduction of synthetics. Now the attention is again on natural pesticides.

As we assess the results of those 40 years, we recognize that effective insect pest control has eluded us. The introduction of synthetics has led to new problems of contamination, resistance, toxicity, and pest resurgence. Despite the achievements of science, it has failed to protect food provision adequately. Food is the only indispensible product produced by man, yet we saw our inadequacies in that pursuit sharply revealed by the tragic famine in the Sahel of Africa in 1984-85. The present loss of rice to pests is estimated to be 46% of the potential crop. If this loss could be reduced to 20%, about another 177 million people could be adequately fed, without bringing any new land into cultivation. Similar, though less dramatic figures can be quoted for wheat, corn, potatoes, soybeans, and many other staple crops. Losses due to pests, even in the U.S., with its intensive use of pesticides, continue at a level of 30% of total sales value.

The incidence of malaria in India and Sri Lanka is rising again, despite near eradication by 1970, due to resistant strains of mosquito and plasmodium.

We now have to reassess our knowledge to devise selective, safe pesticides. "Natural" methods of biological control have in some notable cases been successful but these are exceptional. Pyrethrum is an effective natural insecticide, relatively nontoxic to mammals, nonpolluting, nonpersistent. From a systematic study of structure and activity, a new generation of pesticides, with special applications has arisen. But since the days of natural pyrethrum and derris, a whole new world of substances has been discovered; substances which affect pest growth, behaviour, feeding, oviposition, aggregation, mating, and so on. Much might be learned from these substances, if only the scattered information can be brought together and organized in a way that will permit careful re-examination of these substances, often in a context or application quite foreign to the original discoveries or investigator's concept of these chemicals.

In a recent survey of plant species which have been recorded as possessing pest-control properties, 1005 species were listed as having insecticidal activity, 384 with antifeeding activity, and 279 with repellent activity. The definition of these activities is not exact, but these figures indicate that a great mass of material is waiting to be checked and evaluated.

This Handbook Series is trying to achieve some order and illumination in the half-explored world of Naturally Occurring Pesticides. The present work is concerned with that great army of pests, the insects; *Insect Growth Regulators* brings together all those substances, which occur naturally, whether in plant or animal, which are known to affect insect growth and development. This beings with a thorough look at those hormones which occur naturally in insects and which regulate their development. Not many such compounds are known, the presence of many more has been inferred or suggested, but the known ones have been studied intensively, there is a great accumulation of literature on them, and many specialized books and reviews. This volume is different in that it attempts to assess our knowledge from the point of view of the reader wishing to know more about pest control and wishing to engender new ideas. As well as the naturally occurring hormones, substances related by structure from other sources are considered as well as materials from all origins which are known to have effects upon growth and development of insects.

The subject is introduced by a brief consideration of the physiological and endocrinological aspects of insect hormones at our present state of knowledge, by two world experts in that subject. They provide a glossary of terms from their subject that may be unclear or unfamiliar to the chemist, entomologist, or agriculturalist.

The chapters which then follow deal with the known insect hormones. The treatment of these necessarily varies. There are over 70 ecdysteroids known but only four juvenile hormones. The knowledge of peptide hormones is still very scanty, and other substances, chiefly from plants, which affect insect growth and development are difficult to categorize under just one heading. The same substance may turn up in more than one place, but the treatment and emphasis of the different authors are such that there is no serious overlap.

We are fortunate in being able to include a chapter on the substances from the neem and chinaberry trees which affect insect feeding and development. The subject has received growing attention in the periodical literature, but this is the first comprehensive introduction to the subject to appear. The substance azadirachtin, obtained from neem seeds would appear to hold great promise, in the form of a crude extract, as a cheap pesticide in the Third World. It is already challenging the ingenuity of chemists to discover and mimic the relationship between its structure and activity.

We wish to thank all the contributors for their great efforts in bringing together all the information contained here. Our thanks are due also to Iris Jones and Margaret Furnival for their help throughout the editorial stage.

E.D.M.
N.B.M.

THE EDITORS

E. David Morgan, D.Phil., is a Chartered Chemist, a Fellow of the Royal Society of Chemistry, and a Fellow of the Royal Entomological Society of London. He received his scientific training in Canada and England, and has worked for the National Research Council of Canada, Ottawa, The National Institute for Medical Research, London, the Shell Group of Companies and is now Reader in Chemistry at the University of Keele, Staffordshire, England. He is co-author of a textbook on aliphatic chemistry with the Nobel prizewinner, Sir Robert Robinson, and with him is a co-inventor of a number of patents. Dr. Morgan has contributed to over 100 papers, most of them on aspects of insect chemistry and has written a number of reviews on insect hormones and pheromones.

N. Bhushan Mandava, holds B.S., M.S., and Ph.D. degrees in chemistry and has published over 120 papers including two patents, several monographs and reviews, and books in the areas of pesticides and plant growth regulators and other natural products. As editorial advisor, he has edited two special issues on countercurrent chromatography for the *Journal of Liquid Chromatography*. He is now a consultant in pesticides and drugs. Formerly, he was associated with the U.S. Department of Agriculture and the Environmental Protection Agency as Senior Chemist. He has been active in several professional organizations, was President of the Chemical Society of Washington, and serves as Councilor of the American Chemical Society.

CONTRIBUTORS

Alexej B. Bořkovec, Ph.D.
Chief
Insect Reproduction Laboratory
U.S. Department of Agriculture
Agricultural Research Center
Beltsville, Maryland

Jules A. Hoffman, Ph.D.
Directeur de Recherche
Laboratoire Biologie Generale
University Louis Pasteur
Strasbourg, France

Caleb W. Holyoke, Ph.D.
Senior Research Chemist
Agricultural Products Department
E.I. DuPont DeNemours & Company
Wilmington, Delaware

W. Mordue, D.Sc.
Professor
Department of Zoology
University of Aberdeen
Aberdeen, Scotland

P. J. Morgan, Ph.D.
Department of Zoology
University of Aberdeen
Aberdeen, Scotland

John C. Reese, Ph.D.
Associate Professor
Department of Entomology
Kansas State University
Manhattan, Kansas

Geoff Richards, Ph.D.
Directeur de Recherche
Laboratoire de Génétique Moléculaire des
 Eukaryotes
Institut de Chimie Biologique
Strasbourg, France

H. Schmutterer, D.Phil.Nat.
Professor
Institute of Phytopathology and Applied
 Zoology
Justus Liebig University
Giessen, Federal Republic of Germany

Nobel Wakabayashi, Ph.D.
Research Chemist
U.S. Department of Agriculture
Agricultural Research Service
Beltsville, Maryland

Rolland M. Waters, Ph.D.
Research Chemist
Science and Educational Administration
Agricultural Research Service
Beltsville, Maryland

Ian D. Wilson, Ph.D.
Safety of Medicines
ICI Pharmaceuticals Division
Macclesfield, England

TABLE OF CONTENTS

Part A

Introduction to the Insect Neuroendocrine System.. 1
Geoffrey Richards and Jules A. Hoffmann

The Ecdysteroids 15
Ian D. Wilson

Juvenile Hormones and Related Compounds 87
N. Wakabayashi and R. M. Waters

The Chemistry and Biology of Selected Insect Peptides............................... 153
W. Mordue and P. J. Morgan

Index 185

Part B

Chemosterilants 1
Alexej B. Bořkovec

Allelochemics Affecting Insect Growth and Development............................... 21
John C. Reese and Caleb W. Holyoke, Jr.

Acute Insect Toxicants from Plants... 67
Caleb W. Holyoke, Jr. and John C. Reese

Insect Growth-Disrupting and Fecundity-Reducing Ingredients from the Neem
and Chinaberry Trees.. 119
Heinrich Schmutterer

Index 171

CHEMOSTERILANTS

Alexej B. Bořkovec

Chemical regulation of reproduction is a cornerstone of modern population management. In insects, best known for their immense reproductive potential and adversary relationship to man, a direct or indirect inhibition of reproduction is the basis of most chemical control procedures, and compounds that directly interfere with reproductive processes are referred to as chemosterilants.[1,2] Although some of the earliest reports on insect chemosterilants concerned plant extracts such as colchicine[3] and pyrethrins,[4] the systematic search for active compounds relied primarily on synthetics with specific cytotoxic or mutagenic activity against the cells comprising the reproductive tissues.[5] In many respects, the development of chemosterilants closely followed the advances in cancer chemotherapy with the inclusion of cytotoxic natural products derived from plants or animals.[6] More recently, advances in insect biochemistry and in the techniques for isolating and identifying natural products led to the realization that insects themselves may furnish valuable clues and model compounds for regulating their own reproductive processes, and that certain plants contain materials identical or closely related to insect reproduction and development regulators.[7] There is little doubt that this trend in the characterization of natural products by their biological activity in insects will continue at increasingly rapid pace in the foreseeable future.

CLASSIFICATION

Chemosterilants are usually classified by their chemical and structural characteristics (alkylating agents and aziridines) by their mode of action (mutagens and chitin synthesis inhibitors), or by the effects they induce in insects, (male or female sterilants).[1] Classification by origin or source serves a useful purpose only in a review such as the present one, in which natural origin is the leading aspect. The great variety and structural complexity of naturally occurring chemosterilants make systematic treatment difficult, but a more serious obstacle is the frequent lack of reliable quantitative data and information on the biochemical and physiological effects of the compounds. The assignment of sterilizing activity usually follows a series of tests in which treated insects are allowed to mate and their reproductive performance is compared to that of a control group. Positive results, i.e., a statistically significant reduction in fecundity or fertility, are the first indications of sterilizing activity. However, such preliminary tests are insufficient for classifying biological activity of a candidate compound that is not a homolog of already known sterilants. Follow-up tests consisting of studies of the dose-response relationship, sex specificity, histopathology, and biochemical parameters are required to determine the compound's physiological effect and possible mode of action. Unfortunately, such detailed studies were performed with only a handful of materials derived from natural products and thus the designation of many of the compounds listed in Tables 1 to 3 as chemosterilants is only tentative.

The largest group of organisms that were searched for constituents with sterilizing activity were the plants. Materials derived from higher plants, specifically from Pteridophyta and Spermatophyta, are listed in Table 1 together with the insect species in which they were tested. Occasionally and in accord with the general practice in economic entomology, members of the order Acarina are included among the insects. Similar arrangement applies to Table 2 that contains materials derived from Thallophyta represented entirely by bacteria and fungi. Table 3 includes all other naturally occurring sterilants, many of them widely distributed in animals and plants. An important group in this table is regulatory substances

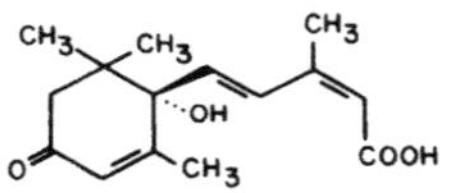

1. Abscisic acid

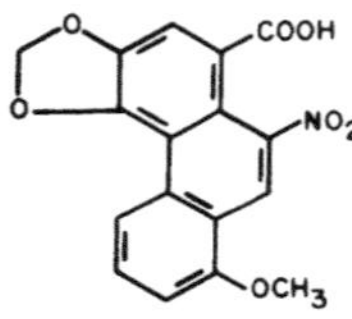

2. Aristolochic acid

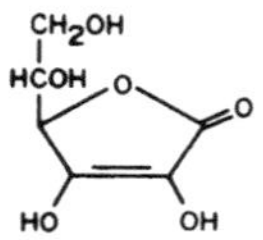

3. Ascorbic acid

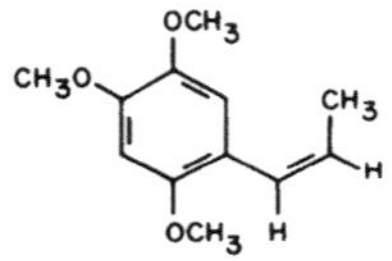

4. β-Asarone

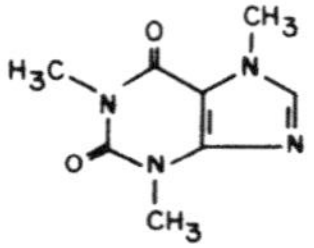

5. Caffeine

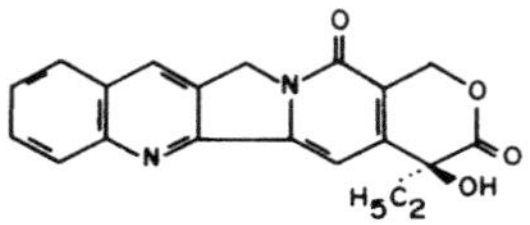

6. Camptothecin

7. L-Canavanine

$$NH_2 \qquad NH_2$$
$$HN=CNHOCH_2CH_2CHCOOH$$

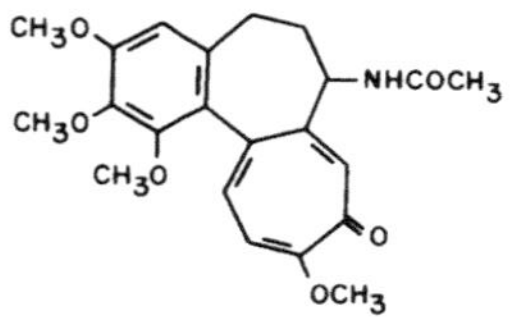

8. Colchicine

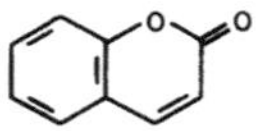

9. Coumarin

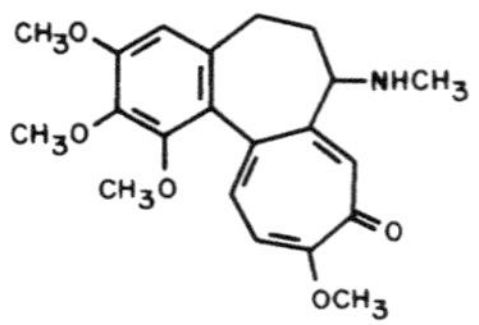

10. Demecolcine

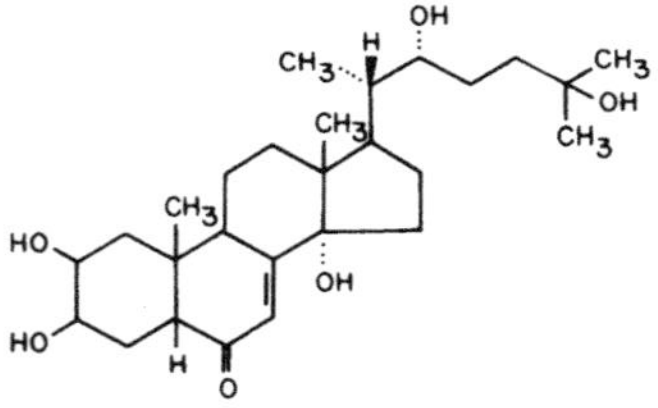

11. α-Ecdysone

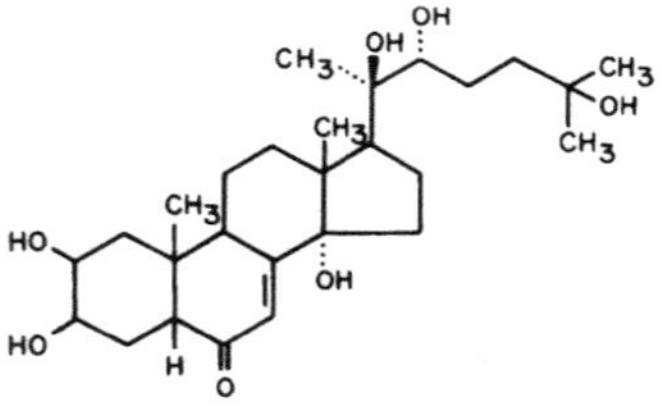

12. β-Ecdysone

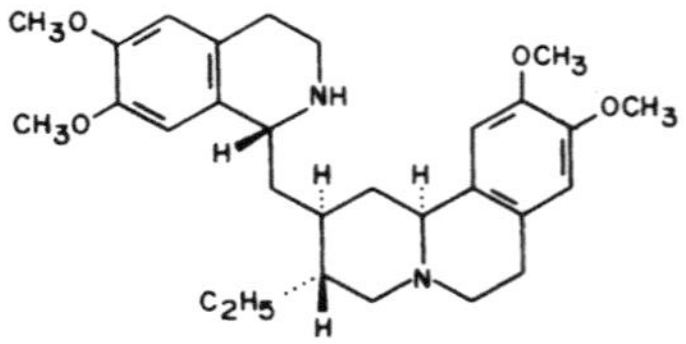

13. Emetine

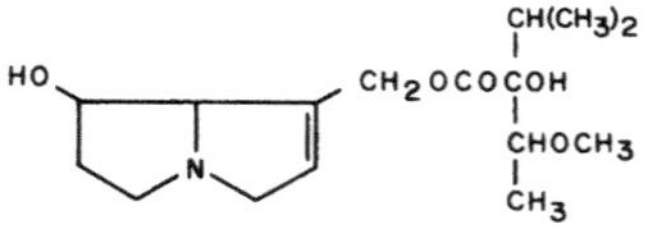

14. Heliotrine

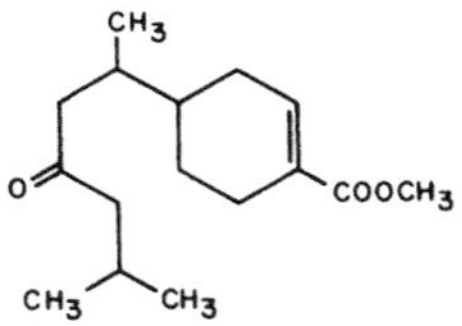

15. Juvabione

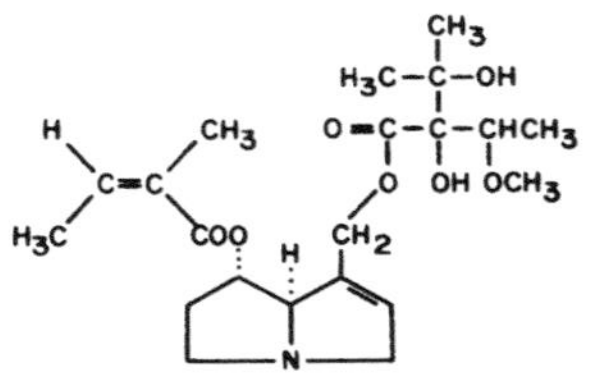

21. Reserpine

24. *m*-Xylohydroquinone

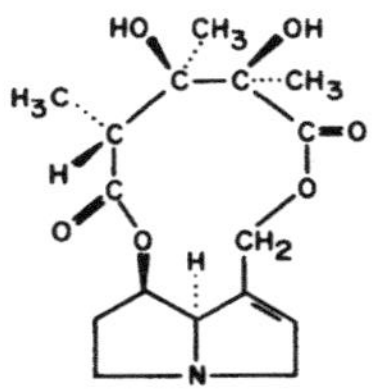

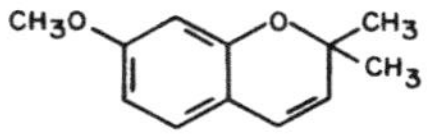

19. Precocene II

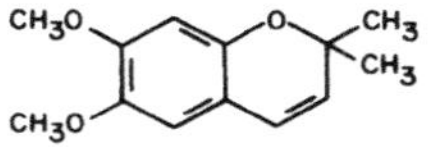

18. Precocene I

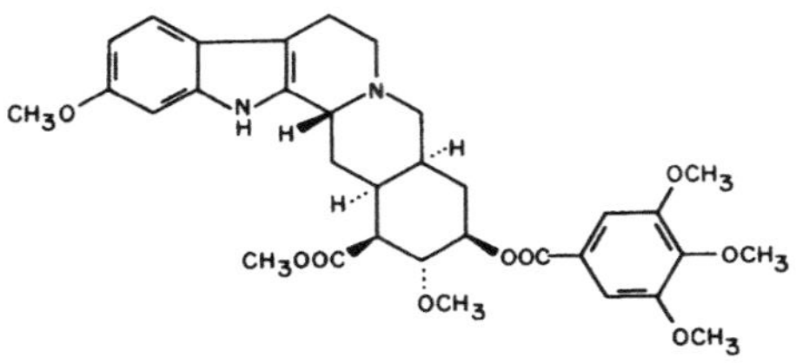

23. Vincristine

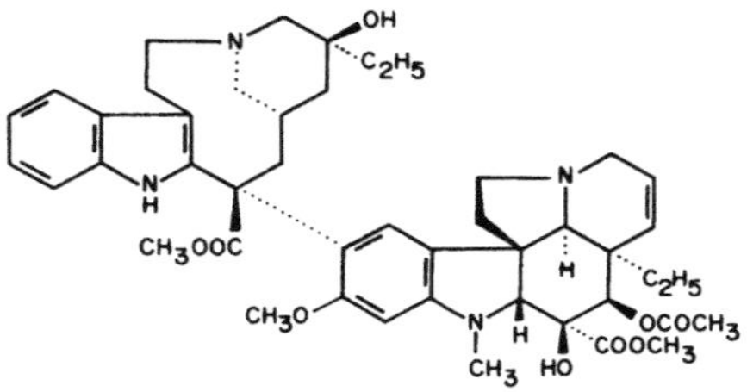

17. Monocrotaline

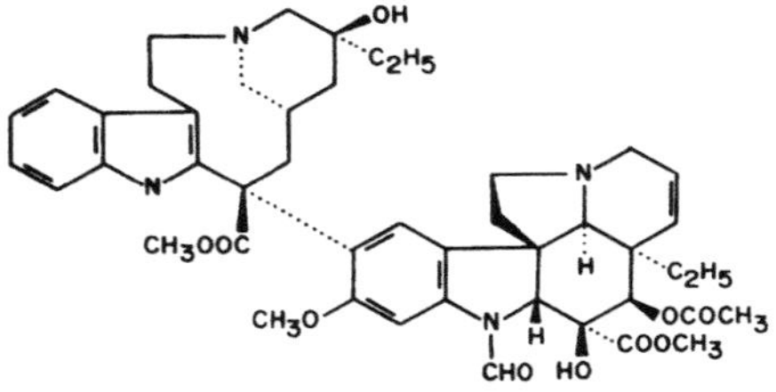

16. Lasiocarpin

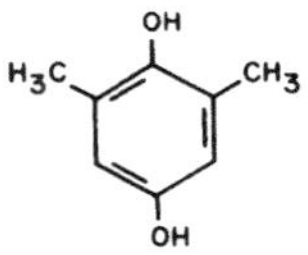

22. Vinblastine

Structures for Table 1.

Table 1
CHEMOSTERILANTS DERIVED FROM HIGHER PLANTS, PTERIDOPHYTA AND SPERMATOPHYTA, AND INSECTS THEY AFFECT

Compound	Plant	Insect	Ref.
Abscisic acid	Variety	*Aulocara elliotti*	21
Aristolochic acid	*Aristolochia bracteata*	*Aedes aegypti*	22
		Dysdercus koenigii	22
		Musca domestica	12
		Tribolium castaneum	22
Ascorbic acid	Variety	*Xylebius ferrugineus*	23
β-Asarone	*Acorus calamus*	*Anthrenus flavipes*	24
		Callosobruchus chinensis	24, 25
		D. koenigii	26—29
		M. domestica	30
		Sitophylus oryzae	24
		Thermobia domestica	31—33
		Trogoderma granarium	24, 34
Caffeine	Variety	*M. domestica nebulo*	13
Camptothecin	*Camptotheca acuminata*	*M. domestica*	35
L-Canavanine	*Canavalia ensiformis*	*Manduca sexta*	36
		Dysdercus koenigii	37
Colchicine	*Colchicum autumnale*	*Chrysomya megacephala*	38
		Ceratitis capitata	39
		Cochliomyia hominivorax	40
		Conotrachelus nenuphar	41
		Dacus dorsalis	39
		D. cucurbitae	39
		Drosophila melanogaster	3, 42, 43
		Musca domestica	44—48
Coumarin	Variety	*M. domestica*	49
Demecolcine	*Colchicum autumnale*	*M. domestica*	50
α-Ecdysone	Variety	*Oncopeltus fasciatus*	51
		Stomoxys calcitrans	52, 53
β-Ecdysone	Variety	*M. domestica*	54, 55
		Oncopeltus fasciatus	51
		Stomoxys calcitrans	52, 53
		Tribolium confusum	54, 55
Emetine	*Uragoga ipecacuanha*	*Cochliomyia hominivorax*	56
Heliotrine	*Heliotropium lasiocarpum*	*D. melanogaster*	57—59
		M. autumnalis	60
Juvabione	*Abies balsamea*	*Pyrrhocoris apterus*	61—63
Lasiocarpin	*H. lasiocarpum*	*D. melanogaster*	58
Monocrotaline	*Crotalaria spectabilis*	*Anthonomus grandis*	64
		D. melanogaster	58
		M. domestica	47
Precocene I	*Ageratum houstonianum*	*Blatella germanica*	65
		Cimex lectularius	66
Precocene II	*A. houstonianum*	*Dysdercus cingulatus*	19
		D. flavidus	67
		D. similis	68
		Epilachna varivestis	19
		Glossina morsitans	69, 70
		Locusta migratoria	71, 72
		Oncopeltus fasciatus	19, 73—75
		Periplaneta americana	76
		Rhagoletis pomonella	19
		Schistocerca gregaria	77
Pyrethrin, unspecified mixture	Variety	*Lasioderma serricorne*	4

Table 1 (continued)
CHEMOSTERILANTS DERIVED FROM HIGHER PLANTS, PTERIDOPHYTA AND SPERMATOPHYTA, AND INSECTS THEY AFFECT

Compound	Plant	Insect	Ref.
Reserpine	*Rauwolfia serpentina*	*Anastrepha ludens*	78
		Anthonomus grandis	79
		Dacus oleae	80
		Diparopsis castanea	81
		Heliothis virescens	82, 83
		M. domestica	84—86
		T. confusum	87
Vinblastine	*Vinca rosea* (syn. *Catharanthus*)	*Dysdercus cingulatus*	88, 89
Vincristine		*Laspeyresia funebrana*	90
		M. domestica	6
m-Xylohydroquinone	*Pisum sativum*	*Drosophila melanogaster*	91
		M. domestica	92—95
		M. domestica vicina	96
Unspecified	*Nephrolepis exaltata*	*Dysdercus cingulatus*	97
	Parthenium hysterophorus	*D. cingulatus*	97
	Podocarpus gracilior	*Spodoptera littoralis*	98
	Santalum album	*Atteva fabriciella*	99, 100
		T. castaneum	100
	Sterculia foetida	*M. autumnalis*	101
		M. domestica	102

that are endogenous to insects, and they are present in extremely low concentrations. However, since the structures of most of them were known, they could be produced synthetically and tested. Several other endogenous reproduction regulators, such as the egg development neurosecretory hormones,[8] are now being studied and will be included in this group in the near future. Chemical names in all tables follow the first listing in the *Merck Index*, 10th edition, and synonyms are included only when such names were mentioned in the references.

MODE OF ACTION

Mutagenesis

In the broadest sense, chemosterilants can be divided into mutagens and nonmutagens.[9,10] Both categories are amply represented among the natural products although clear evidence for such designation is available in only a few cases. A distinctive feature of some mutagens is their capacity for inducing the formation of dominant lethal mutations in sperm without affecting motility or the ability of the sperm to form a zygote.[11] In insects, male sterility induced without the impairment of mating and sperm transfer is usually a good indication of a mutagenic mode of action of the sterilant. In females, on the other hand, dominant or recessive mutations can be expressed in many different ways during ovarian maturation and oogenesis. Unfortunately, no simple rule exists by which a chemically induced female sterility could be unequivocally designated as mutagenic in origin. In some cases that will be mentioned, mutagenicity demonstrated in other organisms may be used for a tentative classification but such assignment should not be considered final.

In Table 1, which contains predominantly plant alkaloids, mutagens are represented in the colchicum alkaloids by colchicine and demecolcine; the pyrrolizidine alkaloids by heliotrine, lasiocarpin, and monocrotaline; the vinca alkaloids by vinblastine and vincristine.

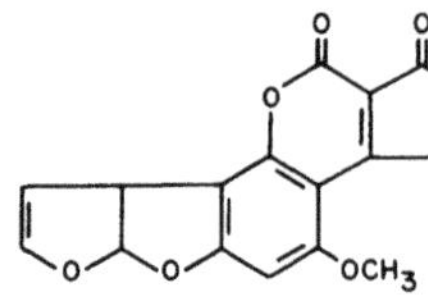

25. Aflatoxin

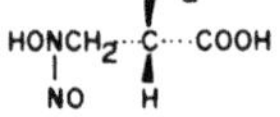

26. Alanosine

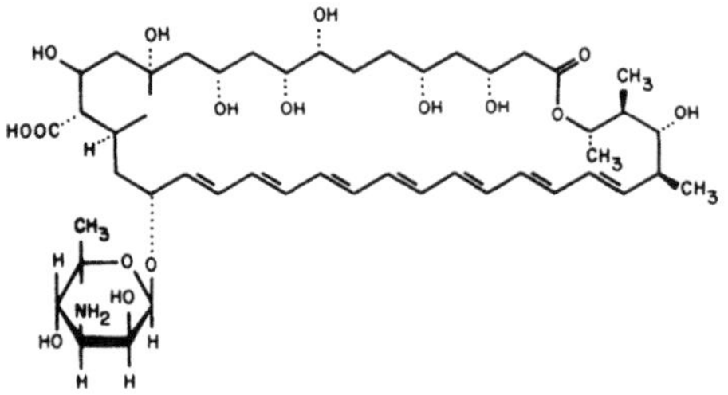

27. Amphotericin A

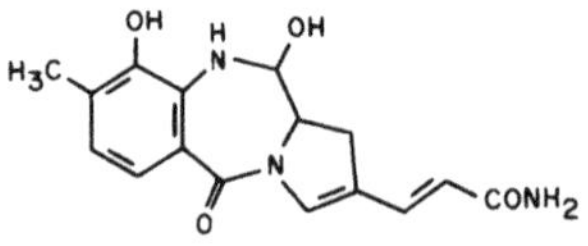

28. Anthramycin

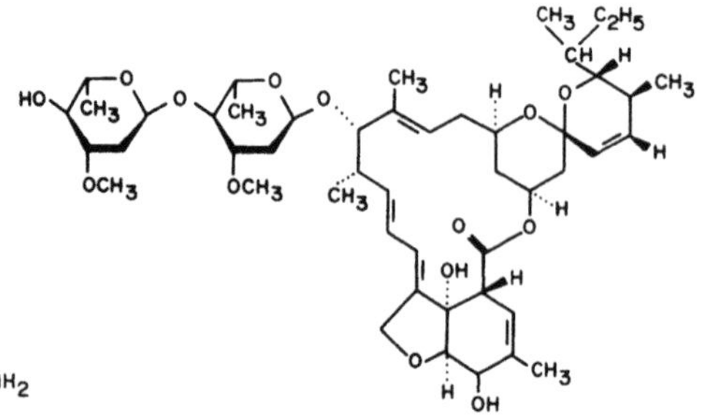

29. Avermectin B₁a

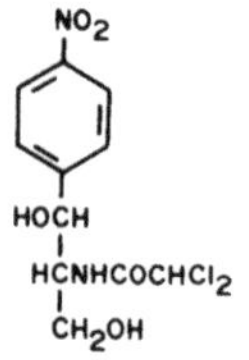

31. Chloramphenicol

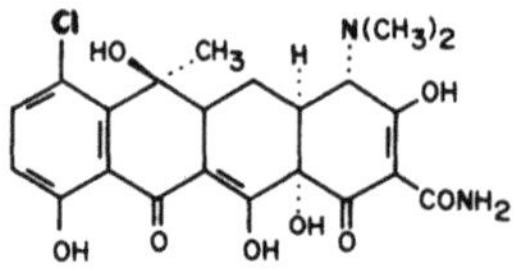

32. Chlortetracycline

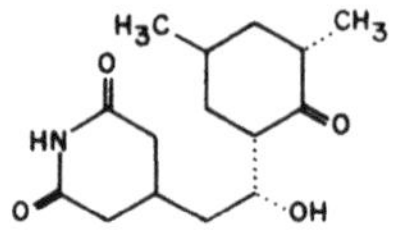

33. Cycloheximide

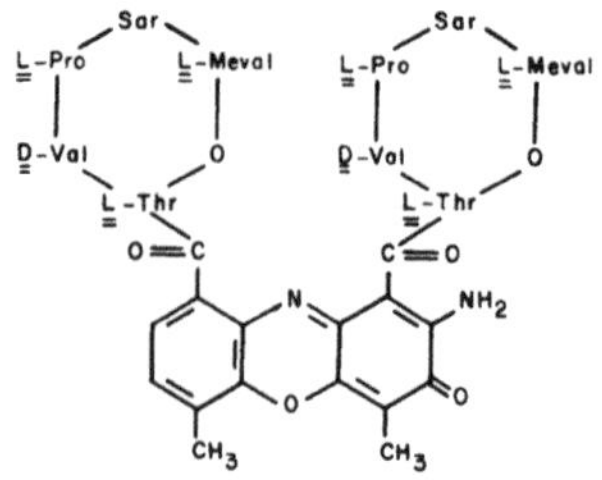

35. Dactinomycin

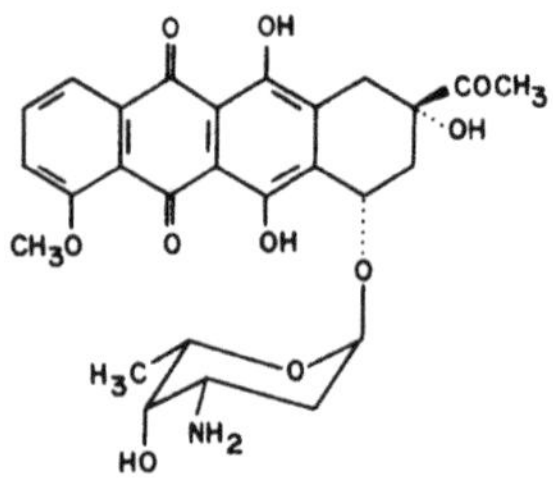

36. Daunorubicin

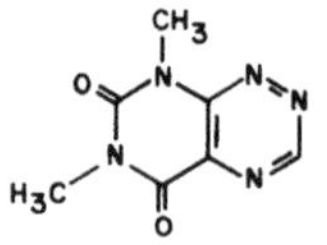

37. Fervenulin

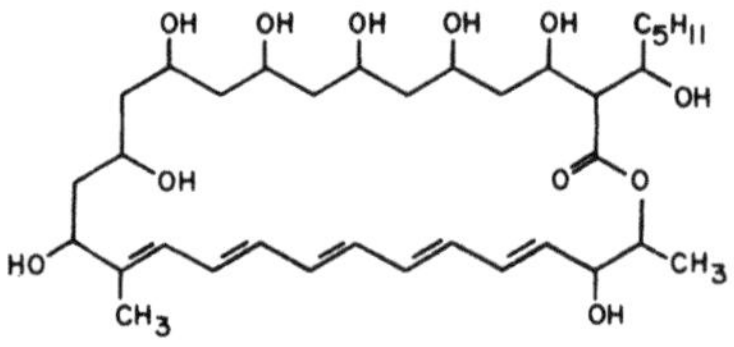

38. Filipin

Structures for Table 2.

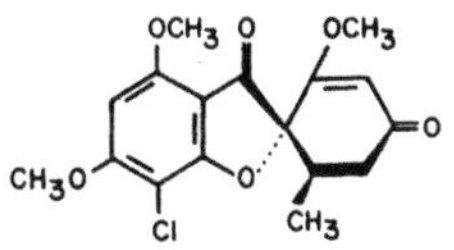

40. Griseofulvin

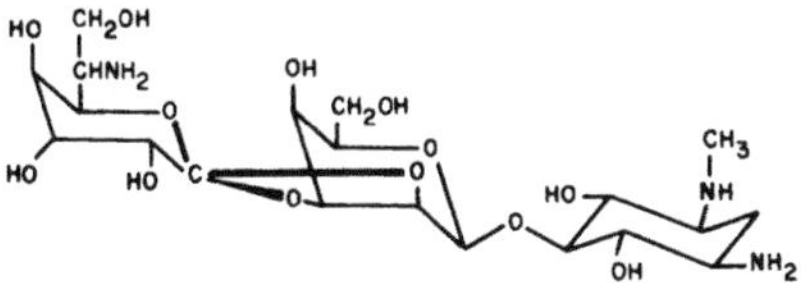

41. Hygromycin B

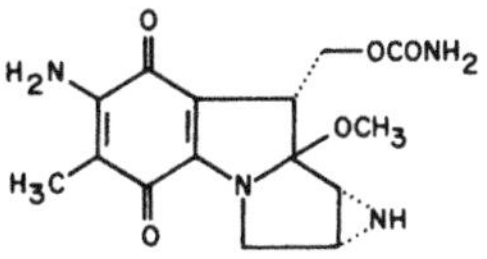

42. Mitomycin C

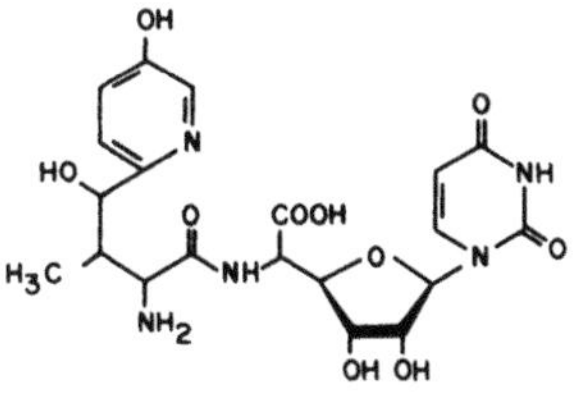

43. Nikkomycin

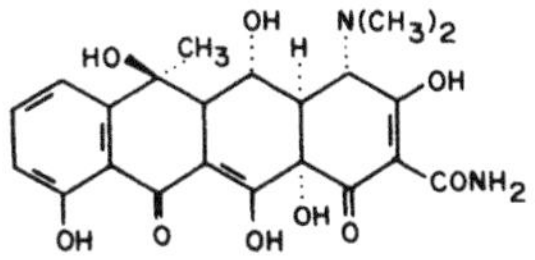

44. Oxytetracycline

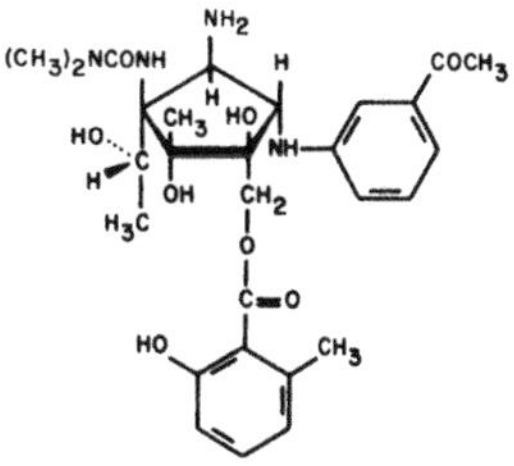

45. Pactamycin

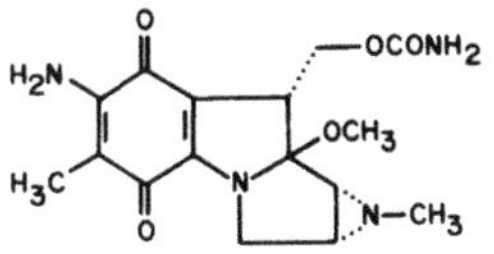

46. Porfiromycin

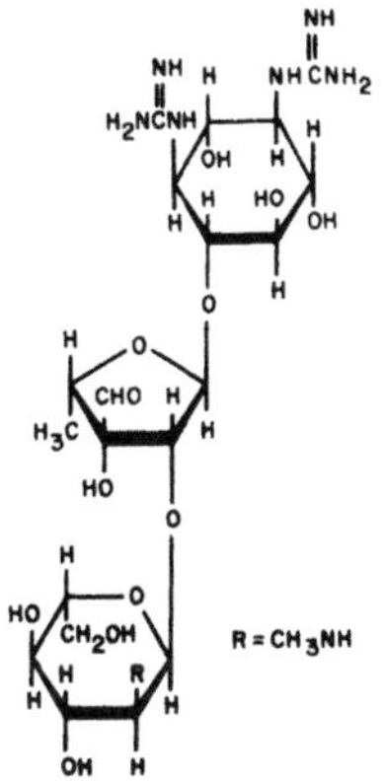

48. Streptomycin

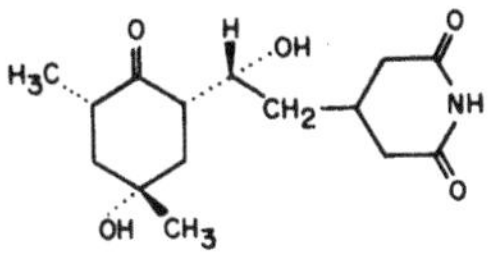

49. Streptovitacin A

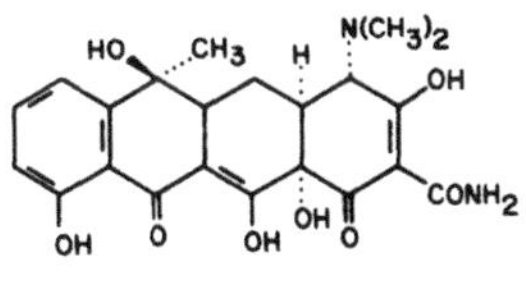

50. Tetracycline

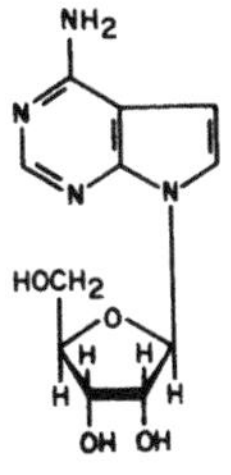

51. Tubercidin

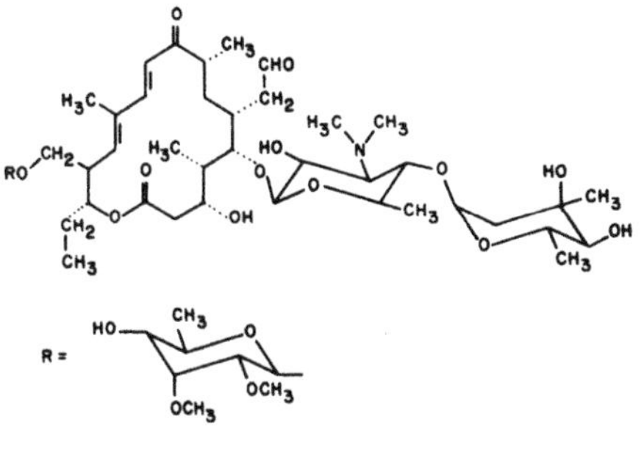

52. Tylosin

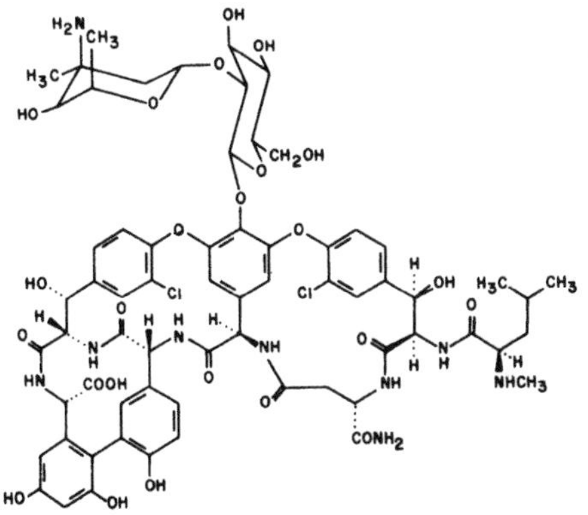

54. Vancomycin

Structures for Table 2 (continued).

Table 2
CHEMOSTERILANTS DERIVED FROM LOWER PLANTS, THALLOPHYTA, AND INSECTS THEY AFFECT

Compound	Source	Insects affected	Ref.
Actidione (see Cycloheximide)			
Actinomycin D (see Dactinomycin)			
Aflatoxin	*Aspergillus flavus, A. parasiticus*	*Aedes aegypti*	103
		Anthonomus grandis	104
		Drosophila melanogaster	103
		Musca domestica	103, 105
Alanosine	*Streptomyces alanosinicus*	*M. domestica*	106
		Phormia regina	107
		Tribolium confusum	106
Amphotericin A	*Streptomycetes* sp.	*Laspeyresia pomonella*	108
Anthramycin	*S. refuineus* v. *thermotolerans*	*D. melanogaster*	109
		M. domestica	110, 111
Aureomycin (see Chlortetracycline)			
Avermectin B₁a	*S. avermitilis*	*Solenopsis invicta*	17, 18
Bambermycin (unspecified mixture)	*S. bambergiensis*	*Aphis fabae*	112
Chloramphenicol	*S. venezuelae*	*A. fabae*	113
Chlortetracycline	*S. aureofaciens*	*A. fabae*	113
Cycloheximide	*S. griseus*	*Aculus cornutus*	114
		Anastrepha ludens	115
		Aphis pomi	116
		Diparopsis castanea	81
		Hippelates collusor	117
		M. domestica	118
		Panonychus citri	119
		P. ulmi	120
		Tetranychus pacificus	119
		T. telarius	114, 120, 121
		T. urticae	122
Cytovirin (structure unknown)	Unidentified	*T. telarius*	121
Dactinomycin	*S.* spp.	*Myzus persicae*	123
		T. urticae	124
Daunomycin (see Daunorubicin)			
Daunorubicin	*S. peucetius*	*M. domestica*	6
Fervenulin	*S. fervens*	*Dysdercus cingulatus*	125
Filipin	*S. filipenensis*	*M. domestica*	126, 127
Filimarisin (see Filipin)			
Flavomycin (see Bambermycin)			
Gibberellin (unspecified mixture)	*Gibberella fijikuroi*	*Aulocara elliotti*	21
		Leptinotarsa decemlineata	128
Griseofulvin	*Penicillium griseofulvum*	*Aphis fabae*	129
Hygromycin B	*S. hygroscopicus*	*Myzus persicae*	123
Likuden (see Griseofulvin)			
Mitomycin C	*S. caespitosus*	*Bombyx mori*	130
		Bracon hebetor	131
		Heliothis zea	132
		M. domestica	133
		Tetranychus urticae	124

Table 2 (continued)
CHEMOSTERILANTS DERIVED FROM LOWER PLANTS, THALLOPHYTA, AND INSECTS THEY AFFECT

Compound	Source	Insects affected	Ref.
Nikkomycin	*S. tendae*	*T. urticae*	134
Oxytetracycline	*S. rimosus*	*A. fabae*	113, 135
Pactamycin	*S. pactum*	*Culex pipiens quinquefasciatus*	117
		Hippelates collusor	117
		M. domestica	136, 137
		T. cinnabarinus	137
Porfiromycin	*S. ardus*	*Anastrepha ludens*	138
		Anthonomus grandis	139
		Cochliomyia hominivorax	138
		Culex pipiens quinquefasciatus	117, 140
		Hippelates collusor	117
		M. domestica	136, 138, 141, 142
		T. atlanticus	138
Ristocetin (unspecified mixture)	*Nocardia lurida*	*Laspeyresia pomonella*	108
Sparsamycin (see Tubercidin)			
Streptomycin	*S. griseus*	*Dacus oleae*	143—145
Streptovitacin A	Unidentified	*Laspeyresia pomonella*	108
Tetracycline	*S. viridifaciens*	*Aphis craccivora*	146
		A. fabae	113
Terramycin (see Oxytetracycline)			
Tubercidin	*S. tubericidus*	*M. domestica*	147
		T. atlanticus	147
Tylosin	*S. fradiae*	*L. pomonella*	113
Tyrothricin (unspecified mixture)	*Bacillus brevis*	*L. pomonella*	108
Vancomycin	*S. orientalis*	*Myzus persicae*	123

Camptothecin and reserpine probably also belong to this category. Aristolochic acid produced chromosomal aberrations in sperm of house flies, *Musca domestica* L., and the subsequent sterilizing effect in males was a good evidence of its mutagenic mode of action.[12] The mutagenicity of caffeine is controversial, nevertheless, Srivasan and Kesavan attributed some of its effects in flies, *Musca domestica nebulo* F., to lethal mutations.[13] Some of the antibiotics listed in Table 2 are known mutagens but only anthramycin, mitomycin C, and porfiromycin had a distinct sterilizing effect in males demonstrating their mutagenic activity. However, aflatoxin and daunorubicin were shown mutagenic in other organisms.[14]

Nutritional Deficiencies

There are numerous nonmutagenic mechanisms by which reproduction can be reduced or prevented but only few of them can be clearly linked with the naturally occurring sterilants. Adequate nutrition is one of the main prerequisites of successful oogenesis and it is not surprising that certain dietary additives can affect female reproduction. In Table 1, ascorbic acid and canavanine belong to this category. In Table 2, filipin, and in Table 3, biotin, cysteine, folic acid, and vitamin E may also be so classified.

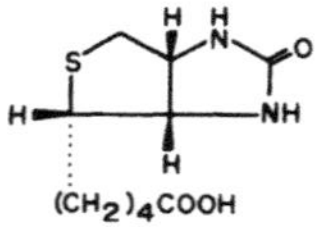

61. D-Glucosamine

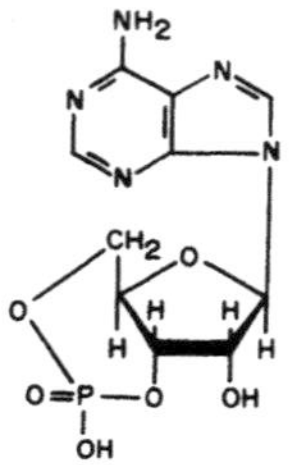

65. Makisterone A

72. Vitamin E

59. Folic acid

HSCH$_2$CH(NH$_2$)COOH

58. Cysteine

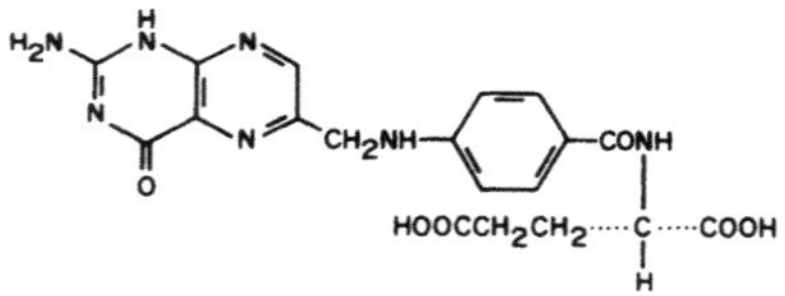

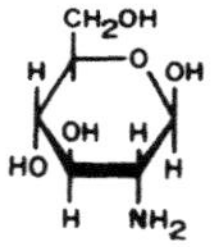

CH$_3$C—CH(CH$_2$)$_2$C=CH(CH$_2$)$_2$C=CHCOOCH$_3$

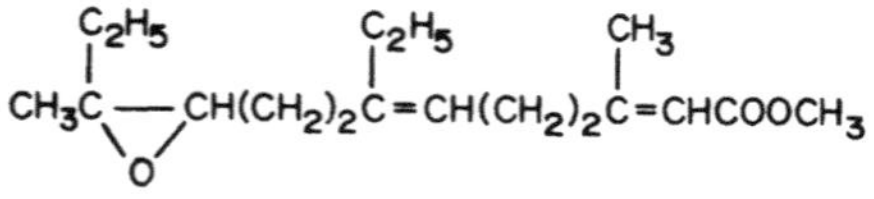

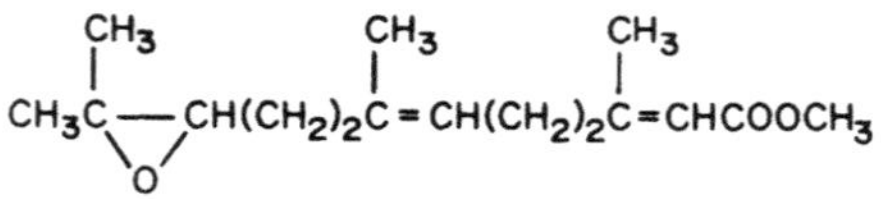

63. JH III

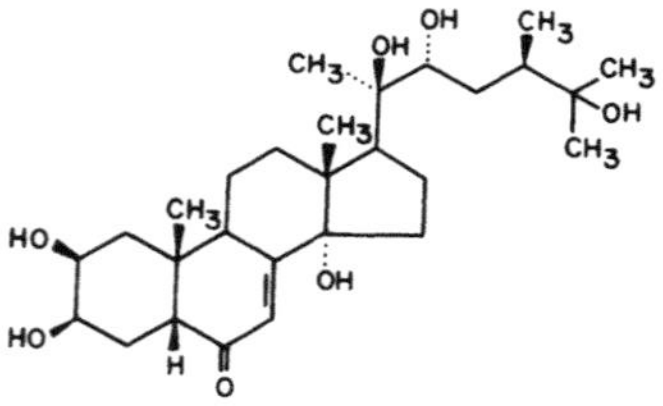

CH$_3$CO(CH$_2$)$_5$

71. Queen substance

57. cAMP

62. JH I

NH$_2$(CH$_2$)$_4$NH$_2$

70. Putrescine

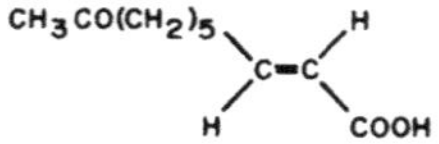

56. Biotin

CH$_3$C—CH(CH$_2$)$_2$C=CH(CH$_2$)$_2$C=CHCOOCH$_3$

62. JH I

CH$_3$SCH$_2$CH$_2$CH(NH$_2$)COOH

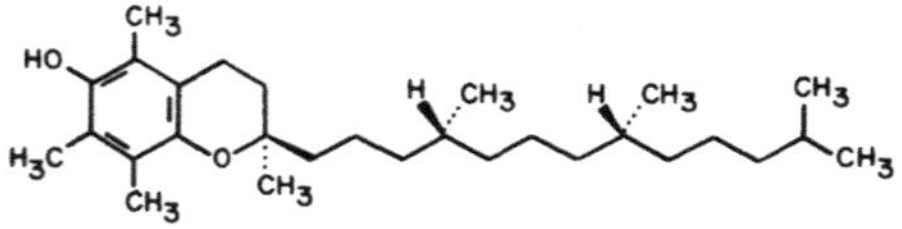

66. Methionine

Structures for Table 3.

Table 3
CHEMOSTERILANTS DERIVED FROM INSECTS OR OTHER NATURAL SOURCES, AND INSECTS THEY AFFECT

Compound	Source	Insect	Ref.
L-Asparaginase (structure unknown)	Animals, plants	*Musca domestica*	6
Biotin	Animals, plants	*Aedes aegypti*	148
		Anastrepha ludens	149
		Dermestes maculatus	150, 151
		Heliothis zea	153
		M. domestica	154
cAMP	Animals, plants	*Dysdercus cingulatus*	125
Cysteine	Animals, plants	*Xyleborus ferrugineus*	23
Folic acid	Animals, plants	*Conotrachelus nenuphar*	41
FSH (structure unknown)	Mammals	*C. nenuphar*	41
D-Glucosamine	Invertebrates	*M. domestica*	49
JH I	Lepidoptera	*Adoxophyes orana*	155
		Aedes aegypti	156
		M. domestica	157
		Spodoptera littoralis	158
JH III	Insects	*H. virescens*	159
LH (structure unknown)	Mammals	*C. nenuphar*	41
Makisterone A	Insects	*Glossina morsitans*	160
		Oncopeltus fasciatus	51
Methionine	Animals, plants	*M. domestica*	161, 162
Monogamy factor	Insects	*A. aegypti*	163—167
		Chironomus riparius	168
		Cochliomyia hominivorax	169
		M. domestica	169, 170
		Phormia regina	169
Oostatic hormone (structure unknown)	*Musca domestica*	*M. domestica*	171
Prostaglandin (unspecified mixture)	Animals	*Dysdercus cingulatus*	125
Putrescine	Animals, plants	*Prodenia litura*	172
Queen substance	*Apis mellifera*	*Acyrtosiphon pisum*	173
		Drosophila melanogaster	174
		D. phalerata	175
		Eurygaster intergriceps	176
		Laspeyresia pomonella	177
		M. domestica	178
Vitamin E	Animals, plants	*X. ferrugineus*	23

Hormonal and Related Regulation

The discovery of two principal classes of insect hormones, juvenile hormones (JHs) and ecdysteroids, opened up new possibilities for regulating reproduction. Although the detailed mechanisms by which either type of hormone affects ovarian development, vitellogenesis, and early embryogenesis are not known, both types of compounds themselves as well as some of their analogs and homologs function as sterilants. Obviously, the insect hormones are natural products but since their natural concentrations in insects are exceedingly low, the compounds became available for broader testing only after they were synthesized in the laboratory. In Table 3, JHs I and III and makisterone A represent this group. However, some of the ecdysteroids are also present in plants in relatively high concentrations and plants are now the most important and least expensive source. Examples are α- and β-ecdysone in Table 1. In addition to the known hormones, insects are believed to contain numerous other regulatory substances of unknown structure and mode of action. Only the oostatic hormone (Table 3) has been mentioned in the literature but this group can be expected to increase substantially in the future. Table 1 contains two other hormonal regulators: precocene I and precocene II. Although the ultimate mode of action of precocenes is the inhibition of JH release, the mechanism by which this effect is achieved is the selective destruction of tissues in the insect's corpora allata.[15] However, since such selective cytotoxicity is a result of an enzymatic conversion of precocene to precocene epoxide, and since epoxides are mutagenic alkylating agents,[14] precocenes could also be classified as mutagens that require biological activation.

Some of the antibiotics listed in Table 2 may also belong to the category of substances that affects regulation of physiological processes through the insect's nervous system. In the boll weevil, *Anthonomus grandis grandis* Boheman, avermectins exhibited several pathological effects including the reduction of pheromone production, and this activity was believed to result from their agonistic action toward a neurotransmitter γ-aminobutyric acid.[16] A similar explanation was advanced for the sterilizing effects of avermectin B_1a in the red imported fire ant, *Solenopsis invicta* Buren,[17,18] but whether any of the other antibiotics listed in Table 2 act by a related mechanism remains to be determined.

In addition to the already mentioned juvenile and oostatic hormones, Table 3 contains several materials that may affect the hormonal and related regulation of reproduction. The two mammalian hormones, FSH (follicle-stimulating hormone) and LH (luteinizing hormone), are glycoproteins involved in reproduction but their activity in insects has not been clarified. The assignation of cAMP and prostaglandin to this category is also entirely tentative.

Undetermined Mode of Action

It was earlier mentioned that most of the data on the mode of action of naturally occurring chemosterilants are tentative, inconclusive, or entirely absent. Consequently, the present category is large and would include a clear majority of the compounds listed in Tables 1 to 3 if rigorous standards for classification were to be employed. One of the reasons for the dearth of data is the fact that most of the chemosterilants mentioned here had a relatively low activity, especially in male insects. The exceptions, such as camptothecin (Table 1) and anthramycin (Table 2), were scarce, complex materials with high mammalian toxicity and little practical development potential. On the other hand, a sterilizing activity produced by an unknown mechanism presents an important challenge and opportunity for exploring previously unsuspected reproductive processes that are susceptible to chemical manipulation.

FUTURE PROSPECTS

There is little doubt that additional natural materials with sterilizing activity will be discovered in the near future. Plant extracts will continue to supply new compounds, some with novel structure and mode of action. Because the chemical taxonomy of plants is still only poorly understood, the search for active extracts is largely empirical and defies attempts for advanced planning. However, since the pharmacological potential of plants is extremely broad and since tests on insects are now common in numerous laboratories it can be expected that new plant products active in insects will continue to appear in the literature. Whether such new compounds will have a serious impact on our understanding of insect physiology and on the development of new insect control procedures will largely depend on the type and quality of data describing their biological activity. Sterility tests conducted with insects whose reproductive biology and physiology are little understood may be adequate in first screening and may help to select potentially interesting materials; nevertheless, only standardized testing with well-known species can furnish reliable information on the compound's specific effects and possible mode of action. The discovery of precocenes may serve as a model for such procedures[19] although in this case the structure of the active material was fairly simple and was determined several years before its biological activity was discovered.

In contrast to plant extracts, materials derived from insects are usually obtained after detailed studies of their activity have been concluded. Research currently underway on peptidic neuroregulators such as the prothoracicotropic hormone and the egg development neurosecretory hormone indicates that both types of materials affect the titer of ecdysone and, therefore, may be expected to disturb reproduction. Availability of larger quantities of such substances for testing can be expected to follow their final isolation and structural identification.

The search for naturally occurring insect chemosterilants has progressed through three philosophically distinct phases. The first phase, between 1940 and 1958, coincided with the first successful attempts for establishing a qualitative biochemical base for reproductive physiology and had no discernible utilitarian motivation. The advent of Knipling's[20] sterility control procedures initiated the second phase, 1958 to around 1975, in which the desired şterilizing properties were clearly identified and the practical utilization potential was focused on control. However, almost imperceptibly and without any organized effort, the concept of chemosterilants and their utilization began to change before and after 1975. Gradually, the need for the development of an exclusive chemosterilant, a substance that would sterilize adults without toxicity and other side effects, became less urgent and more emphasis was being placed on the whole spectrum of physiological effects induced in all developmental stages. Insect control procedures and projections based on hormone agonists and antagonists, chitin synthesis inhibition, and the various pheromone concepts have shown that a multifarious activity at different stages of the life cycle as well as the disruption of several vital physiological processes were not only possible but that they could become more powerful and efficient than a single, highly specific activity. It would be a mistake, however, to relegate reproduction disruption or for that matter its enhancement into a secondary or unimportant category. In all economically important insects, the regulation of reproduction is still the central point of population management and its chemical control remains a most desirable and attainable objective to which the present, third phase of the search for naturally occurring chemosterilants and reproduction regulators will undoubtedly contribute.

REFERENCES

1. **Borkovec, A. B.,** *Insect Chemosterilants,* Interscience, New York, 1966.
2. **LaBrecque, G. C. and Smith, C. N., Eds.,** *Principles of Insect Chemosterilization,* Appleton-Century-Crofts, New York, 1968.
3. **Gelei, G. V. and Csik, L.,** Effect of colchicine on *Drosophila melanogaster, Biol. Zentralbl.,* 60, 275, 1940.
4. **Tenhet, J. N.,** Effect of pyrethrum on oviposition of the cigarette beetle, *J. Econ. Entomol.,* 40, 910, 1947.
5. **Borkovec, A. B.,** Sexual sterilization of insects by chemicals, *Science,* 137, 1034, 1962.
6. **Chang, S. C., Borkovec, A. B., and Braun, B. H.,** Chemosterilant activity of antineoplastic agents, *Trans. N.Y. Acad. Sci.,* 36(2), 101, 1974.
7. **Bowers, W. S.,** Phytochemical disruption of insect development and behavior, presented at Am. Chem. Soc. Div. Pest. Chem. Conf. New Concepts in Pesticide Chemistry, Snowbird, Utah, June 25, 1984.
8. **Masler, E. P., Hagedorn, H. H., Petzel, D., and Borkovec, A. B.,** Partial purification of egg development neurosecretory hormone with reverse-phase liquid chromatographic techniques, *Life Sci.,* 33, 1925, 1983.
9. **Borkovec, A. B.,** Insect chemosterilants as mutagens, in *Chemical Mutagens,* Vol. 3, Hollander, A., Ed., Plenum Press, New York, 1973, 259.

10. **Borkovec, A. B.,** Chemical and synthetic aspects of insect sterilants, in A Rational Evaluation of Pesticidal vs. Mutagenic/Carcinogenic Action, Hart, R. W., Ed., DHEW Publ. No. (NIH) 78-1306, Department of Health, Education, and Welfare, Washington, D.C., 1978, 2.

11. **LaChance, L. E., North, D. T., and Klassen, W.,** Cytogenic and cellular basis of chemically induced sterility, in *Principles of Insect Chemosterilization,* LaBrecque, G. C. and Smith, C. N., Eds., Appleton-Century-Crofts, New York, 1968, 99.

12. **Mathur, A. C., Sharma, A. K., and Verma, V.,** Cytopathological effects of aristolochic acid on male house flies causing sterility, *Experientia,* 36, 245, 1980.

13. **Srinivasan, A. and Kesavan, P. C.,** Reproductive toxicity of caffeine in *Musca domestica, J. Toxicol. Environ. Health,* 5, 765, 1979.

14. **Fishbein, L., Flamm, W. G., and Falk, H. L.,** *Chemical Mutagens,* Academic Press, New York, 1970, 236.

15. **Feyereisen, R., Johnson, G., Koener, J., Stay, B., and Tobe, S. S.,** Precocenes as pro-allatocidins in adult female *Diploptera punctata:* a functional and ultrastructural study, *J. Insect Physiol.,* 27, 855, 1981.

16. **Wright, J. E.,** Biological activity of avermectin B_1 against the boll weevil (Coleoptera: Curculionidae), *J. Econ. Entomol.,* 77, 1029, 1984.

17. **Glancey, B. M., Lofgren, C. S., and Williams, D. F.,** Avermectin B_1a: effects on the ovaries of red imported fire ant queens (Hymenoptera:Formicidae), *J. Med. Entomol.,* 19, 743, 1982.

18. **Lofgren, C. S. and Williams, D. F.,** Avermectin B_1a: highly potent inhibitor of reproduction by queens of red imported fire ant (Hymenoptera:Formicidae), *J. Econ. Entomol.,* 75, 798, 1982.

19. **Bowers, W. S., Ohta, T., Cleere, J. S., and Marsella, P. A.,** Discovery of insect anti-juvenile hormones in plants, *Science,* 193, 542, 1976.

20. **Knipling, E. F.,** Possibilities of insect control or eradication through the use of sexually sterile males, *J. Econ. Entomol.,* 48, 459, 1955.

21. **Visscher, S. N.,** Regulation of grasshopper fecundity, longevity and egg viability by plant growth hormones, *Experientia,* 36, 130, 1980.

22. **Saxena, B. P., Koul, O., Tikku, K., Atal, C. K., Suri, O. P., and Suri, K. A.,** Aristolochic acid — an insect chemosterilant from *Aristolochia bracteata,* Retz., *Indian J. Exp. Biol.,* 17, 354, 1979.

23. **Bridges, J.,** Some Aspects of the Reproduction of the Ambrosia Beetle, *Xyleborus ferrugineus* (F.), Ph.D. thesis, University of Wisconsin, Madison, 1975.

24. **Saxena, B. P., Koul, O., and Tikku, K.,** Non-toxic protectant against the stored grain insect pests, *Bull. Grain Technol.,* 14, 190, 1976.

25. **Tikku, K., Saxena, B. P., and Koul, O.,** Oogenesis in *Callosobruchus chinensis* and induced sterility by *Acorus calamus* L. oil vapours, *Ann. Zool. Ecol. Anim.,* 10, 545, 1978.

26. **Saxena, B. P. and Mathur, A. C.,** Loss of fecundity in *Dysdercus koenigii* F. due to vapours of *Acorus calamus* L. oil, *Experientia,* 32, 315, 1976.

27. **Saxena, B. P., Koul, O., and Tikku, K.,** Inhibition of embryonic development by oil of *Acorus calamus* L. treated males in *Dysdercus koenigii* F., *Indian Perfumer,* 21, 73, 1977.

28. **Saxena, B. P., Koul, O., Tikku, K., and Atal, C. K.,** A new insect chemosterilant isolated from *Acorus calamus* L., *Nature (London),* 270, 512, 1977.

29. **Koul, O., Tikku, K., and Saxena, B. P.,** Mode of action of *Acorus calamus* L. oil vapours on adult male sterility in red cotton bug, *Experientia,* 33, 29, 1977.

30. **Mathur, A. C. and Saxena, B. P.,** Induction of sterility in male houseflies by vapors of *Acorus calamus* L. oil, *Naturwissenschaften,* 62, 576, 1975.

31. **Saxena, B. P. and Rohdendorf, E. B.,** Morphological changes in *Thermobia domestica* under the influence of *Acorus calamus* oil vapours, *Experientia,* 30, 1298, 1974.

32. **Motl, O. and Rohdendorf, E. B.,** *Acorus calamus* oil fractions affecting the ovarioles of *Thermobia domestica, Gen. Comp. Endocrinol.,* 29(Abstr.), 151, 1976.

33. **Rohdendorf, E. B.,** Response of *Thermobia domestica* (Packard) (Thysanura, Insecta) gonads and embryos to foreign substances, *Zh. Obshch. Biol.,* 38, 71, 1977.

34. **Koul, O., Tikku, K., and Saxena, B. P.,** Follicular regression in *Trogoderma granarium* due to sterilizing vapours of *Acorus calamus* L. oil, *Curr. Sci.,* 46, 724, 1977.

35. **DeMilo, A. B. and Borkovec, A. B.,** Camptothecin, a potent chemosterilant against the house fly, *J. Econ. Entomol.,* 67, 457, 1974.

36. **Palumbo, R. E. and Dahlman, D. L.,** Reduction of *Manduca sexta* fecundity and fertility by L-canavanine, *J. Econ. Entomol.,* 71, 674, 1978.

37. **Koul, O.,** L-Canavanine, an antigonadal substance for *Dysdercus koenigii, Entomol. Exp. Appl.,* 34, 297, 1983.

38. **Shukla, G. S. and Singh, R. N.,** Effect of chemosterilant (colchicine) to find out the sterilizing dose on *Chrysomya megacephala* Fab. (Diptera:Calliphoridae), *Z. Angew. Zool.,* 66, 423, 1979.

39. **Mitlin, N., Butt, B. A., and Shortino, T. J.,** Effect of mitotic poisons on house fly oviposition, *Physiol. Zool.,* 30, 133, 1957.

40. **Keiser, I., Steiner, L. F., and Kamasaki, H.,** Effect of chemosterilants against oriental fruit fly, melon fly, and Mediterranean fruit fly, *J. Econ. Entomol.,* 58, 682, 1965.

41. **Chamberlain, W. F. and Hopkins, D. E.,** Effect of colchicine on screw-worms, *J. Econ. Entomol.,* 53, 1133, 1960.

42. **Fye, R. L.,** Screening of chemosterilants against house flies, *J. Econ. Entomol.,* 60, 605, 1967.

43. **Fye, R. L. and LaBrecque, G. C.,** Sterility in house flies offered a choice of untreated diets and diets treated with chemosterilants, *J. Econ. Entomol.,* 60, 1284, 1967.

44. **Jacob, J.,** A study of colchicine-induced sterility in the female fruit fly *Drosophila melanogaster, Growth,* 22, 17, 1958.

45. **Mitlin, N., Butt, B. A., Shortino, T. J.,** Effect of mitotic poisons on house fly oviposition, *Physiol. Zool.,* 30, 133, 1957.

46. **Mitlin, N. and Baroody, A. M.,** Use of the housefly as a screening agent for tumor-inhibiting agents, *Cancer Res.,* 18, 708, 1958.

47. **Piquett, P. G. and Keller, J. C.,** A screening method for chemosterilants of the house fly, *J. Econ. Entomol.,* 55, 261, 1962.

48. **Gouck, H. K. and LaBrecque, G. C.,** Chemicals affecting fertility in adult house flies, *J. Econ. Entomol.,* 57, 663, 1964.

49. **Tsao, T. P. and Chang, J. T. P.,** Studies on insect chemosterilants. IV. Screening of insect chemosterilants, *Acta Entomol. Sin.,* 15, 13, 1966.

50. **Mitlin, N. and Baroody, A. M.,** The effect of some biologically active compounds on growth of house fly ovaries, *J. Econ. Entomol.,* 51, 384, 1958.

51. **Landa, V. and Rezabova, B.,** The effect of chemosterilants on the development of reproductive organs in insects, *Proc. 12th Int. Cong. Entomol. London,* 1965, 516.

52. **Dorn, A. and Buhlmann, K. J.,** Ovicidal activity of maternally applied ecdysteroids in the large milkweed bug (Heteroptera:Lygaeidae), *J. Econ. Entomol.,* 75, 935, 1982.

53. **Wright, J. E. and Kaplanis, J. N.,** Ecdysone and ecdysone analogues: effects on fecundity of the stable fly, *Stomoxys calcitrans, Ann. Entomol. Soc. Am.,* 63, 622, 1970.

54. **Wright, J. E., Chamberlain, W. F., and Barret, C. C.,** Ovarian maturation in stable flies: inhibition by 20-hydroxyecdysone, *Science,* 172, 1247, 1971.

55. **Robbins, W. E., Kaplanis, J. N., Thompson, M. J., Shortino, T. J., Cohen, C. F., and Joyner, S. C.,** Ecdysones and analogs: effects on the development and reproduction of insects, *Science,* 161, 1158, 1968.

56. **Robbins, W. E., Kaplanis, J. N., Thompson, M. J., Shortino, T. J., and Joyner, S. C.,** Ecdysones and synthetic analogs: Molting hormone activity and inhibitive effects on insect growth, metamorphosis and reproduction, *Steroids,* 16, 105, 1970.

57. **Crystal, M. M.,** Antifertility effects of anthelmintics in insects, *J. Econ. Entomol.,* 57, 606, 1964.

58. **Clark, A. M.,** Mutagenic activity of the alkaloid heliotrine in *Drosophila, Nature (London),* 163, 731, 1959.

59. **Clark, A. M.,** Sterilization of *Drosophila* by some pyrrolizidine alkaloids, *Z. Vererbungsl.,* 91, 74, 1960.

60. **Brink, N. G.,** Somatic and teratogenic effects induced by helioitrine in *Drosophila, Mutat. Res.,* 104, 105, 1982.

61. **Zapanta, H. M. and Wingo, C. W.,** Preliminary evaluation of heliotrine as a sterility agent for face flies, *J. Econ. Entomol.,* 61, 330, 1968.

62. **Slama, K. and Williams, C. M.,** "Paper factor" as an inhibitor of the embryonic development of the European bug, *Pyrrhocoris apterus, Nature (London),* 210, 329, 1966.

63. **Slama, K. and Williams, C. M.,** The juvenile hormone. V. The sensitivity of the bug, *Pyrrhocoris apterus,* to a hormonally active factor in American paper pulp, *Biol. Bull.,* 130, 235, 1966.

64. **Williams, C. M. and Slama, K.,** The juvenile hormone. VI. Effects of the "paper factor" on the growth and metamorphosis of the bug *Pyrrhocoris apterus, Biol. Bull.,* 130, 247, 1966.

65. **Klassen, W., Norland, J. F., and Borkovec, A. B.,** Potential chemosterilants for boll weevils, *J. Econ. Entomol.,* 61, 401, 1968.

66. **Belles, X. and Messeguer, A.,** Sterilizing effects of 6,7-dimethoxy-2,2-dimethylchromene (precocene 2) and 6,7-dimethoxy-2,2-dimethylchroman on *Blatella germanica* L., *Dev. Endocrinol. Juv. Horm. Biochem.,* 15, 421, 1981.

67. **Feldlaufer, M. F., Eberle, M. W., and McClelland, G. A. H.,** Development and teratogenic effects of precocene 2 on the bed bug, *Amex lectularius* L., *Insect Sci. Appl.,* 1, 389, 1981.

68. **Fagoonee, I. and Umrit, G.,** Anti-gonadotropic hormones from the goatweed, *Ageratum conyzoides, Insect Sci. Appl.,* 11, 373, 1981.

69. **Judson, P., Rao, B. K., Thakur, S. S., and Revathy, D.,** Effect of precocene II on reproduction of the bug *Dysdercus similis* (F.) (Heteroptera), *Indian J. Exp. Biol.,* 17, 947, 1979.

70. **Samaranayaka-Ramasamy, M. and Chaudhury, M. F. B.,** Precocene treatment of the female tsetse fly *Glossina morsitans morsitans* sterilises her female offspring, *Experientia,* 37, 1027, 1981.

71. **Samaranayaka-Ramasamy, M. and Chaudhury, M. F. B.,** Precocene-induced sterility in F₁ generation of *Glossina morsitans morsitans, J. Insect. Physiol.,* 28, 559, 1982.

72. **Pener, M. P., Orshan, L., and De Wilde, J.,** Precocene II causes atrophy of corpora allata in *Locusta migratoria, Nature (London),* 272, 351, 1978.

73. **Pratt, G. E., Jennings, R. C., Hamnett, A. F., and Brooks, G. T.,** Lethal metabolism of precocene-I to a reactive epoxide by locust corpora allata *in vitro, Nature (London),* 284, 320, 1980.

74. **Bowers, W. S. and Martinez-Pardo, R.,** Antiallatotropins: inhibition of corpus allatum development, *Science,* 197, 1369, 1977.

75. **Rankin, M. A.,** Effects of precocene I and II on flight behavior in *Oncopeltus fasciatus,* the migratory milkweed bug, *J. Insect Physiol.,* 26, 67, 1980.

76. **Bowers, W. S., Evans, P. H., Marsella, P. A., Soderlund, D. M., and Bettarini, F.,** Natural and synthetic allatoxins: suicide substrates for juvenile hormone biosynthesis, *Science,* 217, 647, 1982.

77. **Pratt, G. E. and Bowers, W. S.,** Precocene II inhibits juvenile hormone biosynthesis by cockroach corpora allata *in vitro, Nature (London),* 265, 548, 1977.

78. **Chenevert, R., Paquin, R., and Perron, J. M.,** Anti-juvenilizing activity of precocene I in *Schistocerca gregaria* (Forsk.), *Nat. Can.,* 105, 425, 1978.

79. **Benschoter, C. A.,** Reserpine as a sterilant for the Mexican fruit fly, *J. Econ. Entomol.,* 59, 333, 1966.

80. **Flint, H. M., Earle, N., Eaton, J., and Klassen, W.,** Chemosterilization of the female boll weevil, *J. Econ. Entomol.,* 66, 47, 1973.

81. **Fytizas, E. and Bacoyannis, A.,** Effects of reserpine on the oviposition of *Dacus oleae* Gme., *Ann. Epiphyties,* 19, 623, 1968.

82. **Campion, D. G.,** Chemosterilants for *Diparopsis castanea, PANS,* 21, 359, 1975.

83. **Guerra, A. A., Wolfenbarger, D. A., and Luckefahr, M. J.,** Effects of substerilizing doses of reserpine and gamma irradiation on reproduction of the tobacco budworm, *J. Econ. Entomol.,* 64, 804, 1971.

84. **Hendricks, D. E., De la Rosa, H. H., and Guerra, A. A.,** Attractiveness of tobacco budworm females altered by oral chemosterilants and dietary additives, *J. Chem. Ecol.,* 3, 127, 1977.

85. **Hays, S. B.,** Some effects of reserpine, a tranquilizer, on the house fly, *J. Econ. Entomol.,* 58, 782, 1965.

86. **Hays, S. B. and Amerson, G. M.,** Reproduction control in the house fly with reserpine, *J. Econ. Entomol.,* 60, 781, 1967.

87. **Wicht, M. C. and Hays, S. B.,** Effect of reserpine on reproduction of the house fly, *J. Econ. Entomol.,* 60, 36, 1967.

88. **Huot, L. and Corrivault, G. W.,** Neuroleptic substances and insect behavior. V. Comparative study of activities of reserpine and some of its derivatives on *Tribolium confusum, Arch. Int. Physiol. Biochem.,* 75, 745, 1967.

89. **Sukumar, K. and Osmani, Z.,** Insect sterilants from *Catharanthus roseus, Curr. Sci.,* 50, 552, 1981.

90. **Sukumar, K., Parvathi, H. R., and Osmani, Z.,** Antifertility activity of certain antineoplastic agents against *Dysdercus cingulatus* F., *Sci. Cult.,* 47, 395, 1981.

91. **Velcheva, N.,** Biological evaluation of the effect of some chemosterilants on the propagation potential of *Laspeyresia funebrana* Tr. (Torticidae: Lepidoptera), *Gradinar. Lozar. Nauka,* 8, 9, 1981.

92. **Mukherjee, M. C.,** Sterility in *Drosophila melanogaster* consequent on using a mammalian oral contraceptive, *Sci. Cult.,* 27, 497, 1961.

93. **Ascher, K. R. S. and Hirsch, J.,** The effect of *m*-xylohydroquinone on oviposition in the housefly, *Entomol. Exp. Appl.,* 6, 337, 1963.

94. **Ascher, K. R. S.,** Oviposition inhibiting agents: a screening of simple model substances, *Int. Pest Control,* 7, 8, 1965.

95. **Ascher, K. R. S. and Avdat, N.,** Stabilization of aqueous solutions of the insect chemosterilant *m*-xylohydroquinone (*m*-XHQ) by vitamin C, *Experientia,* 23, 679, 1967.

96. **Ascher, K. R. S. and Avdat, N.,** Sterilising the male housefly with *m*-xylohydroquinone. II., *Int. Pest Control,* 9, 8, 1967.

97. **Ascher, K. R. S. and Avdat, N.,** Sterilising the male housefly with *m*-xylohydroquinone, *Int. Pest Control,* 8, 16, 1966.

98. **Rajendran, B. and Gopalan, M.,** Juvenile-hormone-like activity of certain plant extracts on *Dysdercus cingulatus* Fabricius (Heteroptera: Pyrrhocoridae), *Indian J. Agric. Sci.,* 50, 781, 1980.

99. **El-Ibrashy, M. T.,** Sterilization of the Egyptian cotton leafworm *Spodoptera littoralis* Boisd. with foliage extract of *Podocarpus gracilior* P., *Z. Angew. Entomol.,* 75, 107, 1974.

100. **Shankaranarayana, K. H., Ayyar, K. S., and Rao, G. S. K.,** Insect growth inhibitor from the bark of *Santalum album, Phytochemistry,* 19, 1239, 1980.

101. **Shankaranarayana, K. H., Shivaramankrishnan, V. R., Ayyar, K. S., and SenSarma, P. K.,** Isolation of a compound from the bark of sandal, *Santalum album* L., and its activity against some lepidopterans and coleopterous insects, *J. Entomol. Res.,* 3, 116, 1979.

102. **Lang, J. T. and Treece, R. E.,** Sterility and longevity effects of *Sterculia foetida* oil on the face fly, *J. Econ. Entomol.,* 64, 455, 1971.

103. **Beroza, M. and LaBrecque, G. C.**, Chemosterilant activity of oils, especially oil of *Sterculia foetida* in the house fly, *J. Econ. Entomol.*, 6, 196, 1967.
104. **Matsumura, F. and Knight, S. G.**, Toxicity and chemosterilizing activity of aflatoxin against insects, *J. Econ. Entomol.*, 60, 871, 1967.
105. **Moore, J. H., Hammond, A. M., and Llewllyn, G. C.**, Chemosterilant and insecticidal activity of mixed aflatoxins against *Anthonomus grandis* (Coleoptera), *J. Invertebr. Pathol.*, 31, 365, 1978.
106. **Al-Adil, K. M., Kilgore, W. W., and Painter, R. R.**, Toxicity and sterilization effectiveness of aflatoxins B_1 and G_1 and distribution of aflatoxin B_1-^{14}C in house flies, *J. Econ. Entomol.*, 65, 375, 1972.
107. **Kenaga, E. E.**, Some hydroxynitrosamino aliphatic acid derivatives as insect reproduction inhibitors, *J. Econ. Entomol.*, 62, 1006, 1969.
108. **Kratsas, R. G.**, Decrease in fertility, fecundity, and egg size after feeding alanosine to *Phormia regina*, *J. Econ. Entomol.*, 68, 581, 1975.
109. **Harries, F. H.**, Fecundity and mortality of female codling moths treated with novobiocin and other antibiotics, *J. Econ. Entomol.*, 60, 7, 1967.
110. **Barnes, J. R., Fellig, J., and Mitrovic, M.**, The chemosterilant effect of anthramycin methyl ether in *Drosophila melanogaster*, *J. Econ. Entomol.*, 62, 902, 1969.
111. **Borkovec, A. B., Chang, S. C., and Horwitz, S. B.**, Chemosterilization of house flies with anthramycin methyl ether, *J. Econ. Entomol.*, 64, 983, 1971.
112. **Horwitz, S. B., Chang, S. C., Grollman, A. P., and Borkovec, A. B.**, Chemosterilant action of anthramycin: a proposed mechanism, *Science*, 174, 159, 1971.
113. **Lal, O. P. and Schmutterer, H.**, Contact and systemic actions of flavomycin along with some surfactants on the development and reproductivity of *Aphis fabae* Scop. on *Vicia faba* L. plants, *Indian J. Entomol.*, 43, 158, 1981.
114. **Ehrhardt, P., Jayraj, S., and Schmutterer, H.**, The effect of some antibiotics on the development and fertility of *Aphis fabae* and *Vicia faba*, *Entomol. Exp. Appl.*, 9, 332, 1966.
115. **Harries, F. H.**, Control of insects and mites on fruit trees by trunk injection, *J. Econ. Entomol.*, 58, 631, 1965.
116. **Shaw, J. G. and Riviello, M. S.**, Use of chemicals as sexual sterilants for the fruit fly, *Ciencia*, 22, 17, 1962.
117. **Harries, F. H. and Mattson, V. J.**, Effects of some antibiotics on three aphid species, *J. Econ. Entomol.*, 56, 412, 1963.
118. **Mulla, M. S.**, Chemosterilants for control of reproduction in the eye gnat *(Hippelates collusor)* and the mosquito *(Culex quinquefasciatus)*, *Hilgardia*, 39, 297, 1968.
119. **LaBrecque, G. C. and Gouck, H. K.**, Compounds affecting fertility in adult house flies, *J. Econ. Entomol.*, 56, 476, 1963.
120. **Jeppson, L. R., Jesser, M. J., and Complin, J. O.**, Cycloheximide derivatives and mite control, with special reference to mites on citrus, *J. Econ. Entomol.*, 59, 15, 1966.
121. **Harries, F. H.**, Effects of some antibiotics and other compounds on fertility and mortality of orchard mites, *J. Econ. Entomol.*, 56, 438, 1963.
122. **Harries, F. H.**, Effects of certain antibiotics and other compounds on the two-spotted spider mite, *J. Econ. Entomol.*, 54, 122, 1961.
123. **Harries, F. H.**, Reproduction and mortality of the two-spotted spider mite on fruit seedlings treated with chemicals, *J. Econ. Entomol.*, 59, 501, 1966.
124. **Harries, F. H. and Wiles, W. G.**, Tests of some antibiotics and other chemosterilants on the green peach aphid, *J. Econ. Entomol.*, 59, 694, 1966.
125. **Harries, F. H.**, Further studies of effects of antibiotics and other compounds on fecundity and mortality of the two-spotted spider mite, *J. Econ. Entomol.*, 61, 12, 1968.
126. **Datta, S. and Banerjee, P.**, Prostaglandins, cyclic AMP, U-7118, and acetic acid as insect growth regulators and sterilants, *Indian J. Exp. Biol.*, 16, 872, 1978.
127. **Sweeley, C. C., O'Connor, J. D., and Bieber, L. L.**, Effect of polyene macrolides on growth and reproduction of *Musca domestica* and on the uptake of cholesterol in *Galleria mellonella* larvae, *Chem. Biol. Interact.*, 2, 247, 1970.
128. **Schroeder, F. and Bieber, L. L.**, Effects of filipin and cholesterol on housefly, *Musca domestica* L. and wax moth, *Galleria mellonella* L., *Chem. Biol. Interact.*, 4, 239, 1971.
129. **Ivanova, A. N.**, Chemosterilant, U.S.S.R. Patent 544,342, 1978.
130. **Jayraj, S. and Schmutterer, H.**, On the use of certain sulphanilamides against the black bean aphid, *Aphis fabae*, Scop., *Z. Pflanzenkr. Pflanzenschutz*, 73, 660, 1966.
131. **Inagaki, E. and Oster, I. I.**, Changes in the mutational response of silkworm spermatozoa exposed to mono- and polyfunctional alkylating agents following storage, *Mutat. Res.*, 7, 425, 1969.
132. **Smith, R. H.**, Induction of mutations in *Habrobracon* sperm with mitomycin C, *Mutat. Res.*, 7, 231, 1969.
133. **Stefanov, S. G.**, Chemical sterilization of *Chloridea obsoleta*, *Plant Sci. (Sofia)*, 10, 171, 1973.

134. **Painter, R. R. and Kilgore, W. W.**, Temporary and permanent sterilization of house flies with chemosterilants, *J. Econ. Entomol.*, 57, 154, 1964.

135. **Zebit, C. P. W.**, Effect of the chitin synthesis inhibitor nikkomycin (AMS 0896, GT 25/76) on the two-spotted spider mite *Tetranychus urticae* Koch (Acari:Tetranychidae), *Z. Pflanzenkr. Pflanzenschutz*, 90, 89, 1983.

136. **Jayraj, S., Ehrhardt, P., and Schmutterer, H.**, The effect of certain antibiotics on reproduction of the black bean aphid, *Aphis fabae* Scop., *Ann. Appl. Biol.*, 59, 13, 1967.

137. **Kohls, R. E., Lemin, A. J., and O'Connell, P. W.**, New chemosterilants against the house fly, *J. Econ. Entomol.*, 59, 745, 1966.

138. **Lemin, A. J. and O'Connell, P. W.**, Pactamycin insecticide, U.S. Patent 3,272,697, 1966.

139. **O'Connell, P. W.**, Profiromycin insecticide, U.S. Patent 3,272,696, 1966.

140. **Haynes, J. W., Mitlin, N., Davich, T. B., and Sloan, C. E.**, Evaluation of candidate chemosterilants for the boll weevil, *USDA-ARS Prod. Res. Rep.*, 120, 24 pp., 1971.

141. **Boston, M. D., Patterson, R. S., and Lofgren, C. S.**, Screening of chemosterilants against the southern house mosquito, *Culex pipiens quinquefasciatus, Fla. Entomol.*, 53, 215, 1970.

142. **Fye, R. L.**, Screening of chemosterilants against house flies, *J. Econ. Entomol.*, 60, 605, 1967.

143. **Fye, R. L. and LaBrecque, G. C.**, Sterility in house flies offered a choice of untreated diets and diets treated with chemosterilants, *J. Econ. Entomol.*, 60, 1284, 1967.

144. **Fytizas, E. and Tzanakakis, M. E.**, Development of larvae of *Dacus oleae* in olives, after their parents received streptomycin incorporated in their diet, *Ann. Epiphyties*, 17, 53, 1966.

145. **Fytizas, E. and Tzanakakis, M. E.**, Possible control of *Dacus oleae* with antibiotics, *Meded. Rijksfac. Landbouwwet. Gent*, 31, 782, 1966.

146. **Fytizas, E. and Tzanakakis, M. E.**, Some effects of streptomycin, when added to the adult food, on the adults of *Dacus oleae* (Diptera:Tephritidae) and their progeny, *Ann. Entomol. Soc. Am.*, 59, 269, 1966.

147. **Kuppuswamy, N. T., Jayraj, S., and Subramaniam, T. R.**, Sterility inducing effect of some antibiotics and sulphanilamides on the legume aphid, *Aphis craccivora* Koch, *Madras Agric. J.*, 9(Suppl. 57), 24, 1970.

148. **Lemin, A. J. and O'Connell, P. W.**, Sparsomycin insecticide, U.S. Patent 3,272,698, 1966.

149. **Pillai, M. K. K. and Madhukar, B. V. R.**, Effect of biotin on the fertility of the yellow fever mosquito, *Aedes aegypti, Naturwissenschaften*, 56, 218, 1969.

150. **Benschoter, C. A. and Paniagua, R. G.**, Reproduction and longevity of Mexican fruit flies, *Anastrepha ludens* (Diptera:Tephritidae), fed biotin in the diet, *Ann. Entomol. Soc. Am.*, 59, 298, 1966.

151. **Cohen, E. and Levinson, H. Z.**, Disrupted fertility of the hidebeetle *Dermestes maculatus* (Deg.) due to dietary overdosage of biotin, *Experientia*, 24, 367, 1968.

152. **Cohen, E. and Levinson, H. Z.**, Studies on the chemosterilizing effect of biotin on the hidebeetle *Dermestes maculatus* (Dermestidae; Coleoptera), *Comp. Biochem. Physiol.*, 43B, 143, 1972.

153. **Levinson, H. Z. and Cohen, E.**, The action of overdosed biotin on reproduction of the hide beetle, *Dermestes maculatus, J. Insect Physiol.*, 19, 551, 1973.

154. **Guerra, A. A.**, Effect of biologically active substances in the diet on development and reproduction of *Heliothis* spp., *J. Econ. Entomol.*, 63, 1518, 1970.

155. **Benschoter, C. A.**, Effect of dietary biotin on reproduction of the house fly, *J. Econ. Entomol.*, 60, 1326, 1967.

156. **Schooneveld, H. and Abdallah, M. D.**, Effects of insect growth regulators with juvenile hormone activity on metamorphosis, reproduction, and egg fertility of *Adoxophyes orana, J. Econ. Entomol.*, 68, 529, 1975.

157. **Judson, C. L. and De Lumen, H. Z.**, Some effects of juvenile hormone and analogues on ovarian follicles of the mosquito *Aedes aegypti* (Diptera:Culicidae), *J. M l. Entomol.*, 13, 197, 1976.

158. **Morgan, P. B. and LaBrecque, G. C.**, Hormones or hormonelike substances and the development and fertility of house flies, *J. Econ. Entomol.*, 64, 1479, 1971.

159. **Abdallah, M. D., Zaazou, M. A., and El-Tantavi, M.**, Reduction in fecundity of adult females and hatchability of egg larvae of *Spodoptera littoralis* Boisd. after exposure of pupae and eggs to juvenile hormone and analogues, *Z. Angew. Entomol.*, 78, 176, 1975.

160. **Guerra, A. A., Wolfenbarger, D. A., and Garcia, R. D.**, Activity of juvenile hormone analogues against the tobacco budworm, *J. Econ. Entomol.*, 66, 833, 1973.

161. **Whitehead, D. L.**, The effect of phytosterols on tsetse reproduction, *Insect Sci. Appl.*, 1, 281, 1981.

162. **Gouck, H. K., Crystal, M. M., Borkovec, A. B., and Meifert, D. W.**, A comparison of techniques for screening chemosterilants of house flies and screw-worm flies, *J. Econ. Entomol.*, 56, 506, 1963.

163. **LaBrecque, G. C., Adcock, P. H., and Smith, C. N.**, Tests with compounds affecting house fly metabolism, *J. Econ. Entomol.*, 53, 802, 1960.

164. **Craig, G. B., Jr.**, Mosquitoes: Female monogamy induced by male accessory gland substance, *Science*, 156, 1499, 1967.

165. **Craig, G. B., Jr.**, Sterilization of female mosquitoes with a hormone derived from male mosquitoes, U.S. Patent 3,531,566, 1970.

166. **Craig, G. B., Jr. and Fuchs, M.,** Sterilization of the female mosquitoes with a hormone derived from male mosquitoes, U.S. Patent 3,450,816, 1969.

167. **Fuchs, M. S., Craig, G. B., and Hiss, E. A.,** The biochemical basis of female monogamy in mosquitoes. I. Extraction of the active principle from *Aedes aegypti, Life Sci.,* 7(Part 2), 835, 1968.

168. **Lowe, M. L. and Leahy, M. G.,** Isolation and characterization of an extract interfering with insemination in female *A. aegypti, Bull. Entomol. Soc. Am.,* 13, 193, 1967.

169. **Downe, A. E. R.,** The development of procedures for the assessment and interpretation of some methods of autocidal insect control, *Proc. Entomol. Soc. Ontario,* 103, 32, 1972.

170. **Nelson, D. R., Adams, T. S., and Pomonis, J. G.,** Initial studies on the extraction of the active substance inducing monocoitic behavior in house flies, black blow flies, and screw-worm flies, *J. Econ. Entomol.,* 62, 634, 1969.

171. **Riemann, J. G. and Thorson, B. J.,** Effect of male accessory material on oviposition and mating by female house flies, *Ann. Entomol. Soc. Am.,* 62, 828, 1969.

172. **Adams, T. S.,** Effect of the oostatic hormone on egg development and neurosecretion in the house fly, *Musca domestica,* in *Advances in Invertebrate Reproduction,* Adiyodi, K. G. and R. G., Eds., Vol. 1, ISIR, Peralam-Kenoth, Karivellur, India, 1977, 380.

173. **El-Ibrashy, M. T.,** Polyamines as selective chemosterilants for the male cotton leafworm, *Naturwissenschaften,* 58, 148, 1971.

174. **Bhalla, O. P. and Robinson, A. G.,** Effects of chemosterilants and growth regulators on the pea aphid fed an artificial diet, *J. Econ. Entomol.,* 61, 552, 1968.

175. **Sannasi, A.,** Inhibition of ovary development of the fruit-fly, *Drosophila melanogaster* by synthetic "queen substance", *Life Sci.,* 8(Part 2), 785, 1969.

176. **Shaposhnikova, N. G. and Maslennikova, V. A.,** Inhibition of the reproduction of *Drosophila phalerata* Meig. induced with trans-9-keto-2-decenoic acid, *Proc. All Union Sci. Res. Inst. Plant Prot.,* 40, 150, 1974.

177. **Shaposhnikova, N. G.,** Suppression of the process of oviposition in the noxious pentatomid, *Zashch. Rast. (Moscow),* 9, 20, 1978.

178. **Borisova, A. E.,** Effect of synthetic decenoic acids on the life cycle of the codling moth, *Proc. All Union Sci. Res. Inst. Plant Prot.,* 40, 146, 1974.

179. **Nayar, J. K.,** Effect of synthetic "queen substance" (9-oxodec-trans-2-enoic acid) on ovary development in the house fly, *Musca domestica* L., *Nature (London),* 197, 923, 1963.

ALLELOCHEMICS AFFECTING INSECT GROWTH AND DEVELOPMENT*

John C. Reese and Caleb W. Holyoke, Jr.

INTRODUCTION

The field of plant-insect interactions has exploded with information, particularly in the area of the chemical or physiological mechanisms of these interactions. While the effects of plant allelochemics on immediate behavioral responses have received a great deal of attention for many years, only more recently have the more chronic effects on growth and development been studied extensively. This chapter is an attempt to compile much of this growth and development literature. The rapid increase in this one small aspect of the chemical basis of plant-insect interactions is illustrated by the 35 references tabulated by Beck and Reese[1] vs. the more than 100 references tabulated in this chapter. The reader is also referred to recent books edited by Rodriguez,[2] Maxwell and Harris,[3] Jermy,[4] Gilbert and Raven,[5] Wallace and Mansell,[6] Hedin,[7] Labeyrie,[8] Marini-Bettolo,[9] Chapman and Bernays,[10] Rosenthal and Janzen,[11] Warthen,[12] Maxwell and Jennings,[13] and Harris.[14] Sizeable portions of the literature on plant-insect interaction and host plant resistance have also been reviewed by Beck,[15] Beck and Reese,[1] Rosenthal,[16] Russell,[17] Burnett et al.,[18] Bernays and Chapman,[18a] Panda,[19] and Waiss et al.[20] Most recently, the literature has been summarized by Rosenthal[21] and Hedin.[22]

The survival of plants through evolutionary time is " . . . due largely to their own defensive strategies . . . "[23] The study of these defenses, both chemical and physical, has provided insight into the interrelationships between plants and insects, but the picture is extremely complex. For example, allelochemics that inhibit growth in some species may be a nutrient for some specialists. One of these, L-canavanine, a structural analog of L-arginine and an important growth inhibitor (see Table 1a) can be detoxified[24] and even used as a nitrogen source by larvae of *Caryedes brasiliensis* Thunberg.[25] Similarly, *Anacridium melanorhodon* (Walker) survived better and grew faster when certain phenols normally thought of as defensive allelochemics were added to a food plant.[26]

The defensive chemistry of a plant is not a static phenomenon. Leaf quality of oaks can be affected significantly by insect infestations.[27] In some plants, the level of proteinase inhibitors increases rapidly after wounding.[28]

Just as the plant is a dynamic organism, so too is the insect in its ability to tolerate or overcome plant defenses. Some foods act as inducers for detoxifying enzyme systems (see review by Brattsten[29]). Other differences occur in relation to feeding preferences through evolutionary time.[30]

BEHAVIORAL ASPECTS

This chapter is on the postingestive effects of plant allelochemics on growth and development, but the separation of behavioral effects from effects on growth and development is somewhat arbitrary. If an insect does not feed, it cannot grow, but compounds affecting feeding behavior in a choice test may not be strong enough antifeedants to reduce growth. For example, an important antifeedant from *Larix* spp. did not reduce the growth rate of *Pristiphora erichsonii* (Hartig) larvae when they were not given a choice.[31] Behavioral aspects of insect dietetics have been reviewed by Chapman[32] and Hedin et al.[33]

* Contribution 85-56-B from the Kansas Agricultural Experiment Station, Manhattan.

COVERAGE OF CHAPTER

As discussed above, we have for the most part limited the literature tabulated in this chapter to clear demonstrations of the reduction of growth or the lengthening of the time of development. Abnormalities not clearly hormonal in nature have also been included. We have excluded correlational evidence. That is, papers dealing with the relative concentration of compound x in different plant varieties and the relative growth-inhibiting properties of those varieties have not been cited.

Such data are interesting, and with no artificial diet or bioassay, may be the best evidence available. However, they do not demonstrate cause and effect. Compound x must be incorporated into a bioassay system (e.g., an artificial diet) in which the concentration of compound x is the only variable. Even with presumably genetically identical lines that are reported to differ only in the concentration of one compound, the evidence is only suggestive. That one compound must come from precursors and it may be metabolized to certain products. Only with extremely detailed biochemical knowledge could we really be confident that these lines differ with respect to a single compound. Given these difficulties, correlational evidence cannot easily separate the possible effects of compound x from those of its precursors or metabolites, since they too may be especially high or low in various lines.

While very exciting, the recent burst of research activity on the plant compounds that appear to mimic or antagonize insect hormones[33-39] is not tabulated in this chapter. Reviews of this subject have been written by Heftmann[40] and Schoonhoven,[41] and a recent update is included in Wilson's chapter.

For the most part, only compounds with known structures have been included. Biologically active extracts of unknown composition were not included. An exception to this was enzyme inhibitors. Many of these materials have been studied in detail and are fairly well characterized, and yet their amino acid sequence is not known. Related to this, there has been some very interesting work on the induction of enzyme inhibitors after wounding,[28] without actually demonstrating growth reduction in an insect herbivore. It is assumed that a proteinase inhibitor would have deleterious effects on insects.[42,43]

Despite the title of this chapter, we, in fact, limited coverage to plant allelochemics with negative effects. There are a number of cases in which compounds that do not appear to be nutrients nevertheless stimulate growth[26,44] or serve as nutrients in rare cases.[25] There are also examples, such as the sequestration of plant compounds for insect defense, in which there is a presumed cost to the herbivore. This is not always the case; cardenolides ingested by *Oncopeltus fasciatus* (Dallas) imposed no apparent physiological cost or adverse effect on growth and development.[45] We have also limited coverage to nonfungal compounds.

BIOASSAYS

Many of the papers tabulated here represent an attempt to answer questions about a particular plant-insect interaction. Thus, the herbivorous insect in question was used in the bioassay. However, some investigators were searching for biologically active compounds so spice extracts were tested using mosquito larvae, etc. In this type of bioassay, what would be an appropriate concentration? Even in bioassays using ecologically relevant organisms, the bioassays have been performed in so many different ways (topical application, injection, ingestion, etc.), the concentrations used have been reported in so many different units (molarity, percent relative to concentration in plant, percent fresh weight, percent dry weight, amount per unit body weight, etc.), and the physiological state of the bioassay organism has differed so much (sensitive neonates, older larvae, polyphagous species, monophagous species, and nutritional status)[46,47] that the reporting of concentrations at which biological activity was observed in any one bioassay is nearly meaningless in a compilation such as this. Also, the data must be viewed with some caution since nearly every nutrient could be deleterious to an organism if ingested in high enough concentrations,[48] and deleterious materials may stimulate growth if ingested at low enough concentrations.[49] Additional discussion of problems in drawing conclusions from bioassays is contained in Isman and Duffey.[50]

The following tables represent a compilation of the current literature dealing with the effects of plant allelochemics on the growth and development of insects.

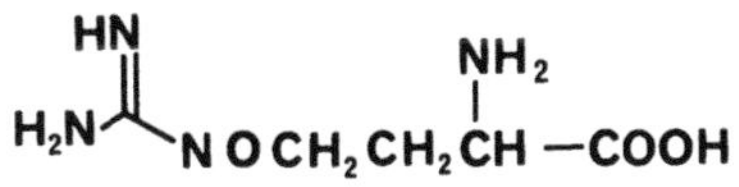

1. L-Canavanine

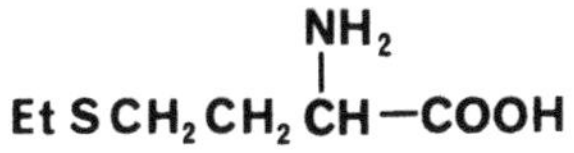

2. Ethionine

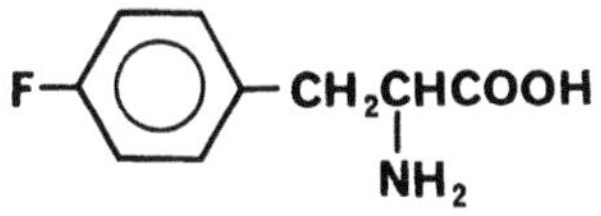

3. 4-Fluorophenyl alanine

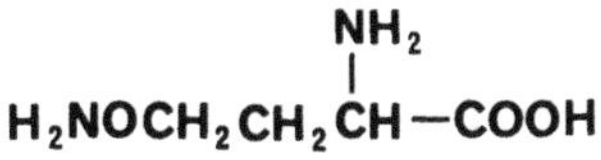

4. Canaline

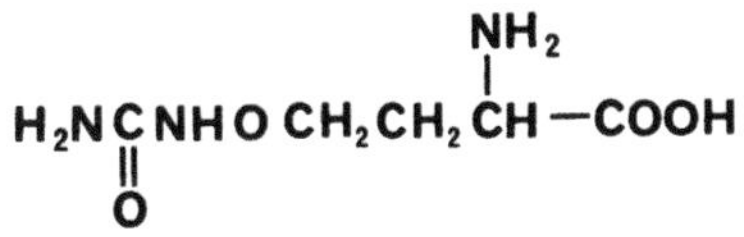

5. *O*-Ureidohomoserine

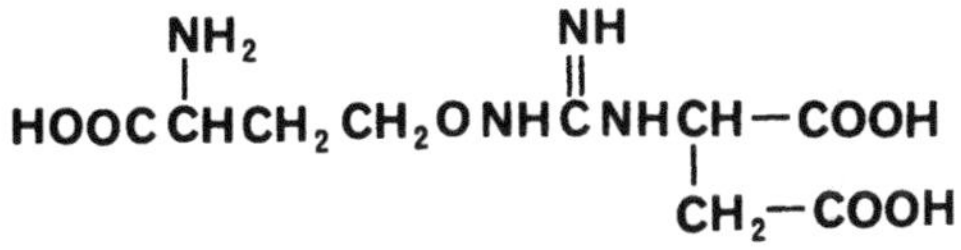

6. Canavanosuccinic acid

7. *S*-(β-Carboxyethyl)-cysteine

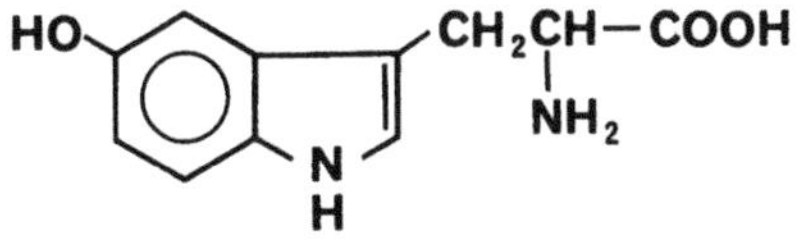

8. *S*-Hydroxytryptophan

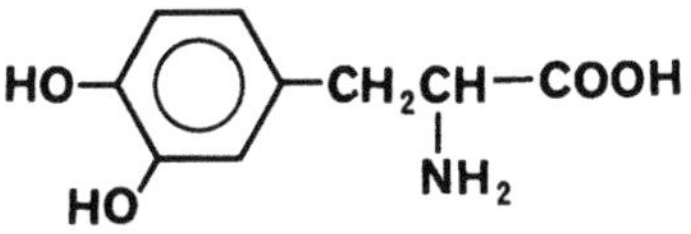

9. L-Dopa[3-(3,4-dihydroxyphenyl)-L-alanine]

Structures for Table 1.

Table 1A
EFFECTS OF AMINO ACID ANALOGS ON INSECT GROWTH AND DEVELOPMENT

Insect	Plant	Compound (conc.) [compound no.]	Effect	Ref.
Manduca sexta (L.)		L-Canavanine (2/25 kg/ha) [1]	Inhibited growth, field	50a
		L-Canavanine (0.5 mg/g body wt) [1]	Inhibited growth when injected	51
		L-Canavanine [1]	Inhibited growth in diet	52
		L-Canavanine (3 mM) [1]	Inhibited growth and caused malformed pupae	53
Tribolium castaneum (Herbst)		Canavanine (sulfate form) (36 mM/kg diet) [1a]	Inhibited growth and arginase activity	54
Pseudosarcophaga affinis (Fallen)		Ethionine (0.07 g/100 mℓ diet) [2]	Inhibited reproduction	55
		p-Fluorophenylalanine (0.4 g/ 100 mℓ diet) [3]	Inhibited reproduction	55
Bombyx mori L.	*Astragalus*	L-Canavanine (5 ppm) [1]	Inhibited metamorphosis	56
Manduca sexta L.		L-Canavanine (5 mM) [1]	Inhibited growth, decreased pupation, and reduced ovarial tissue	57
		L-Canaline (10 mM) [4]	Inhibited growth, decreased pupation, and reduced ovarial tissue	57
		o-Ureido-L-homoserine (5 mM) [5]	Inhibited growth, decreased pupation, and reduced ovarial tissue	57
		L-Canavaninosuccinate (10 mM) [6]	Inhibited growth, decreased pupation, and reduced ovarial tissue	57
Anthonomous grandis Boheman		L-Canavanine (100 mg/100 mℓ diet) [1]	Inhibited growth	58
M. sexta L.		L-Canavanine (2.5 mM) [1]	Inhibited growth, efficiency of conversion of food, and assimilated food	59
Prodenia eridania (Cramer)	*Albizia julibrissin*	S-(β-carboxyethyl)-cysteine [7]	Caused deformed adults	60
	Griffonia simpliciplia	5-Hydroxytryptophan [8]	Caused deformed adults	60
Musca autumnalis DeGeer		L-Canavanine [1]	Reduced pupal weight and some pupal deformation	61
Haematobia irritans L.		L-Canavanine [1]	Reduced pupal weight and some pupal deformation	61
Manduca sexta (L.)		L-Canavanine (0.75 mM) [1]	Reduced fecundity and fertility	62
Agrotis ipsilon (Hufnagel)		L-DOPA (10^{-3} M) (dihydroxyphenylalanine) [9]	Inhibited growth and pupation	63
Spodoptera eridania (Cramer) *Mucuna*		L-DOPA (5%) (dihydroxyphenylalanine) [9]	Abnormal pupae	64

Table 1B

CHEMICAL DATA FOR AMINO ACID ANALOGS, NONPROTEIN AMINO ACIDS, AND AMINES AFFECTING INSECT GROWTH AND DEVELOPMENT

Name [compound no.]	Chem. Abstr. reg. no.	Formula	Physical constants[a]	Synthesis	Comments
L-Canavanine [1]	543-38-4	$C_5H_{12}N_4O_3$	mp 184° $[\alpha]_D^{20} +7.9°$ (H$_2$O)	65 67 68	Distribution review[66]
L-Canavanine sulfate [1a]	2219-31-0	$C_5H_{14}N_4O_7S$	mp (dec) 172° $[\alpha]_D^{17} +19.4°$		
Ethione (S-ethyl-L-homocysteine) [2]	(Racemic) 67-21-0	$C_6H_{13}NO_2S$		69 71	Carcinogenic[70]
4-Fluorophenyl-alanine [3]	(Racemic) 60-17-1 (S) 1132-68-9	$C_9H_{10}FNO_2$	mp 240—243° mp 250—255° $[\alpha]_D^{20} -25.1°$	72 74	Protein synthesis
L-Canaline (O-amino-L-homo serine) [4]	496-93-5	$C_4H_{10}N_2O_3$	mp 214° (dec) $[\alpha]_D^{21} -8.31°$	75 77	Review[76]
O-Ureido-homoserine [5]	1566 18-05-4	$C_5H_{11}N_3O_4$	mp 204°	78	
Canavano-succinate [6]	24764-65-6	$C_9H_{16}N_4O_7$	Not isolated	79	Distribution[80]
S-(β-carboxyethyl)-cysteine [7]	4033-46-9	$C_6H_{11}NO_4S$	mp 218° $[\alpha]_D^{20} -9.33°$	81	
5-Hydroxytryptophan [8]	(Racemic) 56-69-9 (L) 4350-09-8	$C_{11}H_{12}N_2O_3$	mp 298—300° (dec) mp 270° (dec) $[\alpha]_D^{20} -32.5°$ (H$_2$O) $[\alpha]_D^{20} +16.0°$ (4 N HCl)	82 84 85 86	Crystal structure[83]
L-DOPA [9][b]	59-92-7	$C_9H_{11}NO_4$	mp 276—278° dec mp 284—286° $[\alpha]_D^{13} -13.1°$	87 89	

a   °C.
b   Biosynthesis — 88.

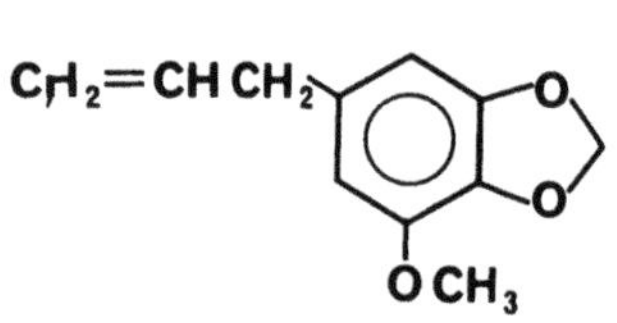

10. Myristicin

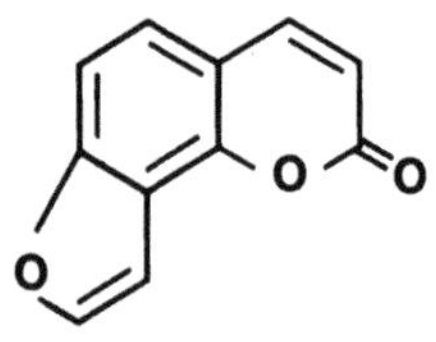

11. Angelicin

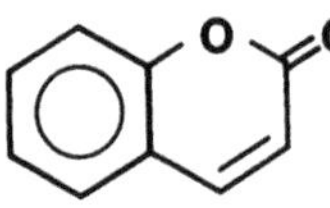

12. Coumarin

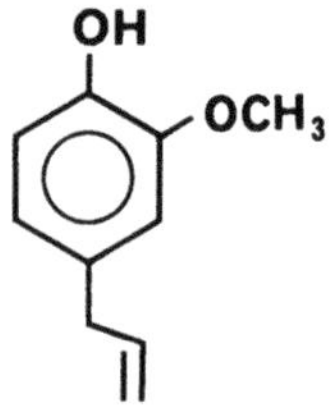

13. Eugenol

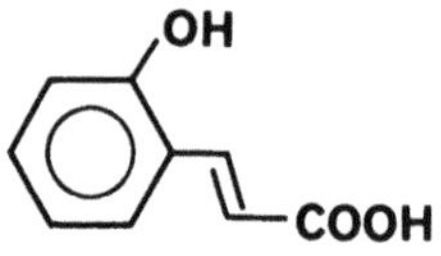

14. *o*-Coumaric acid

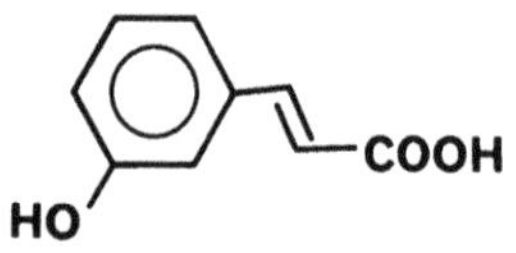

15. *m*-Coumaric acid

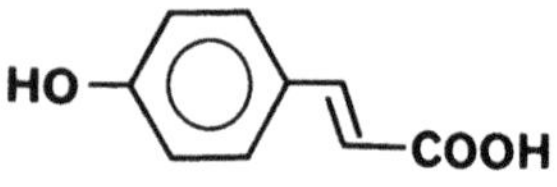

16. *p*-Coumaric acid

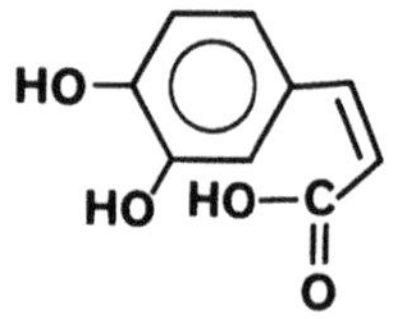

17. *cis*-Caffeic acid

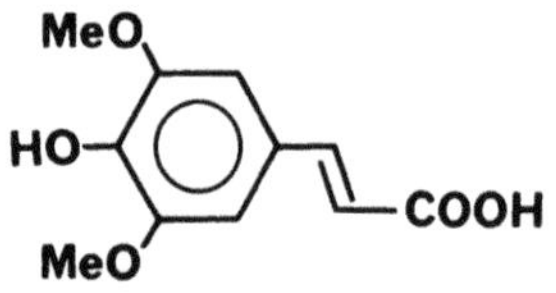

18. Sinapic acid

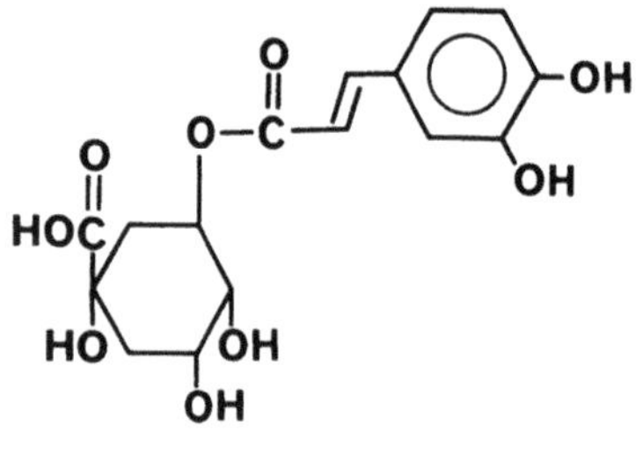

19. Chlorogenic acid

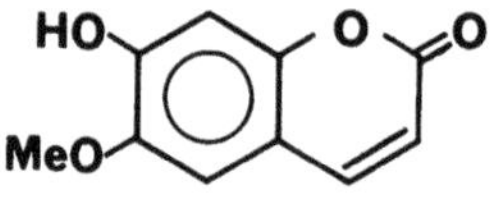

20. Hydrocoumarin

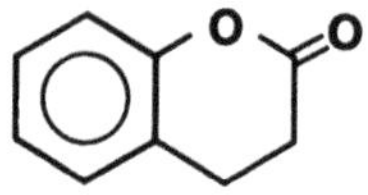

21. Scopoletin

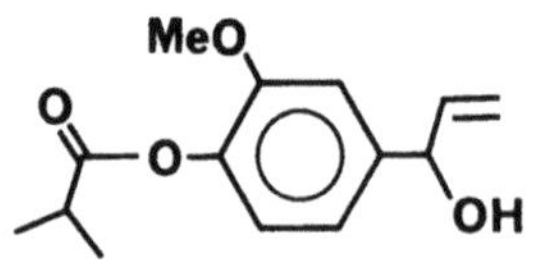

22. 3-(4-isobutyrloxy-3-methoxyphenyl)-1-propen-3-ol

Structures for Table 2.

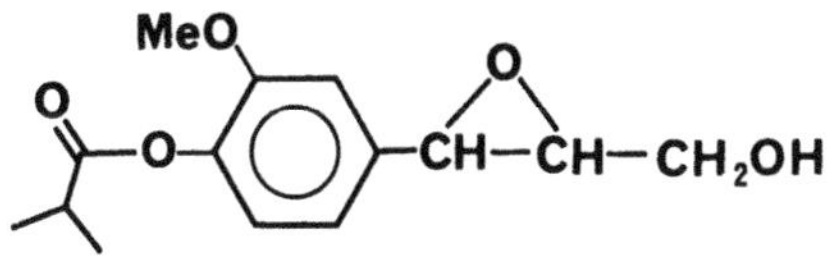

23. 3-(4-isobutyrloxy-3-methoxyphenyl)-2,3-epoxypropan-1-ol

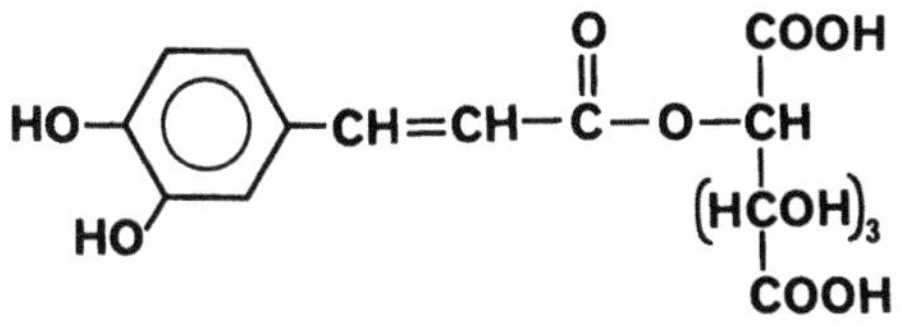

24. Caffeyl aldaric acid

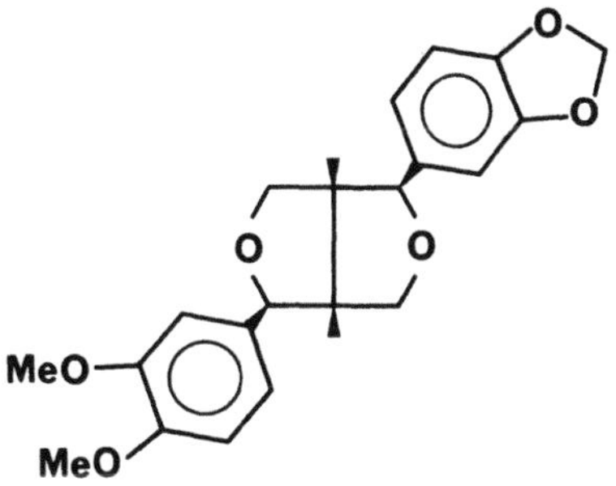

25. Sesamin

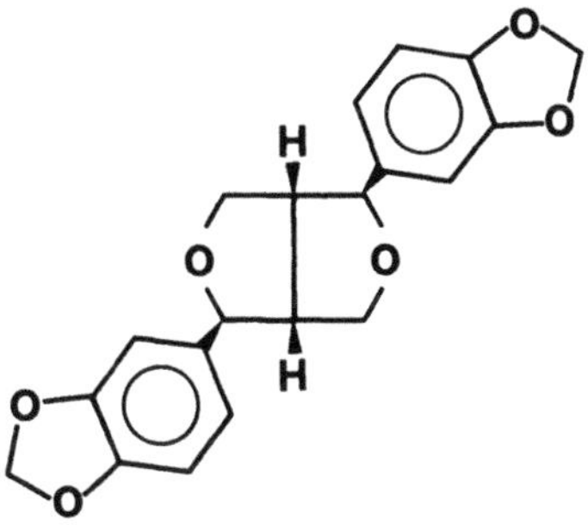

26. Kobusin

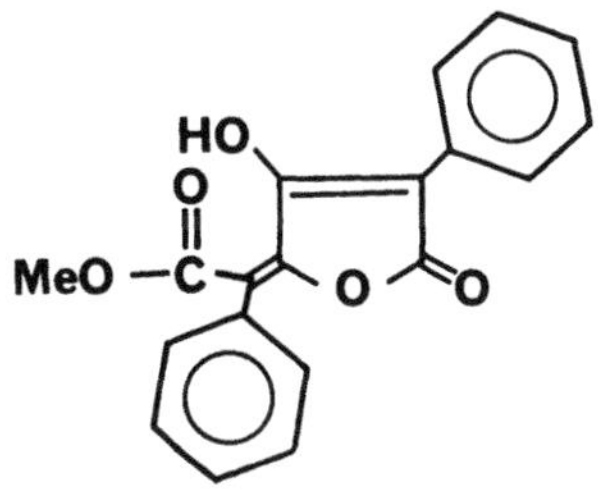

27. Vulpinic acid

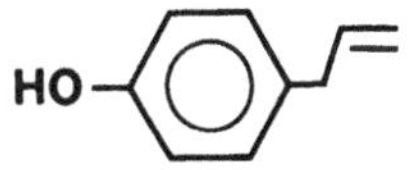

27a. Chavicol

Structures for Table 2 continued.

Table 2A
EFFECTS OF PHENYLPROPANE DERIVATIVES ON INSECT GROWTH AND DEVELOPMENT

Insect	Plant	Compound (conc.) [compound no.]	Effect	Ref.
Bombyx mori L.	*Myristica fragrans* Houtt.	Myristicin [10]	Inhibited growth	90
Papilio polyxenes asterius Stoll	Umbelliferae	Angelicin (0.05% dry wt) [11]	Inhibits growth and fecundity	91
Macrosiphum euphorbiae (Thomas)		Coumarin [12]	Inhibited growth	92
		Coumarin [12]	Prolonged development	92
Aedes aegypti (L.)		Eugenol (0.02 mg/cm^2 water surface) [13]	Inhibited embryonic development	93
Schizaphis graminum (Rondani)		*o*-Coumaric acid (10^{-4} M) [14]	Inhibited growth and reproduction	94
		m-Coumaric acid (10^{-4} M) [15]	Inhibited growth and reproduction	94
		p-Coumaric acid (10^{-4} M) [16]	Inhibited growth and reproduction	94
		cis-Caffeic acid (10^{-4} M) [17]	Inhibited growth and reproduction	94
		Sinapic acid (10^{-4} M) [18]	Inhibited reproduction	94
		Cholorogenic acid (10^{-4} M) [19]	Inhibited growth and reproduction	94
		Hydrocoumarin (10^{-4} M) [20]	Inhibited reproduction	94
		Scopoletin (10^{-4} M) [21]	Inhibited growth and reproduction	94
Drosophila melanogaster Meigen	*Coleopsis lanceolata* L.	3-(4-isobutyryloxy-3-methoxyphenyl)-1-propen-3-ol (1.3 mg/2 g diet) [22]	Inhibited growth	95
	C. lanceolata L.	3-(4-isobutyloxy-3-methoxyphenyl)-2,3-epoxypropan-1-ol (0.6 mg/g diet) [23]	Inhibited growth	95
Heliothis zea (Boddie)	*Lycopersicon*	Chlorogenic acid (2.5 mg/g fresh wt) [19]	Inhibited growth	96
	Lycopersicon	Caffeyl aldaric acid (2.5 mg/g fresh wt) [24]	Inhibited growth	96
Bombyx mori L.	*Magnolia kobus* DC	Sesamin [25]	Inhibited growth	97
	M. kobus DC	Kobusin [26]	Inhibited growth	97
	M. kobus DC	Sesamin (400 ppm) [25]	Inhibited growth	98
	M. kobus DC	Kobusin (400 ppm) [26]	Inhibited growth	98
Spodoptera ornithogalli Quenee	*Letharia vulpina* (L.) Hue	Vulpinic acid (0.6%) [27]	Inhibited growth	99
Drosophila melonogaster Meigen	*Viburnum japonicum* Sprong	Chavicol (1,700 ppm) [27a]	Inhibited growth	100

Table 2B
CHEMICAL DATA FOR PHENYLPROPANE DERIVATIVES AFFECTING INSECT GROWTH AND DEVELOPMENT

Name [compound no.]	Chem. Abstr. reg. no.	Formula	Physical constants[a]	Synthesis	Biosynthesis
Myristicin [10]	607-91-0	$C_{11}H_{12}O_3$	bp 760; 376—377°	101	102
			bp 21; 157°	103	
				104	
Angelicin [11]	523-50-2	$C_{11}H_6O_3$	mp 138—139.5°		105
Coumarin [12]	91-64-5	$C_9H_6O_2$	mp 78—70°	106	107
			bp 760; 297—299°		
			bp 5; 139°		
Eugenol [13]	97-53-0	$C_{10}H_{12}O_2$	bp 255°	108	102
			mp −9.2° to −9.1°		
			n_D^{20} 1.5410		
o-Coumaric acid [14]	583-17-5	$C_9H_8O_3$	mp 207—208° dec	109	110
m-Coumaric acid [15]	588-30-7	$C_9H_8O_3$	mp 191° (194°)	111	
p-Coumaric acid [16]	7400-08-0	$C_9H_8O_3$	mp 210—213°	112	110
Caffeic acid [17]	331-39-5	$C_9H_8O_4$	Monohydrate	113	110
			mp 223—225° (dec)	114	
Sinapic acid [18] (sinapinic acid)	530-59-6	$C_{11}H_{12}O_5$	mp 192°	115	110
Chlorogenic acid [19]	327-97-9	$C_{16}H_{18}O_9$	mp 208°	116	110
			$[\alpha]_D^{26}$ −35.2° (C = 2.8)		
Hydrocoumarin [20]	119-84-6	$C_9H_8O_2$	mp 25°	117	102
			bp 272°	118	
			bp 13; 145°		
Scopoletin [21]	92-61-5	$C_{10}H_8O_4$	mp 204°	119	102
3-(4-Isobnutyryloxy-3-methoxyphenyl)-1-propen-3-ol [22]	75052-10-2	$C_{14}H_{18}O_4$	$[\alpha]_D^{15}$ −23° (EtOH) oil		
3-(4-Isobutyryloxy-3-methoxyphenyl)-2,3-epoxypropan-1-ol [23]	75052-11-8	$C_{14}H_{18}O_5$	$[\alpha]_D^{15}$ −19° (EtOH) oil		
Caffeylaldaric acid [24]	80145-06-8	$C_{15}H_{16}O_{11}$	Not reported		
Sesamin[b] (+)-form [25]	607-80-7	$C_{20}H_{18}O_6$	mp 123—124°	120	
			$[\alpha]_D^{20}$ +68.2° (CHCl₃)		
(−)-form	13079-95-3		mp 122—124°		
			$[\alpha]_D^{17}$ −68° (CHCl₃)		
Kobusin [26]	36150-23-9	$C_{21}H_{22}O_6$	$[\alpha]_D^{27}$ +63.9° (CHCl₃)		
Vulpinic acid [27]	521-52-8	$C_{19}H_{14}O_5$	mp 148—140°		
Chavicol [27a]	501-92-8	$C_9H_{10}O$	mp 15.8°	122	102
			bp 760; 238°		
			bp 16; 123°		
			n_D^{18} 1.5441		

[a]   °C.
[b]   See Reference 121.

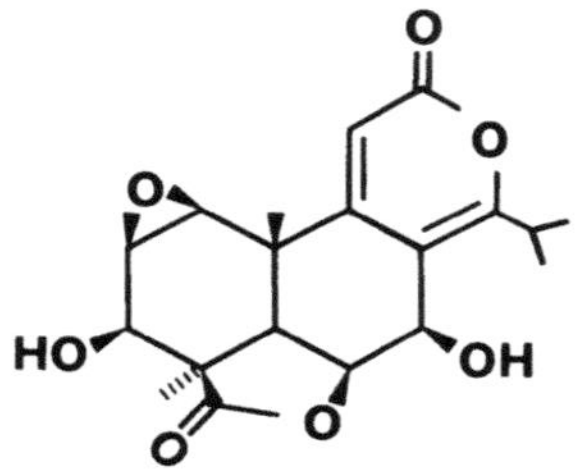

31. Podolactone C

30. Podolactone A

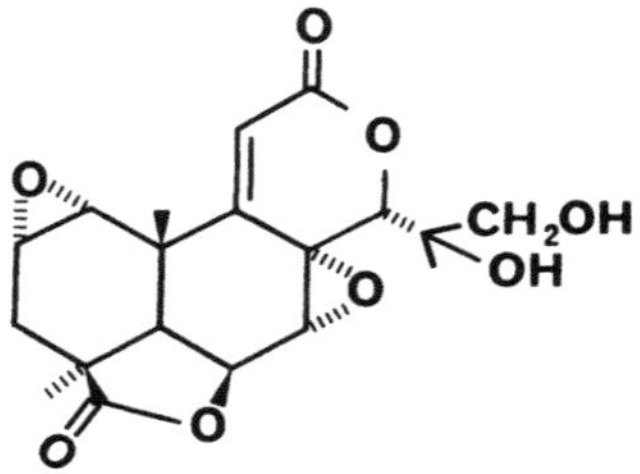

29. Podolactone E

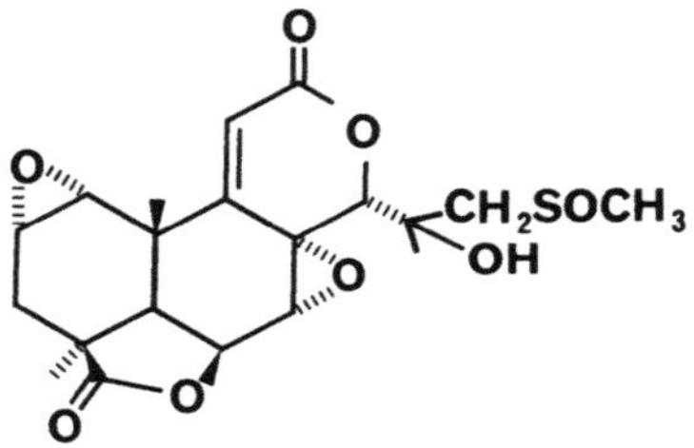

28. Nagilactone C

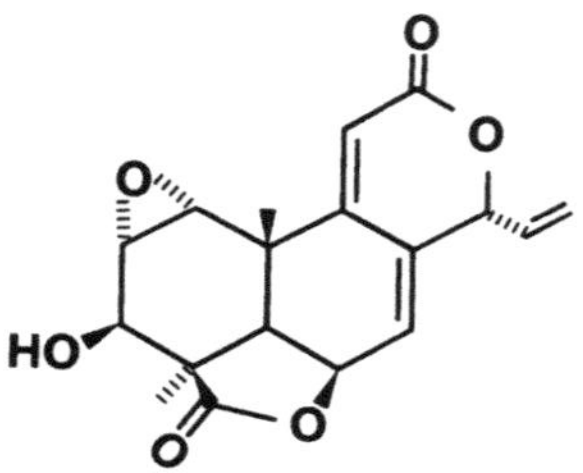

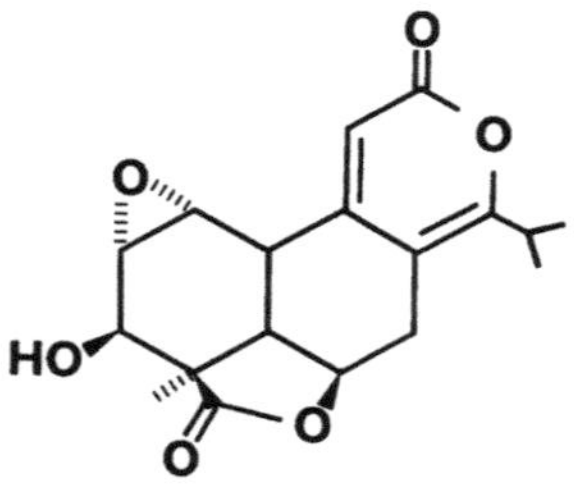

33. Sellowin A

32. Hallactone A

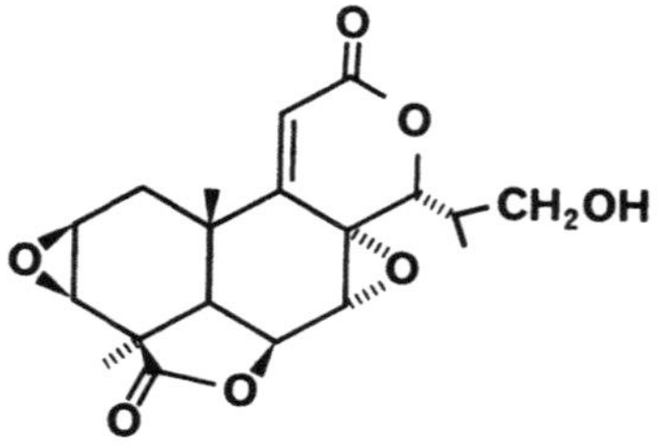

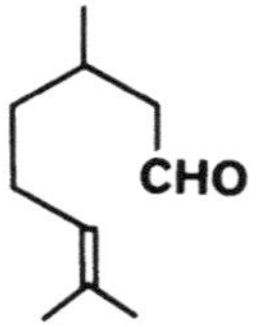

36. Farnesol

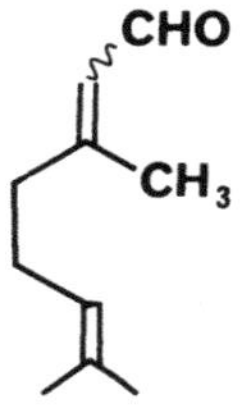

35. Citral; mixture of geranial and neral

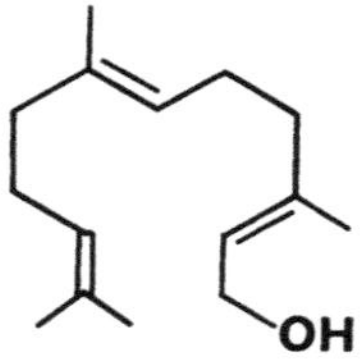

34. Citronellal

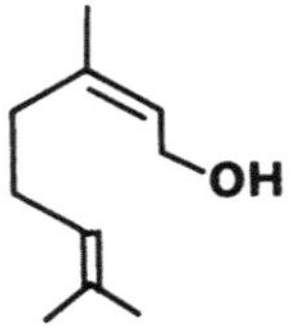

38. (−) Carvone

37. Geraniol

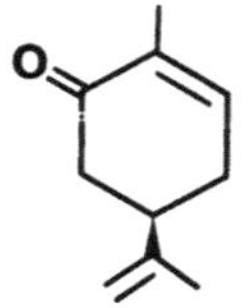

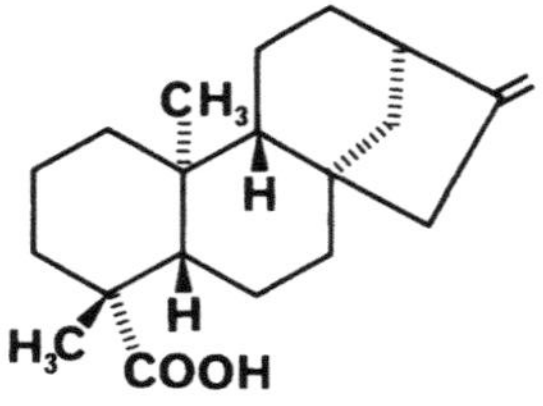

40. (−) Kaur-16-en-19-oic acid

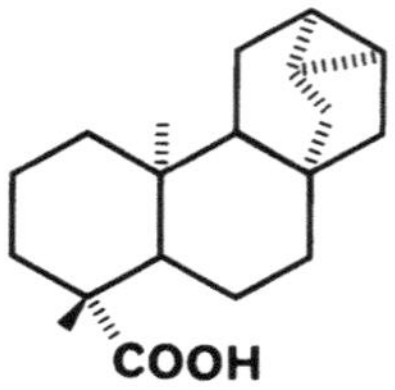

39. Trachyloban–19-oic acid

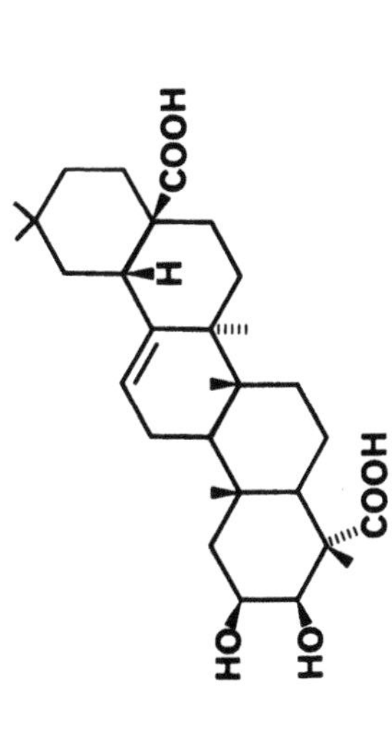

44. Sciadin

50. Caryophyllene

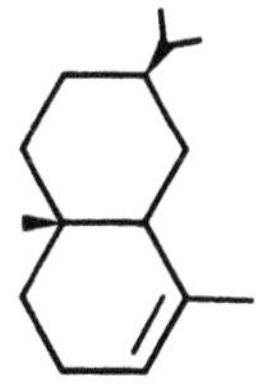

48. α-Selinene

49. β-Selinene

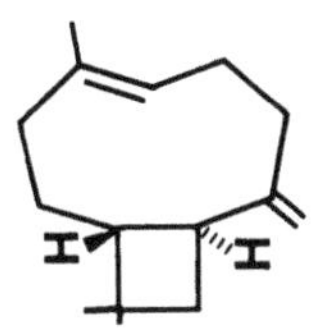

43. Gibberellic acid

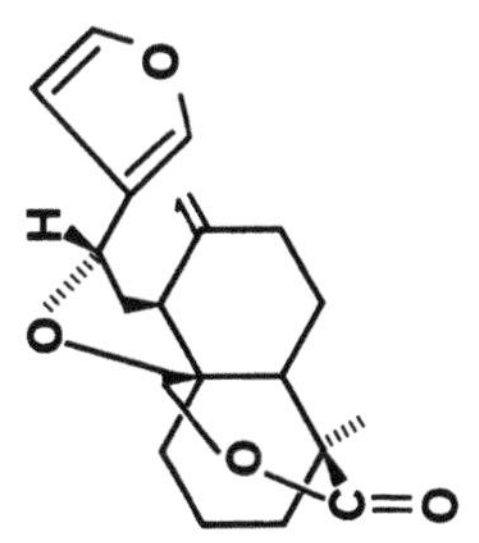

47. Euponin

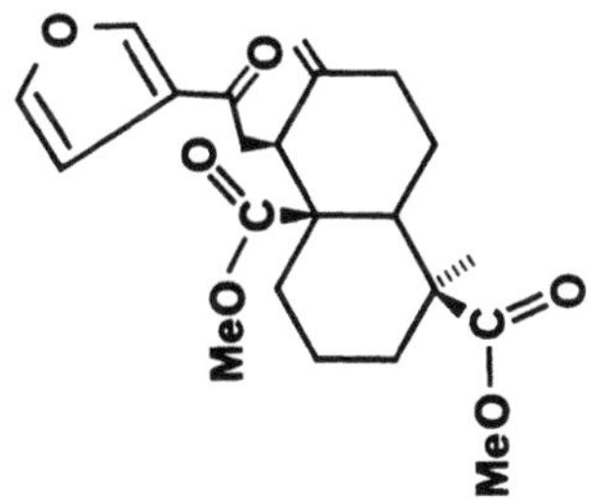

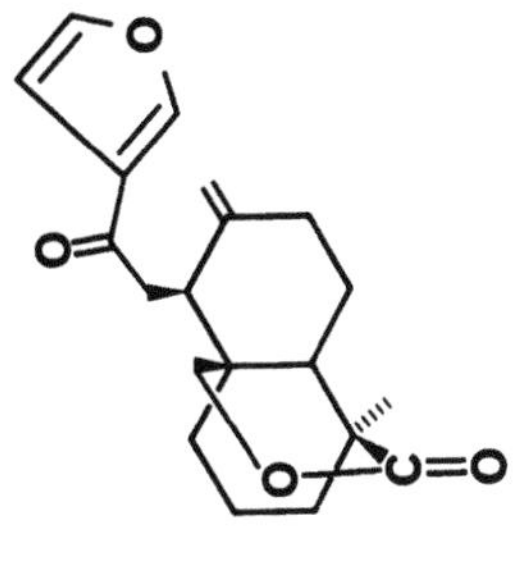

42. Dimethyl Sciadinonate

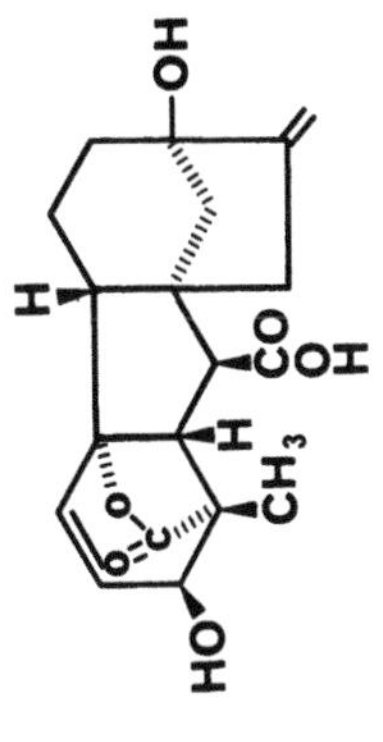

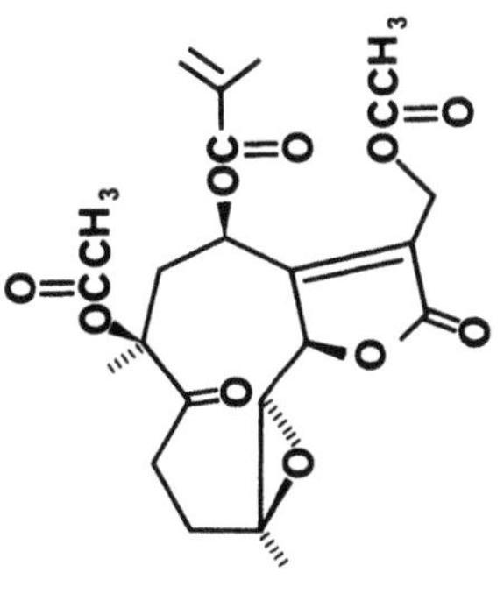

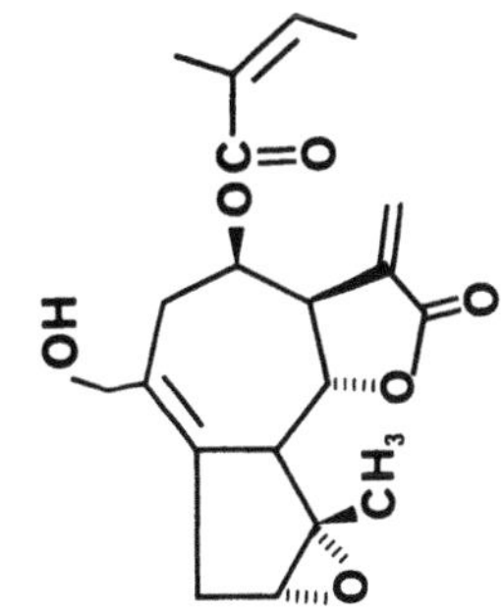

46. Glaucolide A

41. Medicagenic acid

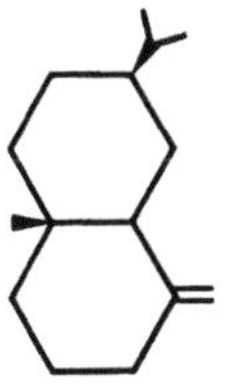

45. Sciadinone

Structures for Table 3.

Table 3A

EFFECTS OF NONSTEROIDAL TERPENOIDS ON INSECT GROWTH AND DEVELOPMENT

Insect	Plant	Compound (conc.) [compound no.]	Effect	Ref.
Musca domestica L.	Podocarpus nivalis Hook	Nagilactone C (12.5 ppm) [28]	Inhibited growth	123
	P. hallii Kirk	Nagilactone C (12.5 ppm) [28]	Inhibited growth	123
	Podocarpus sp.	Podolactone E (5 ppm) [29]	Suppressed development	124
		Podolactone A (25 ppm) [30]	Suppressed development	124
		Podolactone C (12.5 ppm) [31]	Suppressed development	124
		Hallactone A (5 ppm) [32]	Suppressed development	124
		Sellowin A (12.5 ppm) [33]	Suppressed development	124
Aedes aegypti (L.)		Citronellal (0.02 mg/cm² water surface) [34]	Inhibited embryonic development	93
		Citral (0.02 mg/cm² water surface) [35]	Inhibited embryonic development	93
		Farnesol (0.02 mg/cm² water surface) [36]	Inhibited embryonic development	93
		Geraniol (0.02 mg/cm² water surface) [37]	Inhibited embryonic development	93
Spodoptera littoralis Boisduval	Anethum	(−)-Carvone (1%) [38]	Inhibited growth	125
Homoeosoma electellum H.	Helianthus annuus L.	Trachyloban-19-oic acid (0.5%) [39]	Inhibited growth	126
		(−)-Kaur-16-en-19-oic acid (0.5%) (kaurenoic acid) [40]	Inhibited growth	126
Tribolium castaneum (Herbst)	Medicago	Medicagenic acid [41]	Inhibited growth	127
Bombyx mori L.	Persea americana	Dimethyl sciadinonate (100 ppm) [42]	Inhibited growth	128
Pectinophora gossypiella (Saunders)	Helianthus	Kaurenoic acid (0.2%) [40]	Inhibited growth	129
Heliothis virescens (F.)		Gibberellic acid (0.01%) [43]	Inhibited growth	130
B. mori L.	P. americana Mill (avocado)	Dimethylsciadinonate (200 ppm) [42]	Inhibited growth	131
		Sciadin (200 ppm) [44]	Inhibited growth	131
		Sciadinone (200 ppm) [45]	Inhibited growth	131
Spodoptera littoralis Boisduval		Gibberellic acid (0.1%) [43]	Prolonged larval duration	132
Spodoptera eridania (Cramer)	Vernonia	Glaucolide-A (0.125%) [46]	Inhibited growth, reduced pupal weight	133
Spodoptera frugiperda (J. E. Smith)	Vernonia	Glaucolide-A (0.125%) [46]	Inhibited growth	133
Spodoptera ornithogalli	Vernonia	Glaucolide-A (0.125%) [46]	Inhibited growth	133
Drosophila melanogaster Meigen	Eupatorium japonicum	Euponin (2.5 mg/2 g medium) [47]	Inhibited growth	134
Spodoptera exigua (Hubner)	Hymenaea	α- and β-Selinene (1% dry wt) [α = 48, β = 49]	Inhibited growth	135
		Carophyllene (1% dry wt) [50]	Inhibited growth	135

Table 3B
CHEMICAL DATA FOR NONSTEROIDAL TERPENOIDS AFFECTING INSECT GROWTH AND DEVELOPMENT

Name [compound no.]	Chem. Abstr. no.	Formula	Physical constants[a]	Synthesis	Biosynthesis	Comments
Nagilactone C [28]	24338-53-2	$C_{19}H_{22}O_7$	mp 290° (dec) $[\alpha]_D$ +111°			Crystal structure[136]
Podolactone E [29]	37070-59-0	$C_{18}H_{18}O_6$	mp 261—262°			
Podolactone A [30]	26804-81-9	$C_{19}H_{22}O_8$	mp 291—293° (dec)			
Podolactone C [31]	35467-31-3	$C_{20}H_{24}O_8S$	mp 288—290° (dec)			
Hallactone [32]	41787-72-8	$C_{19}H_{22}O_6$	mp 266—268° (dec)			
Sellowin A [33]	34198-79-3	$C_{19}H_{22}O_7$	mp 298° (dec) $[\alpha]_D$ +16° (P_4)			
Citronella [34]	106-22-9	$C_{10}H_{18}O$	bp 47° n^{20} 1.4460 $[\alpha]_D^{25}$ +11.50°		102	
Citral [35]	5392-40-5	$C_{10}H_{16}O$	Geranial bp 26; 92—93° n_D^{20} 1.48982 Neral bp 26; 91—92° n_D^{20} 1.48690	137	102	
Farnesol [36]	4602-84-0	$C_{15}H_{26}O$	*trans-trans* Farnesol liquid bp 0.35; 111° n_D^{15} 0.8871	138 139	102	
Geraniol [37]	106-24-1	$C_{10}H_{12}O$	bp 757; 229—230° bp 12; 114—115° n_D^{20} 1.4766	104	102	
(−)-Carvone [38]	6485-40-1	$C_{10}H_{14}O$	bp 230—231° $[\alpha]_D^{20}$ −62.46° n_D^{20} 1.4988	141	142	Crystal structure[143]
Trachyloban-19-oic acid [39]	26263-39-8	$C_{20}H_{30}O_2$	Methyl ester: mp 98—100° $[\alpha]_D$ −70.5° ($CHCl_3$)			
(−)-Kaur-16-en-19-oic acid [40]	6730-83-2	$C_{20}H_{30}O_2$	mp 179—181° $[\alpha]_D$ −110° ($CHCl_3$)			

Table 3B (continued)
CHEMICAL DATA FOR NONSTEROIDAL TERPENOIDS AFFECTING INSECT GROWTH AND DEVELOPMENT

Name [compound no.]	Chem. Abstr. no.	Formula	Physical constants[a]	Synthesis	Biosynthesis	Comments
Medicagenic acid [41]	599-07-5	$C_{30}H_{46}O_6$	mp 352—353°; $[\alpha]_D^{23} +106°$ (EtOH)	144	145	
Dimethyl sciadiononate [42]	6813-10-1	$C_{22}H_{28}O_6$	mp 122°; $[\alpha]_D -45°$ (CHCl$_3$)			
Gibberellic acid [43]	77-06-5	$C_{19}H_{22}O_6$	mp 233—235°; $[\alpha]_D^{19} +86°$	146		Review[147]
Sciadin [44]	6813-08-7	$C_{20}H_{24}O_4$	mp 160°; $[\alpha]_D +10.3°$ (CHCl$_3$)			
Sciadinone [45]	6704-58-1	$C_{20}H_{24}O_4$	mp 207°; $[\alpha]_D -59.9$ (CHCl$_3$)			
Glaucolide A [46]	11091-29-5	$C_{23}H_{28}O_{10}$	mp 153—154°; $[\alpha]_D^{25} -29°$ (CHCl$_3$)			Crystal structure[148,149]
Euponin [47]	70469-59-9	$C_{20}H_{24}O_6$	mp 148—150°; $[\alpha]_D^{24} -69°$ (EtOH)			
α-Selinene [48]	43-13-2	$C_{15}H_{24}$	bp 268—270°; $[\alpha]_D +31°$	150		
β-Selinene (+)-form [49]	17066-67-0	$C_{15}H_{24}$	bp 6; 121—122°; $[\alpha]_D +61°$	151		
(−)-form	473-12-1		bp 1-2; 70—115°; $[\alpha]_D -46°$ (CHCl$_3$)	152		
β-Caryophyllene [50]	87-44-5	$C_{15}H_{24}$	bp 14; 129—130°; $[\alpha]_D^{15} -5.2°$; n_D^{17} 1.5009	153	154	Review[155]

[a] °C.

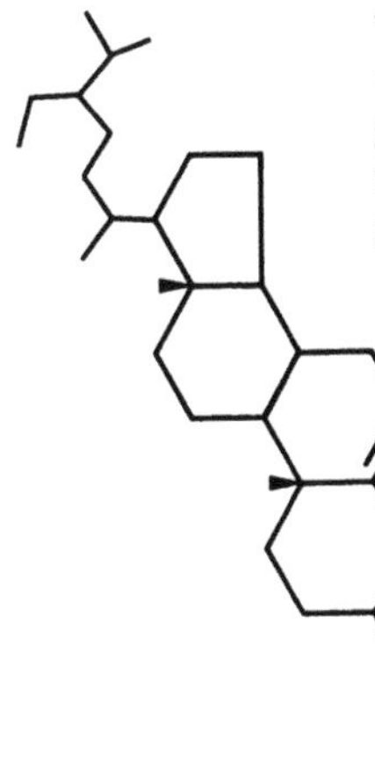

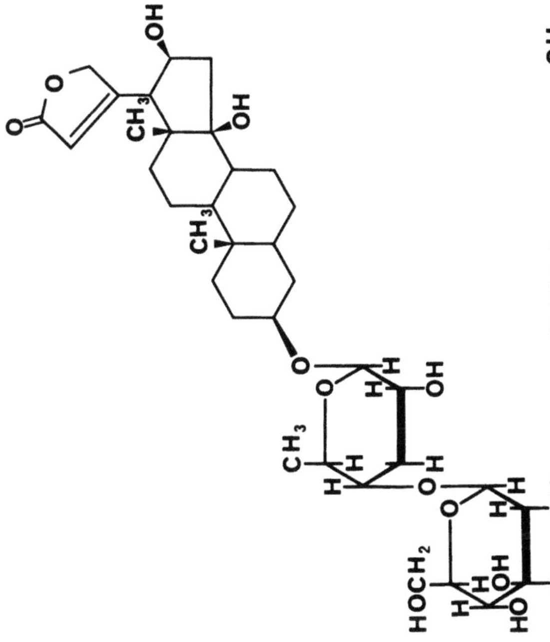

54. Digitalin

55. Digitoxin

51. β-Sitosterol

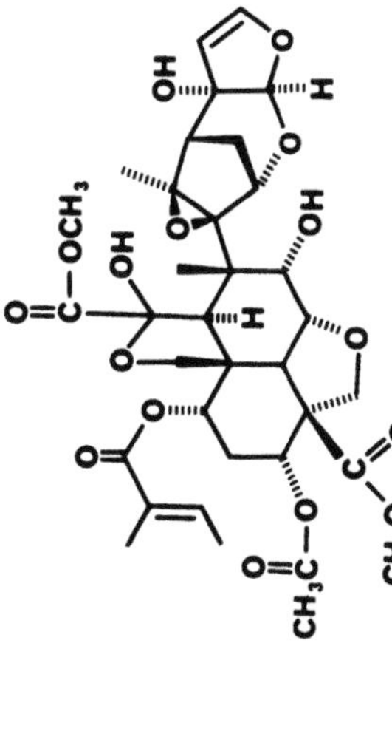

52. Azadirachtin

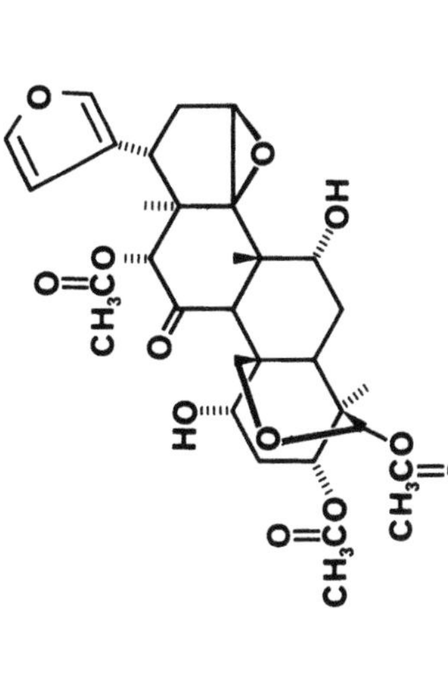

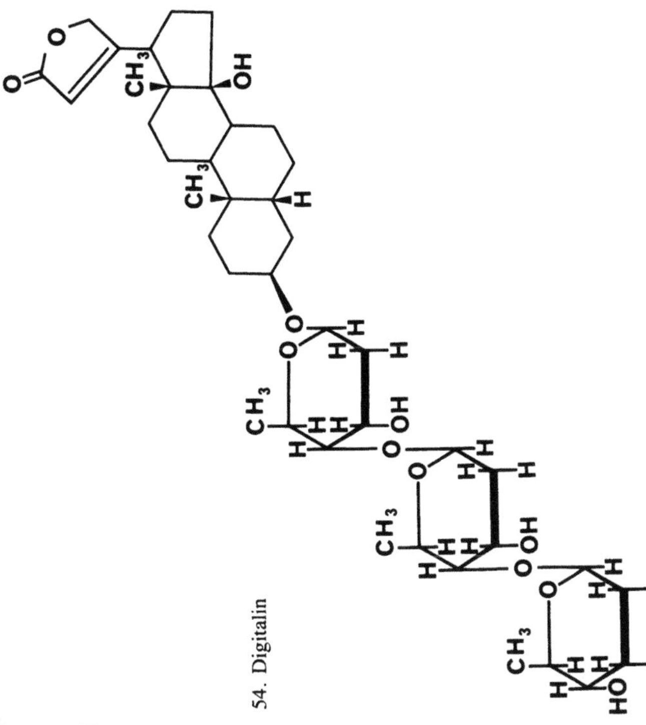

53. Sendanin

Structures for Table 4.

Table 4A

EFFECTS OF STEROIDAL TRITERPENOIDS ON INSECT GROWTH AND DEVELOPMENT

Insect	Plant	Compound (conc.) [compound no.]	Effect	Ref.
Heliothis zea (Boddie)		β-Sitosterol (0.01%) [51]	Inhibited growth	130
Spodoptera littoralis Boisduval		β-Sitosterol (0.1%) [51]	Prolonged larval duration	132
Epilachna varivestis Muls.	*Azadirachta indica* A. Juss.	Azadirachtin (25 ppm) [52]	Disruption of larval-pupal ecdysis	156
Acheta domesticus (L.)	*Azadirachta indica* A. Juss.	Azadirachtin (1 ppm) [52]	Reduced growth rate and increased development time	157
Plutella xylostella (L.)	*Azadirachta indica* A. Juss.	Azadirachtin (0.025 mg) [52]	Reduced growth rate	158
Pieris brassicae	*Azadirachta indica* A. Juss.	Azadirachtin (0.025 mg) [52]	Reduced growth rate	158
Heliothis virescens (Fabricius)	*Azadirachta indica* A. Juss.	Azadirachtin (0.025 mg) [52]	Reduced growth rate	158
Dysdercus fascistus	*Azadirachta indica* A. Juss.	Azadirachtin (0.025 mg) [52]	Reduced growth rate	158
S. frugiperda (J. E. Smith)	*Azadirachta indica* A. Juss.	Azadirachtin (0.2 ppm) [52]	Reduced growth rate	159
Oncopeltus fasciatus Dallas	*Azadirachta indica* A. Juss.	Azadirachtin (0.2 ppm) [52]	Inhibited molting	159
		Azadirachtin (0.355 μg/insect) [52]	Inhibited molting	160
Pectinophora gossypiella (Saunders)	*Trichilia roka*	Sendanin (9 ppm) [53]	Reduced growth rate	161
Heliothis zea (Boddie)	*Trichilia roka*	Sendanin (55 ppm) [53]	Reduced growth rate	161
H. virescens (Fabricius)		Sendanin (60 ppm) [53]	Reduced growth rate	161
S. frugiperda (J. E. Smith)		Sendanin (11 ppm) [53]	Reduced growth rate	161
Bombyx mori L.	*Digitalis purpurea*	Digitalin (100 ppm) [54]	Atonic effects	98
		Digitoxin (25 ppm) [55]	Atonic effects	98

Table 4B

CHEMICAL DATA FOR STEROIDAL TRITERPENOIDS AFFECTING INSECT GROWTH AND DEVELOPMENT

Name [compound no.]	Chem. Abstr. reg. no.	Formula	Physical constants[a]	Synthesis	Biosynthesis
β-Sitosterol [51]	83-46-5	$C_{29}H_{50}O$	mp 140° $[\alpha]_D^{25} -37°$	162	102 163
Azadirachtin [52]	11141-17-6	$C_{35}H_{44}O_{16}$	mp 155—158° $[\alpha]_D^{20} -53°$		
Sendanin [53]	62078-28-8	$C_{32}H_{40}O_{12}$	Cryst. mp 251—252° $[\alpha]_D^{16} +4.3°$		
Digitalin [54]	752-61-4	$C_{36}H_{56}O_{14}$	mp 240—243° $[\alpha]_D^{20} -1.1°$ (MeOH)		102
Digitoxin [55]	71-63-6	$C_{41}H_{64}O_{13}$	mp 256—257° $[\alpha]_D^{20} +4.8°$ (dioxane)		

[a]   °C.

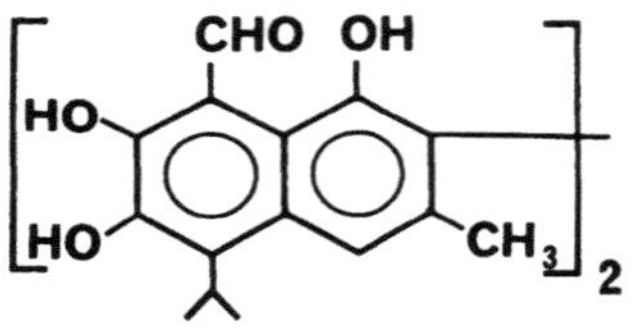

56. Gossypol

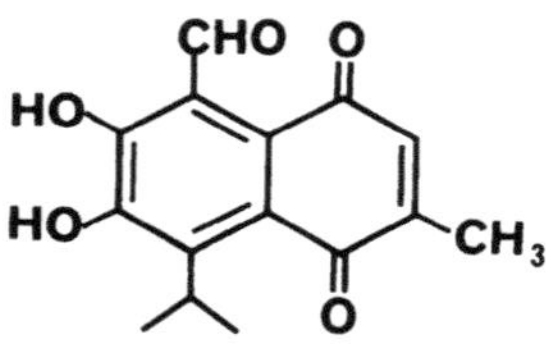

57. Hemigossypolone

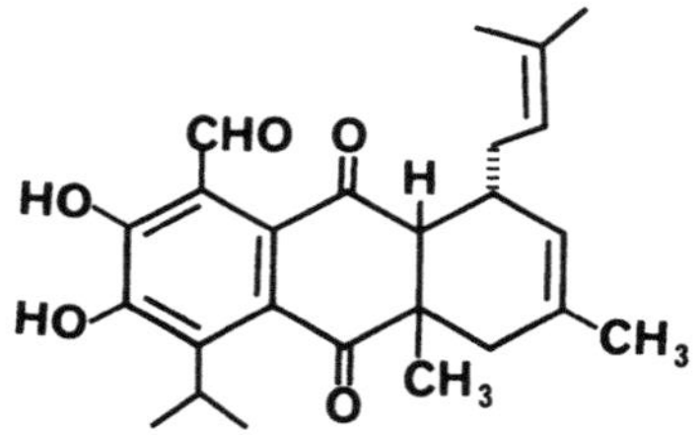

58. Heliocide H₁

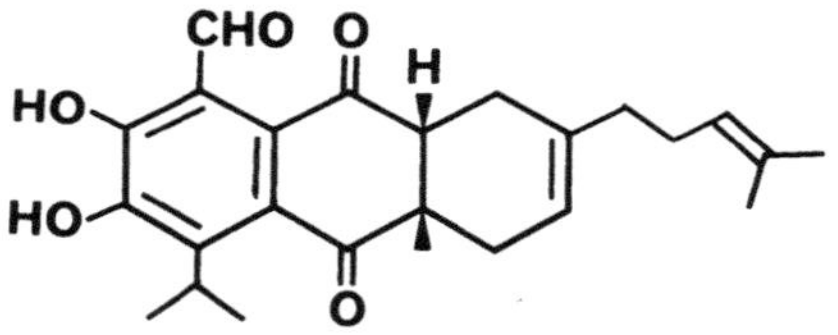

59. Heliocide H₂

Table 5A
EFFECTS OF GOSSYPOL AND RELATED COMPOUNDS ON INSECT GROWTH AND DEVELOPMENT

Insect	Plant	Compound (conc.) [compound no.]	Effect	Ref.
Spodoptera exigua (Hubner)	*Gossypium hirsutum*	Gossypol (0.1%) [56]	Inhibited growth	164
Heliothis zea (Boddie)	*G. hirsutum*	Gossypol (0.1%) [56]	Inhibited growth	164
	G. hirsutum	Gossypol (0.2%) [56]	Inhibited growth	165
Trichoplusia ni (Hubner)	*G. hirsutum*	Gossypol (0.1%) [56]	Inhibited growth	164
Estigmene acrea (Drury)	*G. hirsutum*	Gossypol (0.025%) [56]	Inhibited growth	164
S. littoralis Boisduval	*G. hirsutum*	Gossypol [56]	Inhibited development and certain enzymes	166
H. virescens (F.)	*G. hirsutum*	Gossypol (0.1%) [56]	Inhibited growth	165
Anthonomous grandis Boheman	*G. hirsutum*	Gossypol (2.0%) [56]	Reduced weight of adults at higher conc.	167
H. zea (Boddie)	*G. hirsutum*	Gossypol (0.15%) [56]	Inhibited growth	168
H. virescens (F.)	*G. hirsutum*	Gossypol (0.15%) [56]	Inhibited growth	168
H. zea (Boddie)	*G. hirsutum*	Gossypol (0.2%) [56]	Inhibited pupal weight and development time	169
H. virescens (F.)	*G. hirsutum*	Gossypol (0.2%) [56]	Inhibited pupal weight and development time	169
Pectinophora gossypiella (Saunders)	*G. hirsutum*	Gossypol (0.1%) [56]	Inhibited pupal weight and development time	169
H. virescens (F.)	*G. hirsutum*	Gossypol [56]	Inhibited growth	170

Species	Source	Compound (dose)	Effect	Ref.
H. punctigera Wallengren	G. hirsutum	Gossypol (0.95%) [56]	Inhibited larval growth, decreased pupal and adult weights, and increased larval development time	171
H. amigera (Hubner)	G. hirsutum	Gossypol (1.0%) [56]	Inhibited larval growth, decreased pupal and adult weights, and increased larval development time	171
H. virescens (F.)	G. hirsutum	Hemigossypolone (212 µmol) [57]	Inhibited growth	172
		Heliocide H$_1$ (49 µmol) [58]	Inhibited growth	172
		Gossypol (17 µmol) [56]	Inhibited growth	172
P. gossypiela (Saunders)	G. hirsutum	Hemigossypolone (28 µmol) [57]	Inhibited growth	172
		Heliocide H$_1$ (16 µmol) [58]	Inhibited growth	172
		Heliocide H$_2$ (20 µmol) [59]	Inhibited growth	172
		Gossypol (18.7 µmol) [56]	Inhibited growth	172
Earias vittella (F.)	G. hirsutum	Gossypol (1.0%) [56]	Inhibited growth and pupal weight	173
H. virescens (F.)	G. hirsutum	Gossypol (2 mM/kg) [56]	Inhibited growth	174
		Heliocide$_1$ (3 mM/kg) [58]	Inhibited growth	174
		Heliocide$_2$ (3 mM/kg) [59]	Inhibited growth	174
		Hemigossypolene (3 mM/kg) [57]	Inhibited growth	174
Pectinophora gossypiella (Saunders)	G. hirsutum	Gossypol (1 mM/kg) [56]	Inhibited growth	174
		Heliocide$_1$ (2 mM/kg) [58]	Inhibited growth	174
		Heliocide$_2$ (2 mM/kg) [59]	Inhibited growth	174
		Hemigossypolone (3 mM/kg) [57]	Inhibited growth	174

Table 5B

CHEMICAL DATA FOR GOSSYPOL AND RELATED COMPOUNDS AFFECTING INSECT GROWTH AND DEVELOPMENT

Name	Chem. Abstr. no.	Formula	Physical constants[a]	Synthesis	Biosynthesis	Ref.
Gossypol [56]	303-45-7	C$_{30}$H$_{30}$O$_8$	mp 184° (ether) mp 199° (CHCl$_3$)	175 177	102	176
p-Hemigossypolone [57]	3568-47-2	C$_{15}$H$_{14}$O$_5$	mp 166.5—169°			
Heliocide H$_1$ [58]	65024-84-2	C$_{25}$H$_{30}$O$_5$				
Heliocide H$_2$ [59]	63525-06-4	C$_{25}$H$_{30}$O$_5$				

[a] °C.

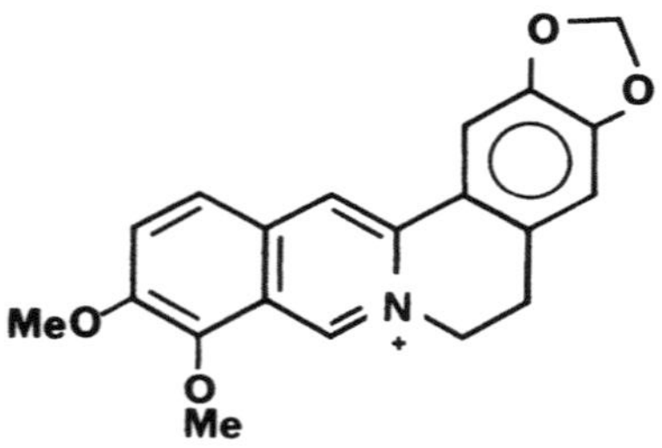

63. Nicotine

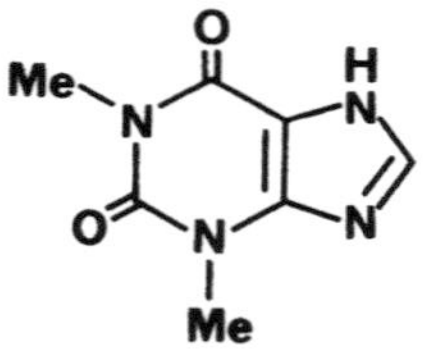

62. Vinblastine

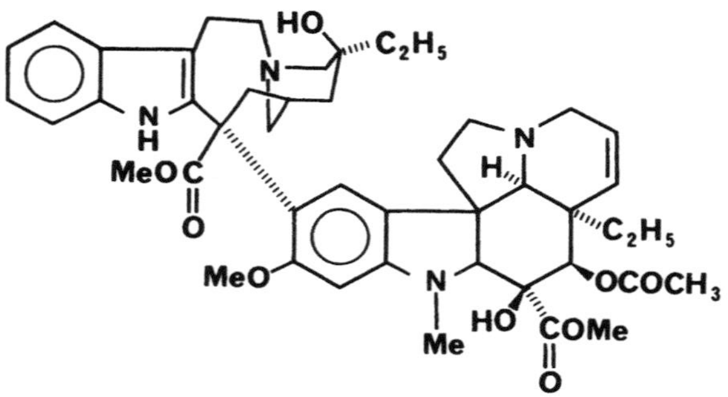

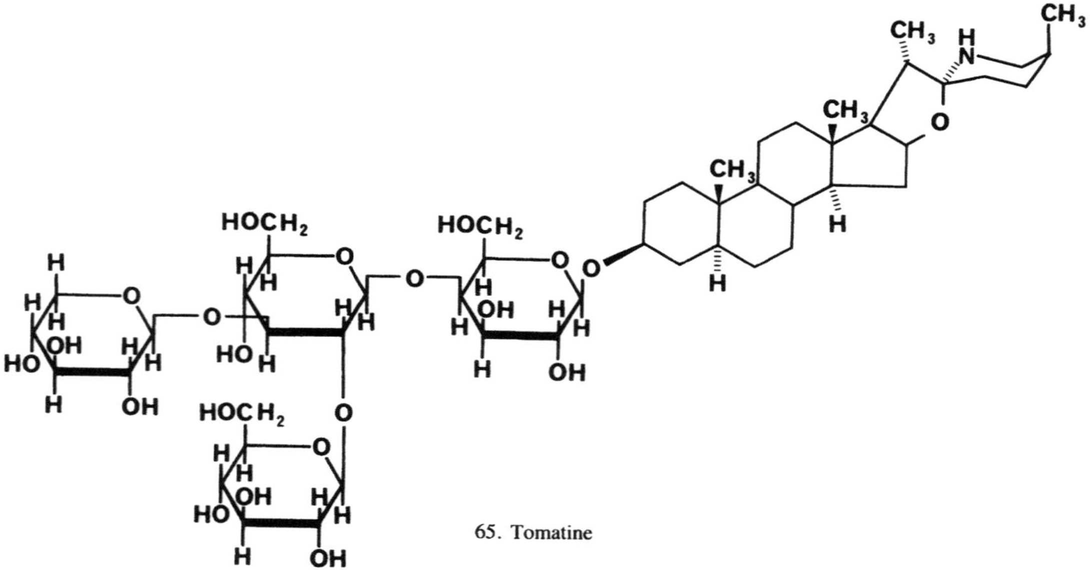

65. Tomatine

61. Caffeine

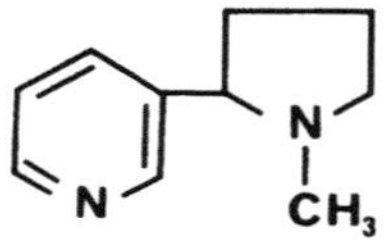

60. Berberine

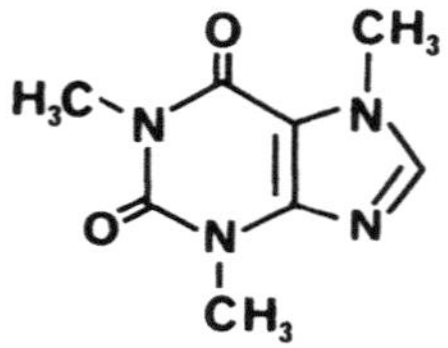

64. Theophylline

Structures for Table 6.

Table 6A

EFFECTS OF ALKALOIDS ON INSECT GROWTH AND DEVELOPMENT

Insect	Plant	Compound (conc.) [compound no.]	Effect	Ref.
Bombyx mori L.	*Phellondendron*	Berberine [60]	Inhibited growth	90
Hyalophora cecropia (L.)		Caffeine (1 mg/g live wt of insect) [61]	Inhibited adult development	177a
Musca domestica L.		Vinblastine (0.1) [62]	Inhibited incorporation of glycine into DNA	178
Manduca sexta (L.)		Nicotine (2%) [63]	Inhibited growth and pupal weight	179
Danaus chrysippus L.		Theophylline (0.1% [64]	Reduced assimilation efficiency of conversion, inhibited metamorphosis	180
		Caffeine (0.1%) [61]	Reduced efficiency of conversion, inhibited metamorphosis	180
Hypogaster exiguae (Viereck) (fed on *Heliothis zea* larvae that had been fed tomatine)	*Lycopersicon*	Tomatine (0.3%) [65]	Lengthened larval period, reduced pupal eclosion and adult weight	181
H. exiquae (Viereck) (fed on *Heliothis zea* larvae that had ingested tomatine)	*Lycopersicon*	Tomatine (20 mol/g dry wt of diet) [65]	Prolonged larval development, disrupted pupal eclosion, caused antennal, abdominal, and genital deformation, and reduced adult weight and longevity	46
Heliothis zea (Boddie)	*Lycopersicon*	Tomatine (0.4 mg/kg)	Inhibited growth	96
Rhodnius prolixus Stal.		Vinblastine [62]	Caused disorganization of microtubular organization of ovary	182

Table 6B

CHEMICAL DATA FOR ALKALOIDS AFFECTING INSECT GROWTH AND DEVELOPMENT

Name [compound no.]	Chem. Abstr. reg. no.	Formula	Physical constants[a]	Synthesis	Biosynthesis	Comments
Berberine [60] (umbellatine)	633-65-8 2086-83-1	$C_{20}H_{18}NO_4$	mp 145°	183	184	Review[185]
Caffeine [61]	58-08-2	$C_8H_{10}N_4O_2$	mp 238°	186 189	187	Crystal structure[188]
Vinblastine [62]	865-21-4	$C_{46}H_{58}N_4O_9$	mp 211—216° $[\alpha]_D^{26}$ +42°		190	
Nicotine [63]	54-11-5	$C_{10}H_{14}N_2$	bp 17; 123—125° n_D^{20} 1.5282 $[\alpha]_D^{20}$ − 169°	191 194	192	Review[193]
Theophylline [64]	58-55-9	$C_7H_8N_4O_2$	Monohydrate mp 270—274°	195 196		
Tomatine [65]	17406-45-0	$C_{50}H_{83}NO_{21}$	mp 263—268° $[\alpha]_D^{20}$ − 18° (Pyridine)		154 (Aglycone)	

[a] °C.

Table 7A
EFFECTS OF TANNINS AND LIGNINS ON INSECT GROWTH AND DEVELOPMENT[a]

Insect	Plant	Compound (conc.)	Effect	Ref.
Heliothis virescens F.	*Gossypium hirsutum*	Condensed tannin (0.2%)	Inhibited growth	197
Operophtera brumata (L.)	*Quercus*	Tannins (11.0%)	Inhibited growth	198
Anthonomous grandis Boheman	*G. hirsutum*	Tannin (0.2%)	Reduced adult weight	167
Chortoicetes terminfera (Walker)		Quebracho (condensed tannin) (10%)	At very high concentration, inhibited growth	199
Locusta migratoria migratorioides (R. & F.)		Quebracho (condensed tannin) (10%)	At very high concentration, inhibited growth and efficiency of conversion of assimilated food	199
Schistocerca gregaria (Forskal)		Quebracho (condensed tannin) (12%)	At very high concentration, inhibited growth and efficiency of conversion of assimilated food	199
Zonocercus variegatus (L.)		Quebracho (condensed tannin) (13%)	At very high concentration, inhibited growth and efficiency of conversion of assimilated food	199
H. virescens F.	*G. hirsutum*	Condensed tannin (0.1%)	Inhibited growth, especially in neonate larvae	200
Drosophila funebris (Fabricius)		Tannic acid (1 ppm)	Induced cytoplasmic alterations in the larval salivary gland cells and nucleolar	201
Schizaphia graminum (Rondani)		Tannic acid (10^{-4})	Inhibited growth and reproduction hypertrophy	94

[a] No chemical data table for these compounds since many are large molecules and not well defined chemically.

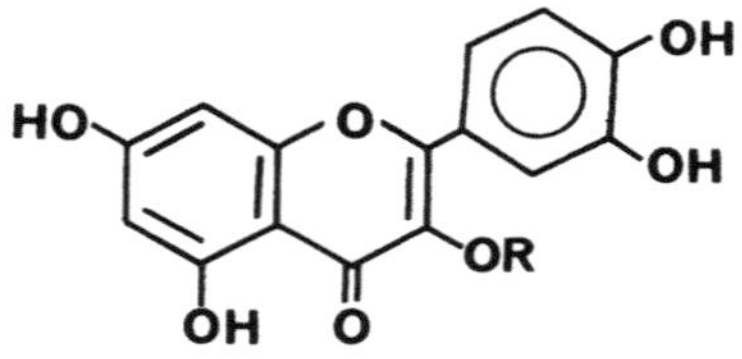

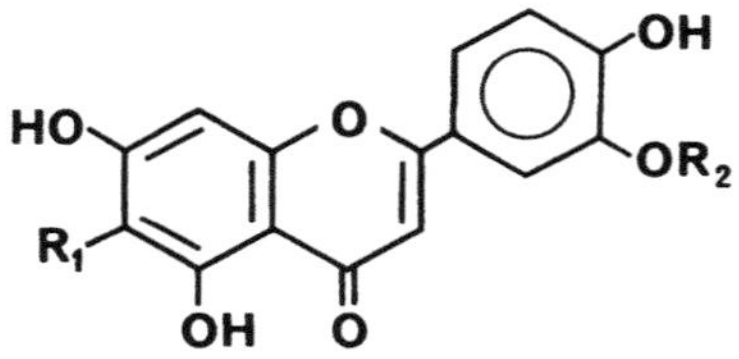

70. Maysin; R₂ = H
76. Maysin-3'-methyl ether; R₂ = CH₃

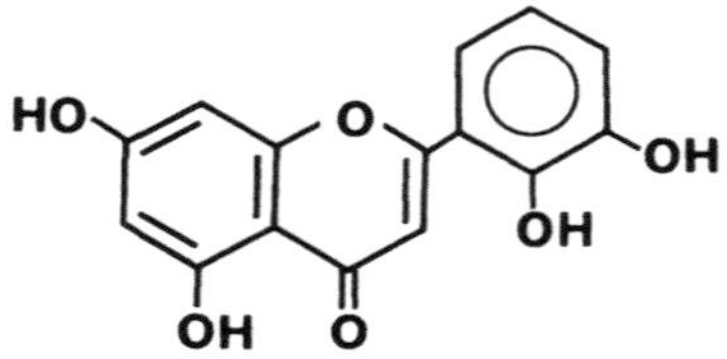

74. Orobol; R = H

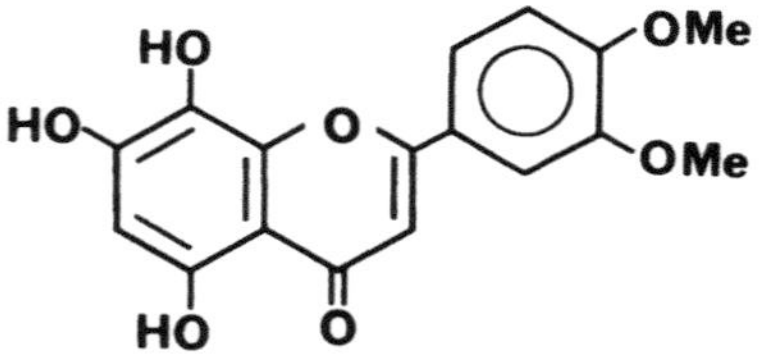

79. Hypolaetin-3',4'-methyl ether

69. Naringenin

73. Eriodictyol

78. Isoscutellarein

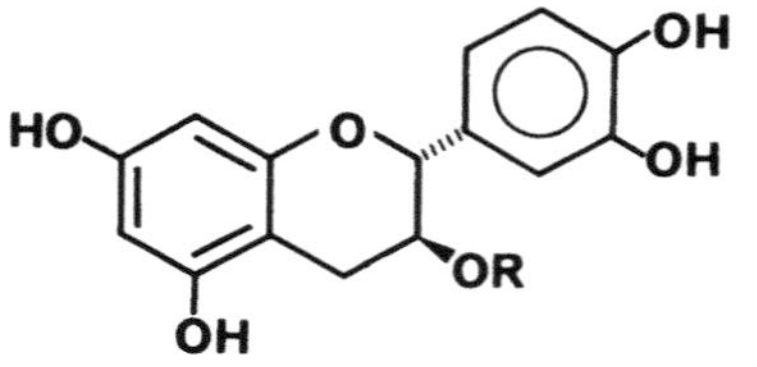

68. D-Catechin; R = H (Dihydroquercetin)
82. Astilbin; R = Rhamnose

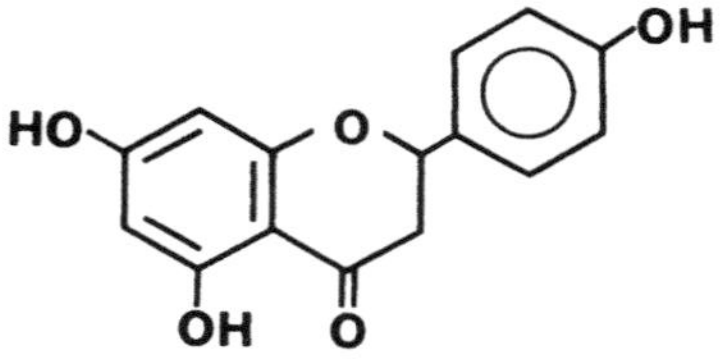

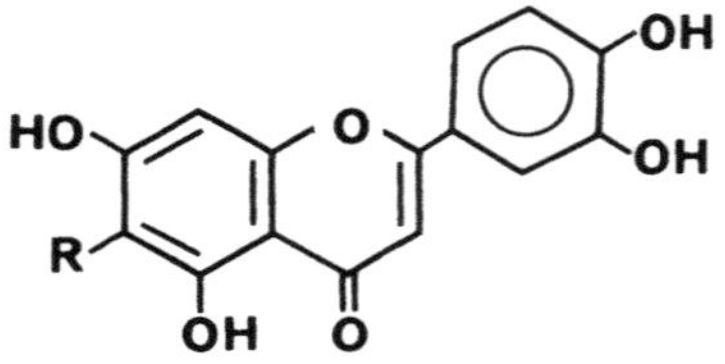

72. Luteolin; R = H
75. Isoorientin; R=C-Glucosyl

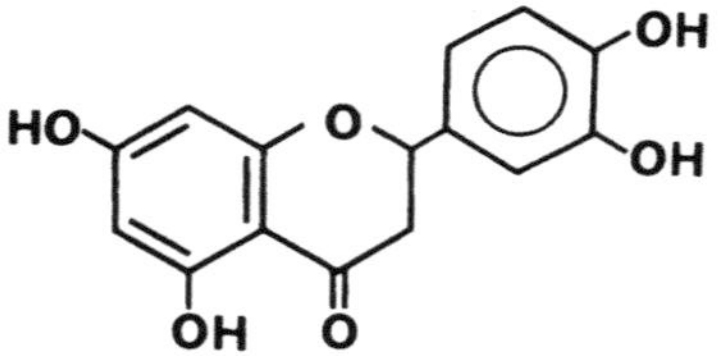

66. Quercetin R = H
67. Rutin R = rutinose

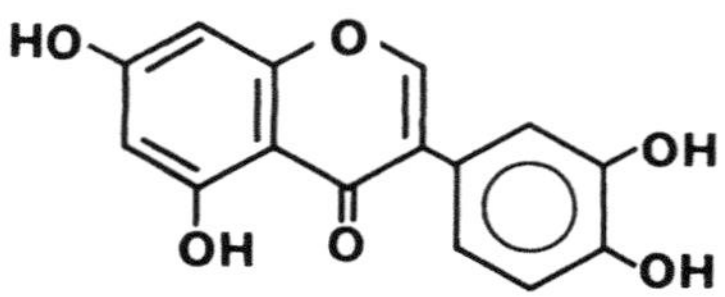

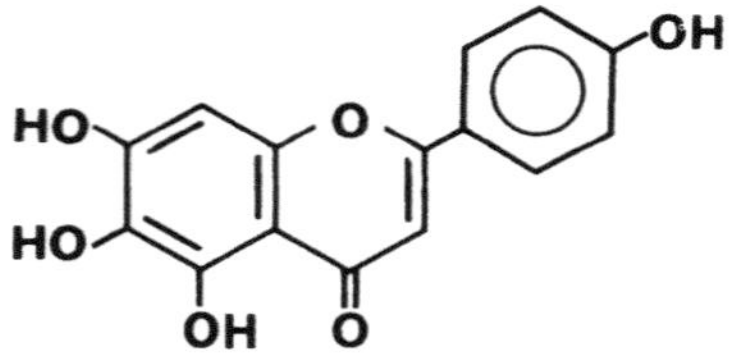

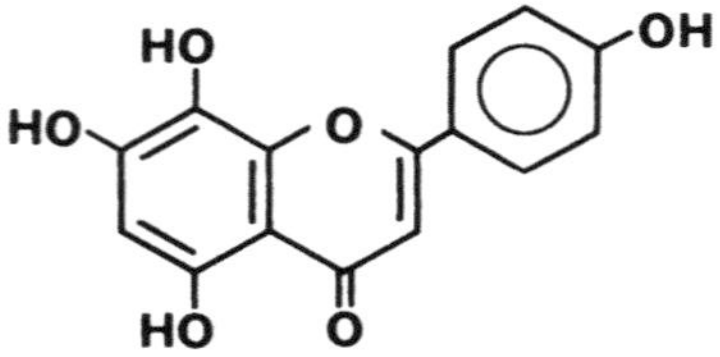

71. 5,7,2',3'Tetrahydroxy Flavone

77. Scutellarein

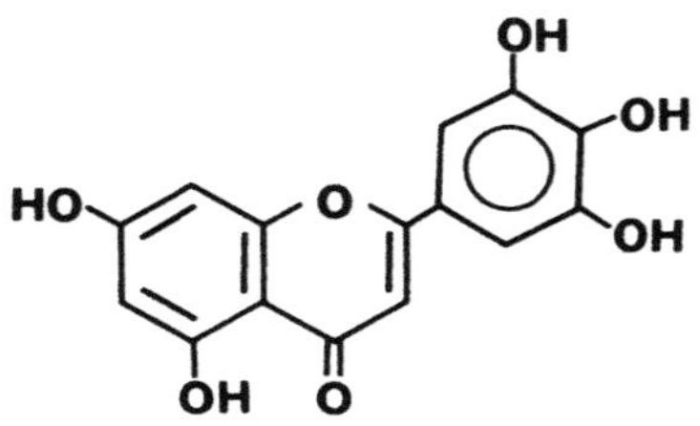

83. Myrcetin

81. Robinetin

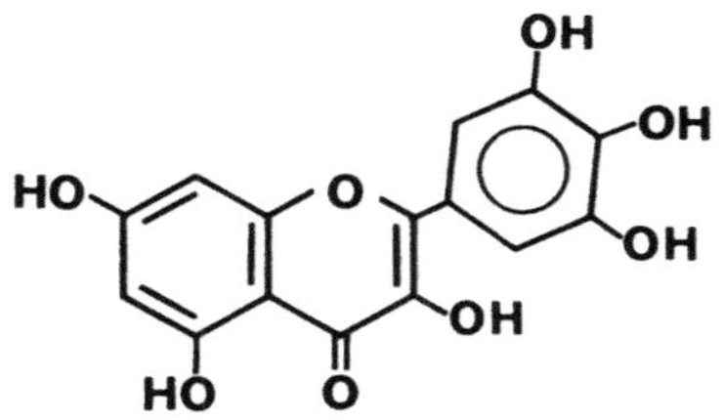

Structures for Table 8.

80. Tricetin

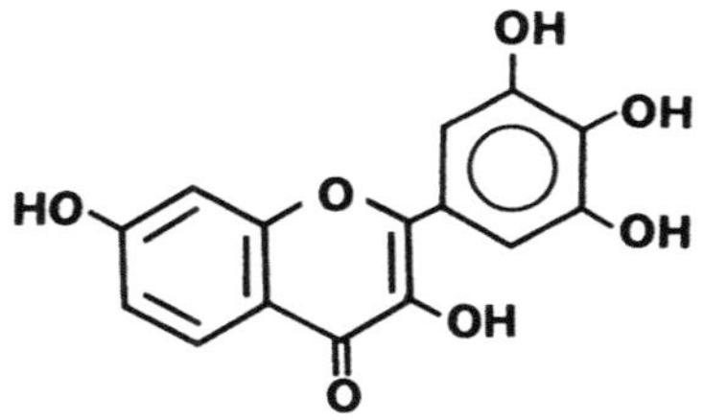

Table 8A
EFFECTS OF FLAVONOIDS ON INSECT GROWTH AND DEVELOPMENT

Insect	Plant	Compound (conc.) [compound no.]	Effect	Ref.
Heliothis zea (Boddie)	*Gossypium hirsutum*	Quercetin (0.05%) [66]	Inhibited growth	165
		Rutin (0.2%) [67]	Inhibited growth	165
Heliothis virescens (F.)	*Gossypium hirsutum*	Quercetin (0.05%) [66]	Inhibited growth	165
		Rutin (0.1%) [67]	Inhibited growth	165
Schizaphis graminum (Rondani)	*G. hirsutum*	Quercetin (10^{-4} M) [66]	Inhibited growth and reproduction	94
		d-Catechin (10^{-4} *M*) (dihydroquercetin) [68]	Inhibited reproduction	94
		Naringenin (10^{-4} *M*) [69]	Inhibited reproduction	94
Heliothis zea (Boddie)	Corn, Zapalote Chico	Maysin (3 mmol/kg diet) [70]	Inhibited growth	202
		5,7,2',3'-Tetrahydroxyflavone (7.5 mmol/kg) [71]	Inhibited growth	203
		Luteolin (5.4 mmol/kg) [72]	Inhibited growth	203
		Eriodictyol (6.2 mmol/kg) [73]	Inhibited growth	203
		Orobol (6.4 mmol/kg) [74]	Inhibited growth	203
		Iso-orientin (3.0 mmol/kg) [75]	Inhibited growth	203
		Maysin (2.6 mmol/kg) [70]	Inhibited growth	203
		Maysin 3'-methyl ether (5.0 mmol/kg) [76]	Inhibited growth	203
		Scutellarein (4.0 mmol/kg) [77]	Inhibited growth	203
		Isoscutellarein (6.2 mmol/kg) [78]	Inhibited growth	203
		Hypolaetin 3',4'-dimethyl ether (7.0 mmol/kg) [79]	Inhibited growth	203
		Tricetin (5.6 mmol/kg) [80]	Inhibited growth	203
		Robinetin (4.0 mmol/kg) [81]	Inhibited growth	203
		Quercetin (3.5 mmol/kg) [66]	Inhibited growth	203
		Dihydroquercetin (3.5 mmol/kg) [68]	Inhibited growth	203
		Quercitrin (4.5 mmol/kg) [66]	Inhibited growth	203
		Astilbin (5.7 mmol/kg) [82]	Inhibited growth	203
		Rutin (4.0 mmol/kg) [67]	Inhibited growth	203
		d-Catechin (5.2 mmol/kg) [68]	Inhibited growth	203
		Myricetin (3.1 mmol/kg) [83]	Inhibited growth	203
	Lycopersicon	Rutin (0.1%) [67]	Inhibited growth	204
		Rutin (2.4 mg/g fresh wt) [67]	Inhibited growth	96

Table 8B

CHEMICAL DATA FOR FLAVONOIDS AFFECTING INSECT GROWTH AND DEVELOPMENT

Name [compound no.]	Chem. Abstr. reg. no.	Formula	Physical constants[a]	Synthesis	Biosynthesis	Comments
Quercetin [66] (dihydrate)	6151-25-3	$C_{15}H_{10}O_7$	mp dec 314°	205	206 208 210a 107	Flavonoid review[207] Isoflavonoid review[209]
Rutin [67]	153-18-4	$C_{27}H_{30}O_{16}$	mp dec 214—215° (ethanol) $[\alpha]_D^{23}$ +13.72° $[\alpha]_D^{23}$ −39.43° (pyridine)	205	102	Review[211]
Catechin [68]	154-23-4	$C_{15}H_{14}O_6$	mp 175—177° $[\alpha]_D^{18}$ +16°—+18.40		212	
Naringenin [69]	480-41-1	$C_{15}H_{12}O_5$	mp 251° $[\alpha]_D^{27}$ +5.9° (acetone)		154	
Maysin [70]	70255-49-1	$C_{27}H_{28}O_{14}$	mp 161—163°		102 110	
5,7,2′,3′-Tetrahydroxy isoflavone [71]	74805-70-2	$C_{15}H_{10}O_6$				
Luteolin [72]	491-70-3	$C_{15}H_{10}O_6$	mp dec 328—330°	213	154	
Eriodictyol [73]	4049-38-1	$C_{15}H_{12}O_6$	Hydrate mp 257° (dec)	214 215	154	
Orobol [74]	480-23-9	$C_{15}H_{10}O_6$	mp 212°	216		Review[218]
Isoorientin [75]	4261-42-1	$C_{21}H_{20}O_{11}$	mp 204—205°	217		
Maysin-3′-methyl ether [76]	74158-05-7	$C_{28}H_{30}O_{14}$				
Scutellarein [77]	529-53-3	$C_{15}H_{10}O_6$	mp 347—349°	219	102	
Isoscutellarein [78]	41440-05-5	$C_{15}H_{10}O_6$		220		
Hypolactin-3′-4′-dimethyl ether [79]	74805-69-9	$C_{17}H_{14}O_7$				
Tricetin [80]	520-31-0	$C_{15}H_{10}O_7$	Hydrate mp 310°	221		
Robinetin [81]	490-31-3	$C_{15}H_{10}O_7$	mp 325—330° dec	222	102	
Astilbin [82]	29838-67-3	$C_{21}H_{22}O_{11}$				
Myricetin [83]	529-44-2	$C_{15}H_{10}O_8$	mp 357°	223	102	

[a] °C.

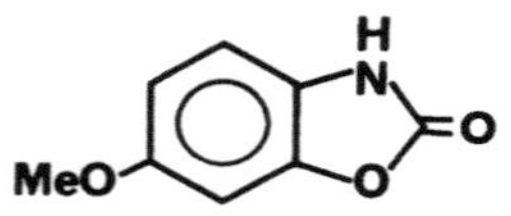

84. 6-Methoxybenzoxazolinone

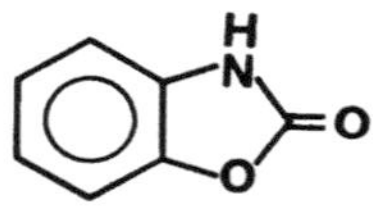

85. Benzoxazolinone

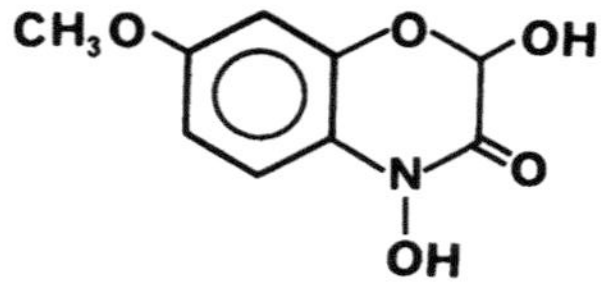

86. 2,4-Dihydroxy-7-methoxy-1,4-benzoxazin-3-one ("DIMBOA")

Table 9A
EFFECTS OF BENZOXAZOLINONES ON INSECT GROWTH AND DEVELOPMENT

Insect	Plant	Compound (conc.) [compound no.]	Effect	Ref.
Ostrinia nubilalis (Hubner)	Zea mays	6-Methoxybenzoxazolinone (0.08 mg/g diet) [84]	Inhibited growth	224
		Benzoxazolinone (10^{-3} M) [85]	Inhibited growth	225
		6-Methoxybenzoxazolinone (0.25 mg/g diet) [84]	Inhibited growth and increased mortality	226
		2,4-Dihydroxy-7-methoxy-1, 4-benzoxazin-3-one (0.32 mg/g diet) [86]	Inhibited growth and increased mortality	227
		6-Methoxybenzoxazolinone (0.5 mg/g diet) [84]	Inhibited growth, effects could be alternated with vitamins	227

Table 9B
CHEMICAL DATA FOR BENZOXAZOLINONES AFFECTING INSECT GROWTH AND DEVELOPMENT

Name [compound no.]	Chem. Abstr. reg. no.	Formula	Physical constant	Synthesis	Ref.
6-Methoxy-2-benzoxazolinone[a] [84]	532-91-2	$C_8H_7NO_3$	mp 154—155°	228 229	231
Benzoxazolin-2-one [85]	59-49-4	$C_7H_5NO_2$	mp 141—142° Monohydrate: mp 97—98°	230	232
2,4-Dihydroxy-7-methoxy-1,4-benzoxazinone (DIMBOA) [86]	15893-52-4	$C_9H_9NO_5$	mp 156—157°	227	233

[a] Biosynthesis — 227.

Table 10A
EFFECTS OF ENZYME INHIBITORS ON INSECT GROWTH AND DEVELOPMENT[a]

Insect	Plant	Compound (conc.)	Effect
Tribolium castaneum (Herbst)	*Glycine*	Trypsin inhibitor (15% soybean meal)	Inhibited growth
Ostrinia nubilalis (Hubner)	*Glycine*	Trypsin inhibitor (Kunitz) (2%)	Inhibited growth and delayed pupation
Oporinia autumnata Packard	*Betula pubescens* spp. *tortuosa*	Trypsin inhibitor (phenolic) (32.3% trypsin inhibition)	Delayed pupation

[a] No chemical data table for these compounds since many are large molecules and not well defined chemically.

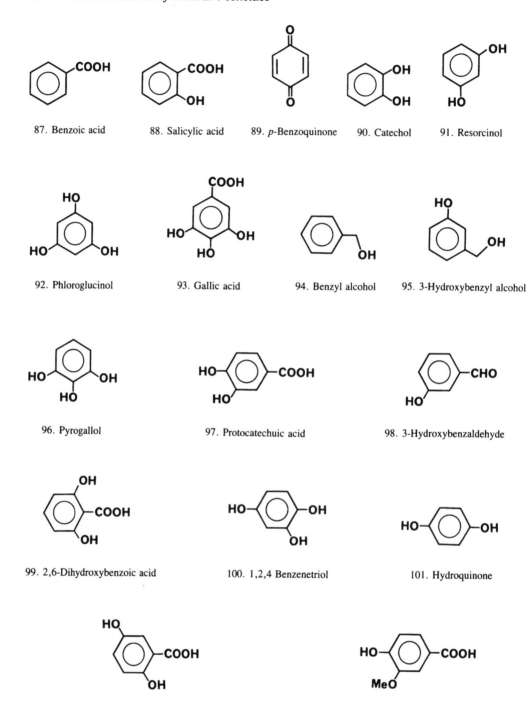

87. Benzoic acid   88. Salicylic acid   89. *p*-Benzoquinone   90. Catechol   91. Resorcinol

92. Phloroglucinol   93. Gallic acid   94. Benzyl alcohol   95. 3-Hydroxybenzyl alcohol

96. Pyrogallol   97. Protocatechuic acid   98. 3-Hydroxybenzaldehyde

99. 2,6-Dihydroxybenzoic acid   100. 1,2,4 Benzenetriol   101. Hydroquinone

102. 2,5-Dihydroxybenzoic acid (gentisic acid)   103. Vanillic acid

Structures for Table 11.

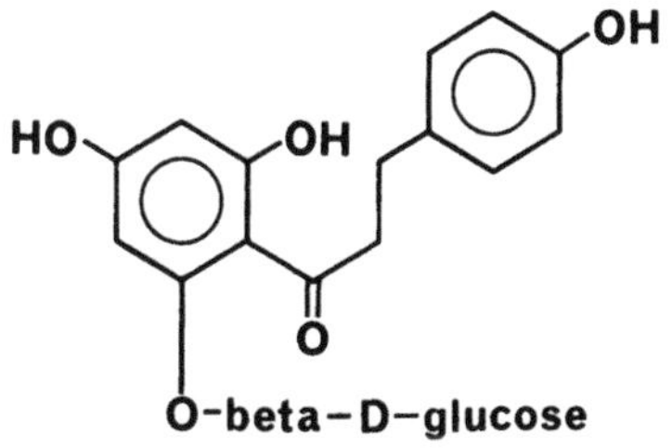

104. Phlorizin

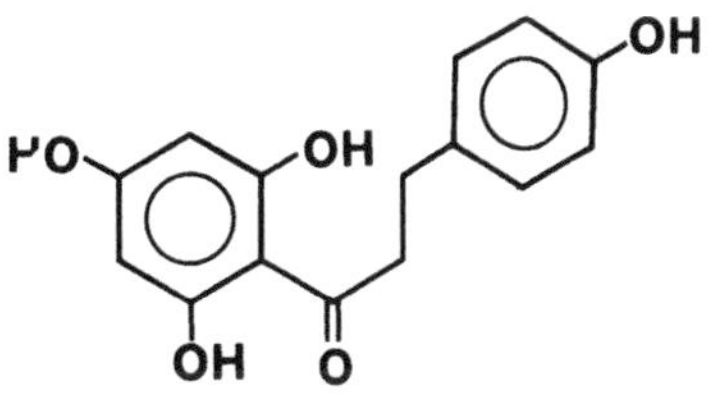

105. Phloretin

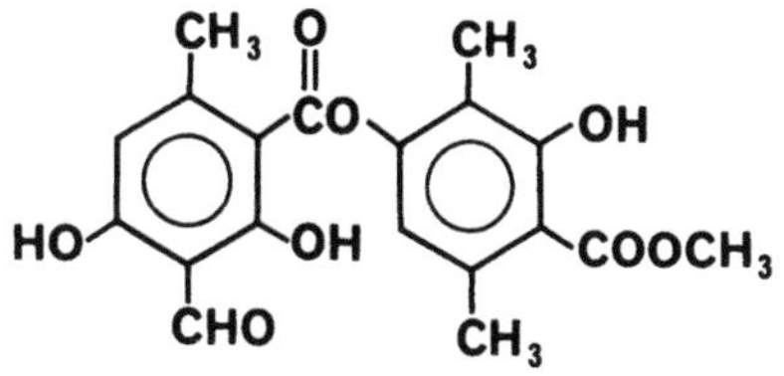

106. Atranorin

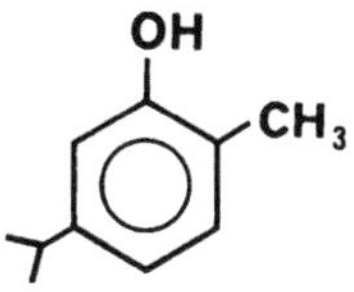

107. Carvacrol

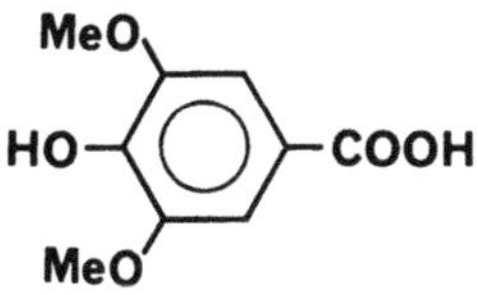

108. Syringic acid

Structures for Table 11 continued.

Table 11A
EFFECTS OF SIMPLE PHENOLICS, BENZOIC ACID, AND BENZYL ALCOHOL DERIVATIVES ON INSECT GROWTH AND DEVELOPMENT

Insect	Compound [compound no.]	Effect	Ref.
Chilo suppressalis Walker[a]	Benzoic acid [87]	Inhibited growth	234
	Salicylic acid [88]	Inhibited growth	234
Agrotis ipsilon (Hufnagel)	p-Benzoquinone (10^{-4} M) [89]	Inhibited growth and pupation	235
	Catechol (10^{-2} M) [90]	Inhibited growth and pupation	63
	Resorcinol (10^{-2} M)[91]	Inhibited growth	236
	Phloroglucinol (10^{-2} M) [92]	Inhibited growth and pupation	236
	Gallic acid (10^{-4} M) [93]	Inhibited growth and pupation	236
	Benzyl alcohol (10^{-2} M) [94]	Inhibited pupation	237
	3-Hydroxybenzyl alcohol (10^{-2} M) [95]	Inhibited pupation and pupal weight (slight)	237
	Pyrogallol (10^{-1} M) [96]	Inhibited growth and pupation	237
	Benzoic acid (10^{-2} M) [87]	Inhibited growth and pupation	237
	Protocatechuic acid (10^{-1} M) [97]	Inhibited growth and pupation	237
	3-Hydroxybenzaldehyde (10^{-2} M) [98]	Inhibited growth and pupation	237
	o-Hydroxybenzoic acid (10^{-2} M) (Salicylic acid) [88]	Inhibited growth and pupation	237
	2,6-Dihydroxybenzoic acid (10^{-1} M) [99]	Inhibited pupation	237
	1,2,4-Benzenetriol (10^{-2} M) [100]	Inhibited growth, pupation, and pupal weight (slight)	237
Schizaphis graminum (Rondani)	Catechol (10^{-1} M) [90]	Inhibited growth and reproduction	94
	Resorcinol (10^{-4} M) [91]	Inhibited growth and reproduction	94
	Hydroquinone (10^{-4} M) [101]	Inhibited growth and reproduction	94
	Pyrogallol (10^{-4} M) [96]	Inhibited growth	94
	Phloroglucinol (10^{-4} M) [92]	Inhibited growth and reproduction	94
	Gentisic acid (10^{-4} M) [102]	Inhibited reproduction	94
	Protocatechuic acid (10^{-4} M) [97]	Inhibited reproduction	94
	Salicylic acid (10^{-4} M) [88]	Inhibited reproduction	94
	Gallic acid (10^{-4} M) [93]	Inhibited reproduction	94
	Vanillic acid (10^{-4} M) [103]	Inhibited reproduction	94
Locusta migratoria migratorioides (R. & F.)	Phlorizin (1 mM/ℓ) [104]	Inhibited glucose reabsorption in malpighian tubes	238
	Phloretin (0.1 mM/ℓ) [105]	Inhibited glucose reabsorption in malpighian tubes	238
Spodoptera ornithogalli Guenee[b]	Atranorin (0.03%) [106]	Inhibited growth	99
Aedes aegypti (L.)	Carvacrol (0.02 mg/cm² water surface) [107]	Inhibited embryonic development	93
Schizaphis graminum (Rondani)	Syringic acid (10^{-4}) [108]	Inhibited growth and reproduction	94

[a]   Plant — *Oryza*.
[b]   Plant — *Letharia vulpina* (L.) Hue.

Table 11B
CHEMICAL DATA FOR SIMPLE PHENOLICS, BENZOIC ACID, AND BENZYL ALCOHOL DERIVATIVES AFFECTING INSECT GROWTH AND DEVELOPMENT

Name [compound no.]	Chem. Abstr. reg. no.	Formula	Physical constants[a]	Synthesis	Biosynthesis
Benzoic acid [87]	65-85-0	$C_7H_6O_2$	mp 122.4°; bp 760; 249.2°; bp 60; 172.8°; bp 10; 132.1°		102
Salicylic acid [88] (2-hydroxybenzoic acid)	69-72-7	$C_7H_6O_3$	mp 157—159°; bp 20; about 211°	239	102 110
1,4-Benzoquinone [89][b]	106-51-4	$C_6H_4O_2$	mp 117°		240
Catechol [90]	120-80-9	$C_6H_6O_2$	mp 105°; bp 240°	Org. synth., CV I, 149	
Resorcinol [91]	108-46-3	$C_6H_6O_2$	mp 111°; bp 227°		
Phloroglucinol [92]	6099-90-7, dihydrate 83-30-7, monohydrate	$C_9H_6O_3$	mp 218°	242 243	154
Gallic acid [93]	149-91-7	$C_7H_6O_5$	mp 235—340°	244 246	245 102
Benzyl alcohol [94]	100-51-6	C_7H_8O	mp −15.19°; bp 760; 240.7°; bp 100; 141.7°; bp 10; 92.6°; bp 1.0; 58.0°; n_D^{20} 1.54035; n_D^{25} 1.53837	247 248	102 110
3-Hydroxybenzyl alcohol [95]	620-24-6	$C_7H_8O_2$	mp 73°	249	102 110
Pyragallol [96]	87-66-1	$C_6H_6O_3$	mp 131—133°; bp 309°	250 246	

Table 11B (continued)
CHEMICAL DATA FOR SIMPLE PHENOLICS, BENZOIC ACID, AND BENZYL ALCOHOL DERIVATIVES AFFECTING INSECT GROWTH AND DEVELOPMENT

Name [compound no.]	Chem. Abstr. reg. no.	Formula	Physical constants[a]	Synthesis	Biosynthesis
Protocatechuic acid [97]	99-50-3	$C_7H_6O_4$	mp about 200° dec	251 Org. synth., CV III, 745	
3-Hydroxybenzaldehyde [98]	100-83-4	$C_7H_6O_2$	mp 108° bp 760; 240° bp 50; 191°	Org. synth., CV III, 453, 564	102
2,6-Dihydroxybenzoic acid [99]	303-07-1	$C_7H_6O_4$	mp 150—170° var.	252	
1,2,4-Benzenetriol [100]	533-73-3	$C_6H_6O_3$	mp 140.5°	253	102
Hydroquinone [101]	123-319	$C_6H_6O_2$	mp 170—171° bp 760; 285—287°		
Gentistic acid [102]	490-79-9	$C_7H_6O_4$	mp 199—200°	254 256	255
Vanillic acid [103]	121-34-6	$C_6H_8O_4$	mp 210°	257	271
Phlorizin [104]	60-81-1	$C_{21}H_{24}O_{10}$	Dihydrate mp 110° $[\alpha]_D^{25} -52°$	258	102
Phloretin [105]	60-82-2	$C_{15}H_{14}O_5$	mp 262° dec	259	
Atranorin [106]	470-20-9	$C_{19}H_{18}O_8$	mp 195°	260	
Carvacrol [107]	499-75-2	$C_{10}H_{14}O$	bp 760; 237—238° bp 18; 118—122° bp 3; 93° n_D 1.52295	261 262	
Syringic acid [108]	530-57-4	$C_9H_{10}O_5$	mp 210.1—210.4°	246	154

a   °C.
b   Crystal structure — 241.

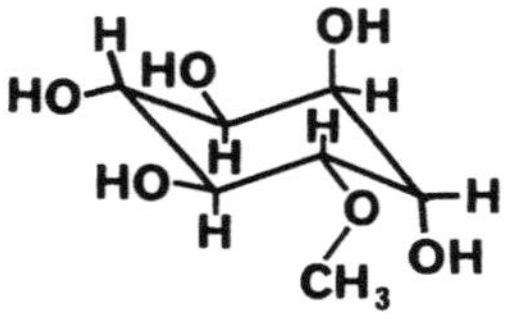

$CH_3(CH_2)_4CH=CHCH_2CH=CH(CH_2)_7\overset{\displaystyle O}{\underset{\|}{C}}CH_2\underset{\underset{\displaystyle OH}{|}}{CH}CH_2OCOCH_3$

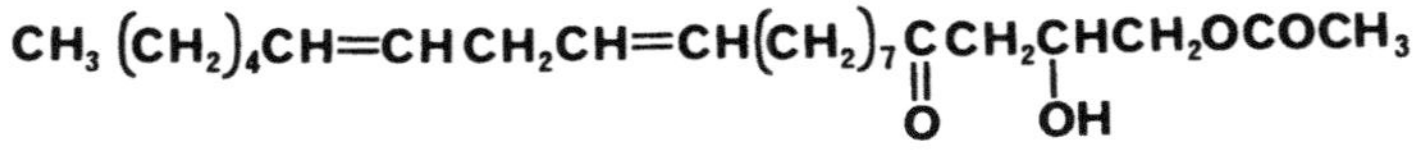

111. 1-Acetoxy-2-hydroxy-4-oxoheneicosa-12,15-diene

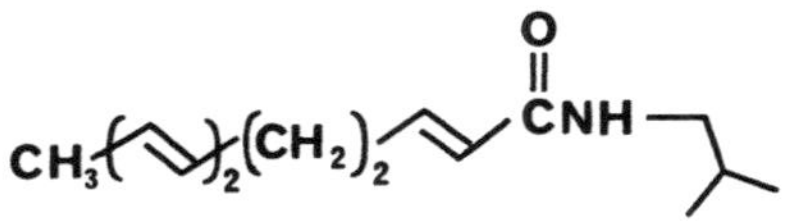

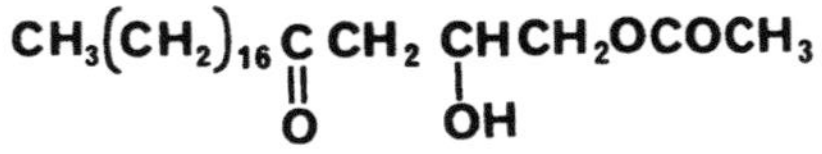

110. Spilanthol (Affinin)

109. Pinitol

$CH_3(CH_2)_{16}\overset{\displaystyle O}{\underset{\|}{C}}CH_2\underset{\underset{\displaystyle OH}{|}}{CH}CH_2OCOCH_3$

112. 1-Acetoxy-2-hydroxy-4-oxo-heneicosane

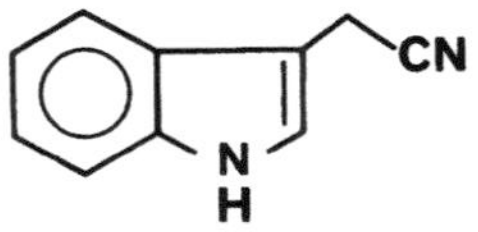

113. Indoleacetonitrile

114. Imidazole

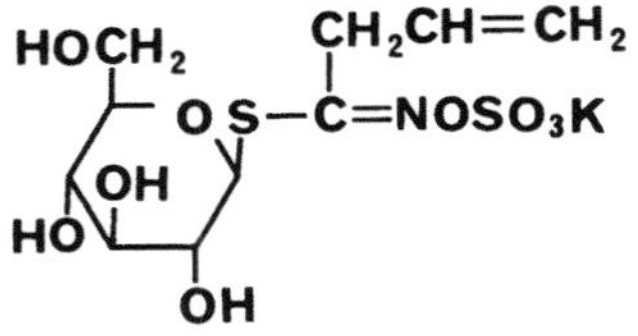

115. Allylglucosinolate

Structures for Table 12.

Table 12A

EFFECTS OF MISCELLANEOUS COMPOUNDS ON INSECT GROWTH AND DEVELOPMENT

Insect	Plant	Compound (conc.) [compound no.]	Effect	Ref.
Heliothis zea (Boddie)	*Glycine*	Pinitol (0.7%) [109]	Inhibited growth	263
Bombyx mori L.	*Spilanthes acmella*	Spilanthol (200 ppm) [110]	Inhibited growth	98
	Persea americana Mill (avocado)	1-Acetoxy-2-hydroxy-4-oxo-heneicosa-12,15-diene (200 ppm) [111]	Inhibited growth	131
		Tetrahydro derivative of above compound (200 ppm) [112]	Inhibited growth	131
Ostrinia nubilalis (Hubner)	*Brassica*	Indole-3-acetonitrile (0.45 mg/g medium) [113]	Inhibited growth	264
Galleria mellonella (L.)	*Brassica*	Indole-3-acetonitrile (0.25 mg/g medium) [113]	Inhibited growth	264
Pseudosarcophaga affinis (Fallen)		Imidazole (0.15 g/100 mℓ diet) [114]	Inhibited reproduction	55
Papilio polyxenes asterius Stoll	Cruciferae	Allylglucosinolate (Sinigrin) (0.68%) [115]	Inhibited growth	265
Spodoptera eridania (Cramer)	Cruciferae	Allylglucosinolate (0.87%) [115]	Inhibited growth	265
P. polyxenes asterius Stoll		Allylglucosinolate (0.1%) [115]	Inhibited growth and assimilation	266

Table 12B
CHEMICAL DATA FOR MISCELLANEOUS COMPOUNDS AFFECTING INSECT GROWTH AND DEVELOPMENT

Name [compound no.]	Chem. Abstr. no.	Formula	Physical constants[a]	Synthesis
	D-(+)-Pinitol [109][b] 10284-63-6	$C_7H_{14}O_6$	mp 185—186° $[\alpha]_D^{20} +67°$ (C = 2.5)	
Spilanthol [110]	25394-57-4		mp 23° bp 0.5; 165° bp 0.001; 120—125°	268
1-Acetoxy-2-hydroxy-4-oxoheneicosa-12,15-diene [111]	60640-59-7	$C_{23}H_{40}O_4$	Oil	
1-Acetoxy-2-hydroxy-4-oxoheneico-sane [112]	60640-58-6	$C_{23}H_{44}O_4$	Oil	
Indoleacetonitrile [113]	771-51-7	$C_{11}H_8N_2$	mp 35—37° bp 0.2; 157—160°	
Imidazole [114]	288-32-4	$C_3H_4N_2$	mp 90—91° bp 760; 257° bp 12; 138.2°	
Allyl glucosinolate (Sinigrin) [115][c]	3952-98-5	$C_{10}H_{16}K$ NO_9S_2	mp 179° mp 147—129° (of monohydrate) $[\alpha]_D^{18} -16.4°$	269

[a] °C.
[b] Biosynthesis — 267.
[c] Crystal structure — 270.

REFERENCES

1. **Beck, S. D. and Reese, J. C.**, Insect-plant interactions: nutrition and metabolism, *Rec. Adv. Phytochem.*, 10, 41, 1976.
2. **Rodriguez, J. G., Ed.**, *Insect and Mite Nutrition*, North-Holland, Amsterdam, 1972.
3. **Maxwell, F. G. and Harris, F. A., Eds.**, *Proceedings of the Summer Institute of Biological Control of Plant Insects and Diseases*, University Press of Mississippi, Jackson, 1974.
4. **Jermy, T., Ed.**, *The Host Plant in Relation to Insect Behaviour and Reproduction*, Plenum Press, New York, 1976; *Int. Symp. Biol. Hung.*, 16, 13, 1976.
5. **Gilbert, L. E. and Raven, P. H., Eds.**, *Coevolution of Animals and Plants*, University of Texas Press, Austin, 1975.
6. **Wallace, J. W. and Mansell, R. L., Eds.**, *Biochemical Interaction Between Plants and Insects*, Rec. Adv. Phytochem. Ser., Vol. 10, Plenum Press, New York, 1976.
7. **Hedin, P. A., Ed.**, *Host Plant Resistance to Pests*, ACS Symp. Ser. 62, American Chemical Society, Washington, D.C., 1977.
8. **Labeyrie, V., Ed.**, *Comportement Des Insectes et Milieu Trophique*, Centre National Recherche Scientifique, Paris, 1977.
9. **Marini-Bettolo, G. B., Ed.**, *Natural Products and the Protection of Plants*, Pontifica Academic Scientiarum, Rome, 1977.
10. **Chapman, R. F. and Bernays, E. A., Eds.**, *Insect and Host Plant*, Nederlandse Entomologische, Vereniging, Netherlands, 1978; *Entomol. Exp. Appl.*, 24, 204, 1978.
11. **Rosenthal, G. A. and Janzen, D. H., Eds.**, *Herbivores: Their Interaction with Secondary Plant Metabolites*, Academic Press, New York, 1979.

12. **Warthen, J. D., Jr.**, *Azadirachta indica:* A Source of Insect Feeding Inhibitors and Growth Regulators, USDA, Agric. Rev. Manuals No. ARM-NE-4, U.S. Department of Agriculture, Washington, D.C., 1979.

13. **Maxwell, F. G. and Jennings, P. E.**, Eds., *Breeding Plants Resistant to Insects*, John Wiley & Sons, New York, 1980.

14. **Harris, M. K.**, Ed., Biology and Breeding for Resistance to Arthropods and Pathogens in Agricultural Plants, Texas Agricultural Experimental Station and University of California Agency for International Development, College Station, Tex., 1980.

15. **Beck, S. D.**, Resistance of plants to insects, *Ann. Rev. Entomol.*, 10, 207, 1965.

16. **Rosenthal, G. A.**, The biological effects and mode of action of L-canavanine, a structural analogue of L-arginine, *Q. Rev. Biol.*, 52, 155, 1977.

17. **Russell, G. E.**, *Plant Breeding for Pest and Disease Resistance*, Butterworths, London, 1978.

18. **Burnett, W. C., Jr., Jones, S. B., Jr., and Mabry, T. J.**, The role of sesquiterpene lactones in plant-animal coevolution, in *Biochemical Aspects of Plant and Animal Coevolution*, Harborne, J. B., Ed., Academic Press, New York, 1978, 233.

18a. **Bernays, E. A. and Chapman, R. F.**, Plant chemistry and acridoid feeding behaviour, in *Biochemical Aspects of Plant and Animal Coevolution*, Harborne, J. B., Ed., Academic Press, New York, 1978, 99.

19. **Panda, N.**, *Principles of Host Plant Resistance to Insect Plants*, Allanheld/Universe, New York, 1979.

20. **Waiss, A. C., Jr., Chan, B. G., Elliger, C. A., Dreyer, D. L., Binder, R. G., and Gueldner, R. C.**, Insect growth inhibitors in plants, *Bull. Entomol. Soc. Am.*, 27, 217, 1981.

21. **Rosenthal, G. A.**, *Plant Nonprotein Amino Acids and Imino Acids*, Academic Press, New York, 1982.

22. **Hedin, P. A.**, Ed., *Plant Resistance to Insects*, ACS Symp. Ser. 208, American Chemical Society, Washington, D.C., 1983.

23. **Feeny, P. P.**, Biochemical coevolution between plants and their herbivores, in *Coevolution of Animals and Plants*, Gilbert, L. E. and Raven, P. H., Eds., University of Texas Press, Austin, 1975, 3.

24. **Rosenthal, G. A., Janzen, D. H., and Dahlman, D. L.**, Degradation and detoxification of canavanine by a specialized seed predator, *Science*, 196, 658, 1977.

25. **Rosenthal, G. A., Hughes, C. B., and Janzen, D. H.**, L-Canavanine, a dietary nitrogen source for the seed predator *Caryedes brasiliensis* (Bruchidae), *Science*, 217, 353, 1982.

26. **Bernays, E. A. and Woodhead, S.**, Plant phenols utilized as nutrients by a phytophagous insect, *Science*, 216, 201, 1982.

27. **Schultz, J. C. and Baldwin, I. T.**, Oak leaf quality declines in response to defoliation by gypsy moth larvae, *Science*, 217, 149, 1982.

28. **Green, T. R. and Ryan, C. A.**, Wound-induced proteinase inhibitor in plant leaves: a possible defense mechanism against insects, *Science*, 175, 776, 1972.

29. **Brattsten, L. B.**, Biochemical defense mechanisms in herbivores against plant allelochemicals, in *Herbivores: Their Interaction with Secondary Plant Metabolites*, Rosenthal, G. A. and Janzen, D. H., Eds., Academic Press, New York, 1979, 199.

30. **Mullin, C. A., Croft, B. A., Stickler, K., Matsumura, F., and Miller, J. R.**, Detoxification enzyme differences between a herbivorous and predatory mite, *Science*, 217, 1270, 1982.

31. **Wagner, M. R. and Benjamin, D. M.**, Allelochemics and nutritional indices for larch sawfly, *Pristiphora erichsonii* (Hartig): a specialist feeding on *Larix* spp., *J. Chem. Ecol.*, 7, 165, 1981.

32. **Chapman, R. F.**, The chemical inhibition of feeding by phytophagous insects: a review, *Bull. Entomol. Res.*, 64, 339, 1974.

33. **Hedin, P. A., Maxwell, F. G., and Jenkins, J. N.**, Insect plant attractants, feeding stimulants, repellents, deterrents, and other related factors affecting insect behavior, in *Proceedings of the Summer Institute of Biological Control of Plant Insects and Diseases*, Maxwell, F. G. and Harris, F. A., Eds., University Press of Mississippi, Jackson, 1974, 494.

33a. **Carlisle, D. B. and Ellis, P. E.**, Bracken and locust ecdysones: their effects on molting in the desert locust, *Science*, 159, 1472, 1968.

34. **Bowers, W. S., Ohta, T., Cleere, J. S., and Marsella, P. A.**, Discovery of insect anti-juvenile hormones in plants, *Science*, 193, 542, 1976.

35. **Bowers, W. S. and Martinez-Pardo, R.**, Antiallatotropins: inhibition of corpus allantum development, *Science*, 197, 1369, 1977.

36. **Brooks, G. T., Pratt, G. E., and Jennings, R. C.**, The action of precocenes in milkweed bugs *(Oncopeltus fasciatus)* and locusts *(Locusta migratoria)*, *Nature (London)*, 281, 570, 1979.

37. **Bowers, W. S. and Aldrich, J. R.**, In vivo inactivation of denervated corpora allata by precocene II in the bug, *Oncopeltus fasciatus*, *Experientia*, 36, 362, 1980.

38. **Bowers, W. S. and Nishida, R.**, Juvocimenes: potent juvenile hormone mimics from sweet basil, *Science*, 209, 1030, 1980.

39. **Bowers, W. S., Evans, P. H., Marsella, P. A., Soderlund, D. M., and Dettarini, F.**, Natural and synthetic allatotoxins: suicide substrates for juvenile hormone biosynthesis, *Science*, 217, 647, 1977.

40. **Heftmann, E.**, Insect molting hormones from plants, *Rec. Adv. Phytochem.*, 3, 211, 1970.

41. **Schoonhoven, L. M.**, Chemical mediators between plants and phytophagous insects, in *Semiochemicals: Their Role in Pest Control*, Nordlund, D. A., Jones, R. L., and Lewis, W. J., Eds., John Wiley & Sons, New York, 1981, 31.
42. **Ryan, C. A.**, Proteolytic enzymes and their inhibitors in plants, *Ann. Rev. Plant Physiol.*, 24, 173, 1973.
43. **Ryan, C. A. and Green, T. R.**, Proteinase inhibitors in natural plant protection, *Rec. Adv. Phytochem.*, 8, 123, 1974.
44. **McFarlane, J. E. and Distler, M. H. W.**, The effect of rutin on growth, fecundity and food utilization on *Acheta domesticus* (L.), *J. Insect Physiol.*, 28, 85, 1982.
45. **Isman, M. B.**, Dietary influence of cardenolides on larval growth and development of the milkweed bug *Oncopeltus fasciatus*, *J. Insect Physiol.*, 23, 1183, 1977.
46. **Campbell, B. C. and Duffey, S. S.**, Alleviation of a-tomatine-induced toxicity to the parasitoid, *Hypogasta oxiguae*, by phytosterols in the diet of the host, *Heliothis zea*, *J. Chem. Ecol.*, 7, 927, 1981.
47. **Dahlman, D. L. and Rosenthal, G. A.**, Potentiation of L-canavanine-induced developmental anomalies in the tobacco budworm, *Manduca sexta* by some amino acids, *J. Insect Physiol.*, 28, 829, 1982.
48. **Manoukas, A. G.**, Effect of excess levels of individual amino acids upon survival, growth and pupal yield of *Dacus oleae* (Gmel.) larvae, *Z. Angew. Entomol.*, 91, 309, 1981.
49. **Reese, J. C.**, Interactions of allelochemicals with nutrients in herbivore food, in *Herbivores: Their Interaction with Secondary Plant Metabolites*, Rosenthal, G. A. and Janzen, D. H., Eds., Academic Press, New York, 1979, 309.
50. **Isman, M. B. and Duffey, S.**, Phenolic compounds in foliage of commercial tomato cultivars as growth inhibitors to the fruitworm *Heliothis zea*, *J. Am. Hortic. Soc.*, 107, 167, 1982.
50a. **Dahlman, D. L.**, Field tests of L-canavanine for control of tobacco hornworm, *J. Econ. Entomol.*, 73, 279, 1980.
51. **Dahlman, D. L. and Rosenthal, G. A.**, Further studies of the effect of L-canavanine on the tobacco hornworm, *Manduca sexta*, *J. Insect Physiol.*, 22, 265, 1976.
52. **Dahlman, D. L. and Rosenthal, G. A.**, L-Canavanine: potential allelochemic in plant defense against insect attack, *Proc. N. Cent. Br. Entomol. Soc. Am.*, 29, 147, 1972.
53. **Dahlman, D. L. and Rosenthal, G. A.**, Non-protein amino acid-insect interactions. I. Growth effects and symptomology of L-canavanine consumption by tobacco hornworm, *Manduca sexta* (L.), *Comp. Biochem. Physiol.*, 51A, 33, 1975.
54. **Harry, P., Dror, Y., and Applebaum, S. W.**, Arginase activity in *Tribolium castaeum* and the effect of canavanine, *Insect Biochem.*, 6, 273, 1976.
55. **Hegdekar, B. M.**, Amino acid analogues as inhibitors of insect reproduction, *J. Econ. Entomol.*, 63, 1950, 1970.
56. **Isogai, A., Murakoshi, S., Suzuki, A., and Tamura, S.**, Isolation from "Astragali Radix" [sic] of L-canavanine as an inhibitory substance to metamorphosis of silkworm, *Bombyx mori* L., *J. Agric. Chem. Soc. Jpn.*, 47, 449, 1973.
57. **Rosenthal, G. A. and Dahlman, D. L.**, Non-protein amino acid-insect interactions. II. Effects of canaline-urea cycle amino acids on growth and development of the tobacco hornworm, *Manduca sexta* L. (Sphingidae), *Comp. Biochem. Physiol.*, 52A, 105, 1975.
58. **Vanderzant, E. S. and Chremos, J. H.**, Dietary requirements of the boll weevil for arginine and the effect of arginine analogues on growth and on the composition of the body amino acids, *Ann. Entomol. Soc. Am.*, 64, 480, 1971.
59. **Dahlman, D. L.**, Effect of L-canavanine on the consumption and utilization of artificial diet by the tobacco hornworm, *Manduca sexta*, *Entomol. Exp. Appl.*, 22, 123, 1977.
60. **Rehr, S. S., Bell, E. A., Janzen, D. H., and Feeny, P. P.**, Insecticidal amino acids in legume seeds, *Biochem. Systemat.*, 1, 63, 1973.
61. **Dahlman, D. L., Herald, F., and Knapp, F. W.**, L-Canavanine effects on growth and development of four species of Muscidae, *J. Econ. Entomol.*, 72, 678, 1979.
62. **Palumbo, R. E. and Dahlman, D. L.**, Reduction of *Manduca sexta* fecundity and fertility by L-canavanine, *J. Econ. Entomol.*, 71, 674, 1978.
63. **Reese, J. C. and Beck, S. D.**, Effects of allelochemics on the black cutworm, *Agrotis ipsilon*: effects of catechol, L-dopa, dopamine, and chlorogenic acid on larval growth, development, and utilization of food, *Ann. Entomol. Soc. Am.*, 69, 68, 1976.
64. **Rehr, S. S., Janzen, D. H., and Feeny, P. P.**, L-DOPA in legume seeds: a chemical barrier to insect attack, *Science*, 181, 81, 1973.
65. **Nyberg, D. D. and Christensen, B. E.**, The synthesis of DL-canaline, DL-canavanine and related compounds, *J. Am. Chem. Soc.*, 79, 1222, 1957.
66. **Turner, B. L. and Harborne, J. B.**, Distribution of canavanine in the plant kingdom, *Phytochemistry*, 6, 863, 1967.
67. **Frankel, M., Knobler, Y., Zuilichovsky, G.**, Synthesis of DL-canavanine, *J. Chem. Soc.*, p. 3127, 1963.

68. **Yamada, Y., Noda, H., and Okada, H.,** An improved synthesis of DL-canavanine, *Agric. Biol. Chem.,* 37, 2201, 1973.

69. **Armstrong, M. D. and Lewis, J. D.,** Thioether derivatives of cysteine and homocysteine, *J. Org. Chem.,* 16, 749, 1951.

70. **Farber, E. L.,** Ethionine carcinogenesis, *Adv. Cancer Res.,* 7, 383, 1963.

71. **Norton, F. H.,** U.S. Patent 2,840,587; Ethionine and its salts, *Chem. Abstr.,* 52, 16206g, 1958.

72. **Bennet, E. L. and Niemann, C.,** The preparation and resolution of the three isomeric nuclear substituted monofluoro-DL-phenylalanines, *J. Am. Chem. Soc.,* 72, 1800, 1950.

73. **Wheatley, D. N. and Henderson, J. Y.,** Fluorophenylalanine and "division-related proteins", *Nature (London),* 247, 281, 1974.

74. **Bosshard, H. R. and Berger, A.,** Synthesis of optically active, ring-substituted *N*-benzyloxycarbonyl-phenylalanines via 2-benzyloxycarbonylamino-2-arylalkylmalonates, *Helv. Chim. Acta,* 56, 1838, 1973.

75. **Gilon, C., Knobler, F., and Sheradsky, T.,** Synthesis of ω-amino oxy-acids by oxygen-alkyl fission of lactones. An improved synthesis of DL-canaline, *Tetrahedron Lett.,* 23, 4441, 1967.

76. **Rosenthal, G. A.,** The biological and biochemical properties of L-canaline, a naturally-occurring analogue of L-ornithine, *Life Sci.,* 23, 93, 1978.

77. **Rahiala, E.-L.,** Canaline: characterization of enzyme-pyridoxal phosphate complex, *Acta Chem. Scand.,* 27, 3861, 1973.

78. **Zvilichovsky, G.,** Some reactions of amino oxy acids. Synthesis of dl-*O*-ureidohomoserine, *Tetrahedron Lett.,* 22, 1445, 1966.

79. **Gmelin, R., Strauss, G., and Hasenmaier,** Isolierung von 2 neuen pflanzlichen aminosauren: S(β-carboxyathyl)-L-cystein und albizzin aus den samen von albizzea, julibrissin durazz (m, mosaceae), *Z. Naturforsch.,* 13B, 252, 1958.

80. **Evans, C. S., Qureshi, M. Y., and Bell, E. A.,** Free amino acids in the seeds of *Acacia* species, *Phytochemistry,* 16, 565, 1977.

81. **Gmelin, R.,** Die freien aminosauren der samen von *Acacia willardiena* Mimosaceae, *Z. Physiol. Chem.,* 316, 164, 1959.

82. **Ek, A. and Witkop, B.,** Synthesis of labile hydroxytryptophan metabolites, *J. Am. Chem. Soc.,* 76, 5579, 1954.

83. **Wakahara, A., Kido, M., Fujiwara, T., and Tomita, K.,** The crystal and molecular structure of 5-hydroxy-D-tryptophan, *Tetrahedron Lett.,* p. 3003, 1970.

84. **Shaw, K. N. F. and Morris, A. G.,** 5-Hydroxy-DL-tryptophan, *Biochem. Prep.,* 9, 92, 1962.

85. **Frangatos, G. and Chubb, F. L.,** A new synthesis of 5-hydroxy tryptophan, *Can. J. Chem.,* 37, 1374, 1959.

86. **Renson, J., Goodwin, F., Weissbach, H., and Undenfriend, S.,** Conversion of tryptophan to 5-hydroxy tryptophan by phenylalanine hydroxylase, *Biochem. Biophys. Res. Commun.,* 6, 20, 1961.

87. **Waser, E. and Lewandowski, M.,** Untersuchungen in der phenylalain-reihe. I. Synthese des 1-3,4 dioxyphenylalanine, *Helv. Chim. Acta,* 4, 657, 1921.

88. **Fowden, L.,** Non protein amino acids, in *The Biochemistry of Plants, A Comprehensive Treatise,* Vol. 7, Stumpf, P. K. and Conn, E. E., Eds., Academic Press, New York, 1981.

89. **Bretschneider, H. and Hohenlohe-Oehringen, K.,** A new synthesis of 3,4-dihydroxyphenyl-L-alanine from L-tyrosine, *Helv. Chim. Acta,* 56, 2857, 1973.

90. **Isogai, A., Murakoshi, S., Suzuki, A., and Tamira, S.,** Growth inhibitory effects of phenylpropanoids in nutmeg on silkworm larvae, *J. Agric. Chem. Soc. Jpn.,* 47, 275, 1973a.

91. **Berenbaum, M. and Feeny, P.,** Toxicity of angular furanocoumarins to swallowtail butterflies: escalation in a coevolutionary arms race, *Science,* 212, 927, 1981.

92. **Chawla, S. S., Perron, J. M., and Cloutier, M.,** Effects of different growth factors on the potato aphid, *Macrosiphum euphorbiae* (Aphididae:Hompotera), fed on artificial diet, *Can. Entomol.,* 106, 273, 1974.

93. **Saxena, K. N. and Sharma, R. N.,** Embryonic inhibition and oviposition induction in *Aedes aegypti* by certain terpenoids, *J. Econ. Entomol.,* 65, 1588, 1972.

94. **Todd, G. W., Getahun, A., and Cress, D. C.,** Resistance in barley to the greenbug, *Schizaphis graminum.* I. Toxicity of phenolic and flavonoid compounds and related substances, *Ann. Entomol. Soc. Am.,* 64, 718, 1971.

95. **Nakajima, S. and Kawazu, K.,** Insect development inhibitors from *Coleopsis lanceolata* L., *Agric. Biol. Chem.,* 44, 1529, 1980.

96. **Ellinger, C. A., Wong, Y., Chan, B. G., and Waiss, A. C., Jr.,** Growth inhibitors in tomato *(Lycopersicon)* to tomato fruitworm *(Heliothis zea), J. Chem. Ecol.,* 7, 753 and 758, 1981.

97. **Kamikado, T., Chang, C.-F., Murakoshi, S., Sakurai, A., and Tamura, S.,** Isolation and structure elucidation of growth inhibitors on silkworm larvae from *Magnolia kobus* DC, *Agric. Biol. Chem.,* 39, 833, 1975.

98. **Murakoshi, S., Kamikado, T., Chang, D.-F., Sakurci, A., and Tamura, S.,** Effects of several compounds from leaves of four species of plants on the growth of silkworm larvae, *Bombyx mori* L. Jap, *J. Appl. Entomol. Zool.,* 20, 26, 1976.

99. **Slansky, F., Jr.,** Effect of the lichen chemicals atranorin and vulpinic acid upon feeding and growth of larvae of yellow-striped armyworm, *Spodoptera ornithogalli, Environ. Entomol.,* 8, 865, 1979.

100. **Ohigashi, H. and Koshimizu, K.,** Chavicol, as a larva-growth inhibitor, from *Viburnum japonicum* Spreng., *Agric. Biol. Chem.,* 40, 2283, 1976.

101. **de Olivera, A. B., de Olivera, G. G., and Shaat, V. T.,** Synthesis and biological activity of eugenol derivatives, *Rev. Latinoam. Quim.,* 13, 8, 1982.

102. **Manitto, P.,** *Biosynthesis of Natural Products,* Halsted Press, New York, 1981.

103. **Fujita, H. and Yamashita, M.,** The methylenation of several allyl-benzene-1,2-diol derivatives in aprotic dipolar solvents, *Bull. Chem. Soc. Jpn.,* 46, 3553, 1973.

104. **Dallacker, F. and Sluysmans, R.,** Derivatives of methylenedioxy-benzenes, 24. Synthesis of O-safrol and myristicins, *Monatsch. Chem.,* 100, 560, 1969.

105. **Steck, W. and Brown, S. A.,** Biosynthesis of angular furanocoumarins, *Can. J. Biochem.,* 48, 872, 1970.

106. **Manimarin, T., Thiruvengadam, T. K., and Ramakrishnan, V. T.,** Synthesis of coumarins, thiacoumarins and carbostyrils, *Synthesis,* 739, 1975.

107. **Brown, S. A.,** Coumarins, in *The Biochemistry of Plants, A Comprehensive Treatise,* Vol. 7, Stumpf, P. K. and Conn, E. E., Eds., Academic Press, New York, 1981, 269.

108. **Claisen, L. and Kremers, F.,** A new synthesis of engenols, *Liebig's Ann. Chem.,* 418, 113, 1919.

109. **Tiemann, F. and Herzfeld, H.,** On the synthesis of coumarins from salicaldehyde, *Chem. Ber.,* 10, 283, 1877.

110. **Gross, G. G.,** Phenolic acids, in *The Biochemistry of Plants, A Comprehensive Treatise,* Vol. 7, Stumpf, P. K. and Conn, E. E., Eds., Academic Press, New York, 1981, 301.

111. **Borsche, W.,** Ueben den einfluss ungestahigter seiten ketten auf des kuppelungsauer mogen von phenolen und die farbung der resultirenden oxyazoverbindungen, *Chem. Ber.,* 37, 4116, 1904.

112. **Eigel, G.,** Bietrag zur kenntniss der paracumersaure, *Chem. Ber.,* 20, 2527, 1887.

113. **Hayduck, F.,** Verusche zur datstellung eines tetraoxyindigos, *Chem. Ber.,* 36, 2930, 1903.

114. **Neish, A. C.,** Preparation of caffeic acid and dihydrocaffeic acids by methods suitable for introduction of C^{14} into the β position, *Can. J. Biochem. Physiol.,* 37, 1431, 1959.

115. **Kung, H. P. and Huang, W.-Y.,** Chemical Investigation of *Draba nemerosa* L. The isolation of sinapine iodide, *J. Am. Chem. Soc.,* 71, 1836, 1949.

116. **Panizzi, L., Scarpati, M. L. and Oriente, G.,** Syntesi dell acido clorogenico, *Gazz. Chim. Ital.,* 86, 913, 1956.

117. **Palfray, L. and Sanetay, S.,** Reduction catalytiques sons hautes pressions 5 communication, *Bull. Soc. Chim. Fr.,* 5, 1423, 1977.

118. **Chatterjee, A., Bhattacharya, S., Benerji, J., and Shosh, P. C.,** A new synthesis of coumarins, *Indian J. Chem. Soc.,* 15, 214, 1977.

119. **Crosby, D. G.,** Improved synthesis of scopoletin, *J. Org. Chem.,* 26, 1215, 1961.

120. **Beroza, M. and Schechter, M. S.,** The synthesis of dl-sesamin and dL-asarinin, *J. Am. Chem. Soc.,* 78, 1242, 1956.

121. **Budowski, P.,** Recent research on sesamin, sesamolin and related compounds, *J. Am. Oil Chem. Soc.,* 41, 280, 1964.

122. **Grignard, V.,** Sur le dedoublement des ethers-oxydes de phenols par les organomagnesiens mixtes, *Compte Rendu,* 151, 322, 1910.

123. **Russell, G. D., Fenemore, P. G., and Singh, P.,** Insect-control chemicals from plants. Nagilactone C, a toxic substance from the leaves of *Podocarpus nivalis* and *P. hallii, Aust. J. Biol. Sci.,* 25, 1025, 1972.

124. **Singh, P., Fenemore, P. G., and Russell, G. B.,** Insect-control chemicals from plants. II. Effects of five natural norditerpene dilactones on the development of the housefly, *Aust. J. Biol. Sci.,* 26, 911, 1973.

125. **Meisner, J., Fleischer, A., and Eizick, C.,** Phagodeterrency induced by (−)-carvone in the larva of *Spodoptera littoralis* (Lepidoptea: Noctuidae), *J. Econ. Entomol.,* 75, 462, 1982.

126. **Waiss, A. C., Jr., Chan, B. G., Elliger, C. A., Garrett, V. H., Carlson, E. C., and Beard, B.,** Larvicidal factors contributing to host-plant resistance against sunflower moth, *Naturwissenschaften,* 64, 341, 1977.

127. **Shany, S., Gestetner, B., Birk, Y., and Bondi, A.,** Lucerne saponins. III. Effect of lucerne saponins on larval growth and their detoxification by various sterols, *J. Sci. Food Agric.,* 21, 508, 1970.

128. **Chang, C.-F., Isogai, A., Kamikado, T., Murakoshi, S., Sakurai, A., and Tamura, S.,** Isolation and structure elucidation of growth inhibitors for silkworm larvae from avocado leaves, *Agric. Biol. Chem.,* 39, 1167, 1975.

129. **Elliger, C. A., Zinkei, D. F., Chan, B. G., and Waiss, A. C., Jr.,** Diterpene acids as larval growth inhibitors, *Experientia,* 32, 1364, 1976.

130. **Guerra, A. A.,** Effect of biologically active substances in the diet on development and reproduction of *Heliothis* spp., *J. Econ. Entomol.,* 63, 1518, 1970.

131. **Murakoshi, S., Isogai, A., Chang, C.-F., Kamikado, T., Sakurai, A., and Tamura, S.,** Effects of two components from avocado leaves *(Persea americana* Mill) and the related compounds on the growth of silkworm larvae, *Bombyx mori* L., *Jpn. J. Appl. Entomol. Zool.,* 20, 87, 1976b.

132. **Salama, H. S. and El-Sharaby, A. M.,** Giberellic acid and β-sitosterol as sterilants of the cotton leafworm *Spodoptera littoralis* Viusdyvak, *Experientia*, 28, 413, 1972.

133. **Jones, S. B., Burnett, W. C., Jr., Coile, N. C., Mabry, T. J., and Betkouski, M. F.,** Sesquiterpene lactones of *Vernonia* — influence of glaucolide-A on growth rate and survival of lepidopterous larvae, *Oecologia (Berlin)*, 39, 71, 1979.

134. **Nakajima, S. and Kawazu, K.,** Coumarin and euponin, two inhibitors of insect development from leaves of *Eupatorium japonicum*, *Agric. Biol. Chem.*, 44, 2893, 1980b.

135. **Langenheim, J. H., Foster, C. E., and McGinley, R. B.,** Inhibitory effects of different quantitative compositions of *Hymenaea* leaf resins on a gemeralist herbivore *Spodoptera exigua*, *Biochem. Syst. Ecol.*, 8, 385, 1980.

136. **Poppleton, B. J.,** Podolactone p-bromobenzoate. Stereochemistry and absolute configuration, *Cryst. Struct. Commun.*, 4, 101, 1975.

137. **Leets, K. V., Shumeiko, A. K., Rozenoer, A. A., Kudryasheva, N. V., Pilyavskaya, A. I.,** A new synthesis of citrol from isoprene, *J. Gen. Chem. U.S.S.R.*, 27, 1584, 1957.

138. **Ruzicka, L. and Firmenich, G.,** Zur kentnis der diterpene synthese des aliphatischen diterpenalkohole geranyl-geraniol, *Helv. Chim. Acta*, 22, 392, 1939.

139. **Popjak, G., Cornforth, J. W., Conforth, R. H., Ryhage, R., and Goodman, D. S.,** Studies in the biosynthesis of cholesterol. XVI. Chemical synthesis of $1-H_2^3-2-C^{14}$- and $1-D_2-2C^{14}$-trans-trans-farnesyl pyrophosphate and their use in squalene biosynthesis, *J. Biol. Chem.*, 237, 56, 1962.

140. **Burrell, J. W. K., Garwood, R. F., Jackman, L. M., Oskay, E., and Weedon, W. C. L.,** Carotenoids and related compounds. XIV. Stereochemistry and synthesis of geraniol, nerol, farnesol and phytol, *J. Chem. Soc. C*, p. 2144, 1966.

141. **Honwad, V. K., Siscovic, E., and Rao, A. S.,** Terpenoids. LXXXIV. Hydrazine reduction of α,β-epoxyketones from carvone and arturmerone and conversion of (−)carvone to its enantiomer, *Indian J. Chem.*, 5, 234, 1967.

142. **Akhita, A., Benthorpe, D. V., and Rowan, M. G.,** Biosynthesis of carvone in *Mentha* spicta, *Phytochemistry*, 19, 1433, 1980.

143. **Oonk, H. A. J. and Kroon, J.,** The carvoxime system. I. X-ray study of dk-carvoxime (mp 92°C), *Acta Cryst. Sect. B*, 32, 500, 1976.

144. **Metzger, J. D., Baker, M. W., and Morris, R. J.,** The synthesis of the A, B and D, E rings of medicagenic acid, *J. Org. Chem.*, 37, 789, 1972.

145. **Nowaki, E., Jurzysta, M., and Dietrych-Szostak, D.,** Biosynthesis of medicagenic acid in germinating alfalfa, *Biochem. Physiol. Pflanz.*, 169, 183, 1976.

146. **Corey, E. J., Brennan, T. M., and Carney, R. L.,** Stereospecific elaboration of the A ring of gibberellic acid by partial synthesis, *J. Am. Chem. Soc.*, 93, 7316, 1971.

147. **Krishnamurthy, H. N., Ed.,** *Gibberellins and Plant Growth*, Wiley/Halsted, New York, 1975.

148. **Watson, W. H. and Wu, I. B., Monti, S. A., David, R. E., Mabry, T. J., and Padolina, W. G.,** Dihidrodesacetoxyglaucolide-A, C_{21},- $H_{28}O_8$, *Cryst. Struct. Comm.*, 3, 697, 1974.

149. **Cox, P. J. and Sim, G. A.,** Sesquiterpenoids. XIX. X-ray crystallographic determination of the stereochemistry and conformation of the germecranolide glaucolide A, *J. Chem. Soc. Trans. Perkin*, 2, 455, 1975.

150. **Ghetty, G. L., Krishna-Rao, G. S., Dev, S., and Bannerjee, D. K.,** Studies in sesquiterpenes. XXV. A synthesis of α-selinene, *Tetrahedron Lett.*, 22, 2311, 1966.

151. **Marshall, J. A., Piuke, M. T., and Carroll, R. D.,** Studies leading to the stereoselective total synthesis of dl-β-eudesmol, dl-β-selinene, dl-costol and related naturally-occurring sesquiterpenes, *J. Org. Chem.*, 31, 2933, 1966.

152. **Vig, O. P., Anand, R., Kumar, B., and Sharma, S. D.,** Terpenoids. XXXVII. Synthesis of dl-β-selinene and dl-sudesmol, *J. Indian Chem. Soc.*, 45, 1033, 1968.

153. **Corey, E. J., Mitra, R. B. and Uda, H.,** Total synthesis of d,l-carophyllene and d,l-isocaryophyllene, *J. Am. Chem. Soc.*, 86, 485, 1964.

154. **Torsell, K. B. G.,** *Natural Product Chemistry, A Mechanistic and Biosynthetic Approach to Secondary Metabolism*, John Wiley & Sons, New York, 1983.

155. **Halsal, T. G. and Theobald, D. W.,** Recent aspects of sesquiterpenoid chemistry, *Q. Rev. (London)*, 16, 101, 1962.

156. **Rembold, H. and Schmutter, H.,** Disruption of insect growth by neem seed components, in *Regulation of Insect Development and Behaviour, International Conference Part II*, Technical University of Poland, Wroclaw, 1981, 1087.

157. **Warthen, J. D., Jr. and Uebel, E. C.,** Effect of azadirachtin on house crickets, *Acheta domesticus*, *Proc. 1st Int. Neem. Conf.*, (Rottach Egem, 1980), 1980, 137.

158. **Ruscoe, C. N. E.,** Growth disruption effects of an insect antifeedant, *Nature (London) New Biol.*, 236, 159, 1972.

159. **Redfern, R. E., Warthen, J. D., Jr., Uebel, E. C., and Mills, G. D., Jr.,** The antifeedant and growth-disrupting effects of azadirachtin on *Spodoptera frugiperda* and *Oncopeltus fasciatus. Proc. 1st Int. Neem. Conf.,* (Rottach Egem, 1980), 1980, 129.

160. **Redfern, R. E., Warthen, J. D., Jr., Mills, G. D., Jr., and Uebel, E. C.,** Molting Inhibitory Effects of Azadirachtin on Large Milkweed Bug, USDA, Agric. Res. Results No. ARR-NE-5, U.S. Department of Agriculture, Washington, D.C., 1979.

161. **Kubo, I. and Klocke, J. A.,** An insect growth inhibitor from *Trichilia roka, Experientia,* 38, 639, 1982.

162. **Fujimoto, Y. and Ikekawa, N.,** Convenient preparation of the C-24 stereoisomers of 24-ethyl and 24-methyl cholesterols, *J. Org. Chem.,* 44, 1011, 1979.

163. **Aexel, R. T., Evans, S., Kelley, M. T., and Nicholas, H. J.,** Biosynthesis and metabolism of β-sitosterol, β-amyrin, and related methyl sterols, *Phytochemistry,* 6, 511, 1967.

164. **Bottger, G. T. and Patana, R.,** Growth, development, and survival of certain Lepidoptera fed gossypol in the diet, *J. Econ. Entomol.,* 59, 1166, 1966.

165. **Lukefahr, M. J. and Martin, D. F.,** Cotton plant pigments as a source of resistance to the bollworm and tobacco budworm, *J. Econ. Entomol.,* 59, 176, 1966.

166. **El-Sebae, A. H., Sherby, S. M., and Mansour, A. M.,** Gossypol suppresses development and inhibits protease and lipid peroxidation activities of *Spodoptera littoralis, Toxicol. Lett.,* 5, 51, 1980.

167. **Maxwell, F. G., Jenkins, J. N., and Parrott, W. L.,** Influence of constituents of the cotton plant on feeding, oviposition, and development of the boll weevil, *J. Econ. Entomol.,* 60, 1294, 1967.

168. **Shaver, T. N., Lukefahr, M. J., and Garcia, J. A.,** Food utilization, ingestion, and growth of larvae of the bollworm and tobacco budworm on diets containing gossypol, *J. Econ. Entomol.,* 63, 1544, 1970.

169. **Shaver, T. N. and Parrott, W. L.,** Gossypol against cotton insects. Relationship of larval age to toxicity of gossypol to bollworms, tobacco budworms, and pink bollworms, *J. Econ. Entomol.,* 63, 1802, 1970.

170. **Shaver, T. N., Dilday, R. H., and Garcia, J. A.,** Interference of gossypol in bioassay for resistance to tobacco budworm in cotton, *Crop Sci.,* 18, 55, 1978.

171. **Kay, I. R., Noble, R. M., and Twine, P. H.,** The effect of gossypol in artificial diet on the growth and development of *Heliothis punctigera* Wallengren and *H. armigera* (Hubner) (Lepidoptera: Noctuidae), *J. Aust. Entomol. Soc.,* 18, 229, 1979.

172. **Lukefahr, M. J., Stipanovic, R. D., Bell, A. A., and Gray, J. B.,** Biological activity of new terpenoid compounds from *Gossypium hirsutum* against the tobacco budworm and pink bollworm, *Proc. Beltwide Cotton Conf.,* Natl. Cotton Council, Memphis, Tenn., 1977, 97.

173. **Dongre, T. K. and Rahalkar, G. W.,** Growth and development of spotted bollworm, *Earias vitella* on glanded and glandless cotton and on diet containing gossypol, *Entomol. Exp. Appl.,* 27, 6, 1980.

174. **Elliger, C. A., Chan, G. B., and Waiss, A. C., Jr.,** Relative toxicity of minor cotton terpinoids compared to gossypol, *J. Econ. Entomol.,* 71, 161, 1978.

175. **Edwards, J. D.,** Total synthesis of gossypol, *J. Am. Chem. Soc.,* 80, 3798, 1958.

176. **Berardi, L. C. and Goldblatt, L. A.,** Gossypol, in *Toxic Constituents of Plants and Food Stuffs,* Liener, I. E., Ed., Academic Press, New York, 1969.

177. **Edwards, J. D.,** Synthesis of gossypol and gossypol derivatives, *J. Am. Oil Chem. Soc.,* 47, 441, 1970.

177a. **McDaniel, C. N. and Berry, S. J.,** Effects of caffeine and aminophylline on adult development of the cecropia silkmoth, *J. Insect Physiol.,* 20, 245, 1974.

178. **Miller, S., Collins, J. M., and Frenkel, L. D.,** The effect of vinblastine sulphate on the incorporation of (2-^{14}C) glycine into housefly DNA, *Insect Biochem.,* 2, 87, 1972.

179. **Parr, J. C. and Thurston, R.,** Toxicity of nicotine in synthetic diets to larvae of the tobacco hornworm, *Ann. Entomol. Soc. Am.,* 65, 1185, 1972.

180. **Muthkrishnan, J., Matharan, S., and Venkatasubbu, K.,** Effect of caffeine and theophylline on food utilization and emergence in *Danaus chrysippus* (Lepidoptera:Danidae), *Entomology,* 4, 309, 1979.

181. **Campbell, B. C. and Duffey, S. S.,** Tomatine and parasitic wasps: potential incompatibility of plant antibiosis with biological control, *Science,* 205, 700, 1979.

182. **Huebner, E. and Anderson, E.,** The effects of vinblastine sulfate on the microtubular organization of the ovary of *Rhodnius prolixus, J. Cell. Biol.,* 46, 191, 1970.

183. **Kametani, T., Noguchi, I., Saito, K., and Kenada, S.,** Studies on the synthesis of heterocyclic compounds. CCCII. Alternative total synthesis of (±) nandinene, (±)-canadine and berberine iodide. *J. Chem. Soc. C,* p. 2036, 1969.

184. **Gear, J. R. and Spencer, I. D.,** The biosynthesis of hydrastine and berberine, *Can. J. Chem.,* 41, 783, 1963.

185. **Hahn, F. E. and Ciak, J.,** Berberine, in *Antibiotics,* Vol. 3, Corcoran, J. W. and Hahn, F. E., Eds., Springer-Verlag, Basel, 1975, 577.

186. **Fischer, E. and Ach, F.,** Synthese des caffeins, *Chem. Ber.,* 28, 3135, 1895.

187. **Suzuki, T. and Takahashi, E.,** Caffeine biosynthesis in *Camellia sinensis, Phytochemistry,* 15, 1235, 1976.

188. **Sutor, D. J.,** The structures of the pyrimidines and purines. VII. The crystal structure of caffeine, *Acta Cryst.*, 11, 453, 1958.

189. **Bredereck, H. and Gotsmann, U.,** Synthesen in der purinreihe. XVI. Uber die dartstllung von 5-alkyl-bzw-5-arylsulfonylamino-4-amino-uracilen, 4-amino-5-alkylamino-uracilen und 4-amino-5-[pyridinio-methylenamino]-uracil-chloriden, *Chem. Ber.*, 95, 1902, 1962.

190. **Scott, A. I., Lee, S.-L., Culver, M. G., Wan, W., Hirata, T., Guirette, F., Baxter, R. L., Nordlov, H., and Dorschel, C. A.,** Indole alkaloid biosynthesis, *Heterocycles*, 15, 1257, 1981.

191. **Pinner, A.,** Ueber nicotin. Die constitution des alkaloids, *Chem. Ber.*, 26, 292, 1893.

192. **Smith, T. A.,** Amines, in *The Biochemistry of Plants, A Comprehensive Treatise*, Vol. 7, Stumpf, P. K. and Conn, E. E., Eds., Academic Press, New York, 1981, 249.

193. **Jackson, K. E.,** Alkaloids of tobacco, *Chem. Rev.*, 29, 123, 1941.

194. **Craig, L. C.,** A new synthesis of nornicotine and nicotine, *J. Am. Chem. Soc.*, 55, 2854, 1933.

195. **Traube, W.,** Der synthetische aufban der hornsaure, des xanthins, theobromins, theophylline und caffeins aus der cyanessigsaure, *Chem. Ber.*, 33, 3035, 1900.

196. **Gebner, B. and Kreps, L.,** A new synthesis of caffeine, theophylline and theobromine, *J. Gen. Chem. U.S.S.R.*, 16, 179, 1946.

197. **Chan, B. G., Waiss, A. C., Jr., and Lukefahr, M.,** Condensed tannin, an antibiotic chemical from *Gossypium hirsutum, J. Insect Physiol.*, 24, 113, 1978.

198. **Feeny, P. P.,** Effect of oak leaf tannins on larval growth of the winter moth, *Operophtera brumata, J. Insect Physiol.*, 14, 805, 1968.

199. **Bernays, E. A., Chamberlain, D. J., and Leather, E. M.,** Tolerance of acridids to ingested condensed tannin, *J. Chem. Ecol.*, 7, 247, 1981.

200. **Waiss, A. C., Jr., Chan, B. G., Elliger, C. A., and Binder, R. G.,** Biologically active cotton constituents and their significance in HPR, *Beltwide Cotton Prod. Res. Conf.*, Natl. Cotton Council, Memphis, Tenn., 1981.

201. **Kreber, R. A. and Einhellig, F. A.,** Effects of tannic acid on *Drosophila* larval salivary gland cells, *J. Insect Physiol.*, 18, 1089, 1972.

202. **Waiss, A. C., Jr., Chan, B. G., Elliger, C. A., Wiseman, B. R., McMillian, W. W., Widstrom, N. W., Zuber, M. S., and Keaster, A. J.,** Maysin, a flavone glycoside from corn silks with antibiotic activity toward corn earworm, *J. Econ. Entomol.*, 72, 256, 1979.

203. **Elliger, C. A., Chan, B. G., and Waiss, A. C., Jr.,** Flavonoids as larval growth inhibitors. Structural factors governing toxicity, *Naturwissenschaften*, 67, 358, 1980.

204. **Duffey, S. S. and Isman, M. B.,** Inhibition of insect larval growth by phenolics in glandular trichomes of tomato leaves, *Experientia*, 37, 574, 1981.

205. **Shakhova, M. K., Samokhvalov, G. I., and Preabrazhenskii, N. A.,** Synthetic investigations in the field of flavonoids. III. Total synthesis of quercetin-3-β-rutinoside, rutin, *J. Gen. Chem. U.S.S.R.*, 32, 382, 1962.

206. **Watkins, J. E., Underhill, E. W., and Neish, A. C.,** Biosynthesis of quercetin in buckwheat. II, *Can. J. Biochem. Physiol.*, 35, 229, 1957.

207. **Hahlbrook, K.,** Flavonoids, in *The Biochemistry of Plants, A Comprehensive Treatise*, Vol. 7, Stumpf, P. K. and Conn, E. E., Eds., Academic Press, New York, 1981.

208. **Patschke, L., Barz, W., and Grisenbach, H.,** Stereospifischer einban von (−) 5.7.4'-trihydroxyflavonon in flavonoids und isoflavons, *Z: Naturforsch.*, 21b, 201, 1966.

209. **Ingham, J. L.,** Naturally occurring isoflavonoids (1855—1981), *Prog. Chem. Org. Nat. Prod.*, 43, 1, 1983.

210. **Grisebach, H.,** The biosynthesis of flavonoids, *Biochem. J.*, 85, 3P, 1962.

210a. **Grisebach, H.,** Lignins, in *The Biochemistry of Plants, A Comprehensive Treatise*, Vol. 7, Stumpf, P. K. and Conn, E. E., Eds., Academic Press, New York, 1981.

211. **Griffith, J. Q., Jr.,** *Rutin and Related Flavonoids*, Mack, Easton, Pa., 1955.

212. **Zaprometov, M. N. and Grisebach, H.,** Dihydrokaempferol as precursor for *Catechinsin* the tea plant, *Z. Naturforsch.*, 28c, 113, 1973.

213. **Hutchins, W. A. and Wheeler, T. S.,** Chalcones: a new synthesis of chrysin, apigenin and luteolin, *J. Chem. Soc.*, 91, 1939.

214. **Reichel, L., Burkart, W., and Muller, K.,** Synthesis of oxy-chalcone and oxy-flavonone, *Liebig's Ann. Chem.*, 550, 146, 1942.

215. **Pew, J. C.,** Conversion of dihydro-quercetin to eriodictyl, *J. Org. Chem.*, 27, 2935, 1962.

216. **Jain, A. C., Tuli, D. K., and Gupta, R. C.,** Synthesis of pomiferin, auriculasin and related compounds, *J. Org. Chem.*, 43, 3446, 1978.

217. **Chopin, J., Durix, A., and Bouillant, M. L.,** C-glucosylation des dihydroxy-5,7 flavones, *Tetrahedron Lett.*, p. 3657, 1966.

218. **Hostettmann, M. and Jacot-Guillarmod, A.,** Xanthones and Flavone-c-glycosides from the genera Gentiana, *Phytochemistry*, 17, 2083, 1978.

219. **Jouanne, M. and Mentzer, C.,** Synthese totale de la scutellareine par un nouveau procede de condensation thermique, *C.R. Acad. Ser. C.,* 263, 1022, 1966.
220. **Farkas, L., Mezey-Vandor, G., and Nogradi, M.,** Die synthese des scutellarins, plantaginins, scutellarein-7-β-rutinoids und de erste harstellung des isoscutillareins, *Chem. Ber.,* 107, 3878, 1974.
221. **Gaydou, E. M. and Bianchini, J.-P.,** Etudes decomposes flavoniques. I. Synthese et proprietes (UV, RMN de 13C) de quelques flavones, *Bull. Soc. Chim. Fr.,* II, 43, 1978.
222. **Whittman, H. and Hehnberger, U.,** Cleavages with diazonium compounds and quinone imidichloride. XIV. Reactions of hydroxyflavones with coupling reagents, *Z. Naturforsch.,* B25, 820, 1970.
223. **Kalff, J. and Robinson, R.,** A synthesis of myricetin and of a galangin monomethyl ether occurring in galanga root, *J. Chem. Soc.,* 127, 181, 1925.
224. **Beck, S. D. and Stauffer, J. F.,** The European corn borer, *Pyrausta nubilalis* (Hubn.), and its principal host plant. III. Toxic factors influencing larval establishment, *Ann. Entomol. Soc. Am.,* 50, 166, 1957.
225. **Beck, S. D. and Smissman, E. E.,** The European corn borer, *Pyrausta nubilalis,* and its principal host plant. IX. Biological activity of chemical analogs of corn resistance factor A (6-methoxybenzoxazolinone), *Ann. Entomol. Soc. Am.,* 54, 53, 1961.
226. **Beck, S. D.,** The European corn borer, *Pyrausta nubilalis* (Hubn.), and its principal host plant. VII. Larval feeding behavior and host plant resistance, *Ann. Entomol. Soc. Am.,* 53, 206, 1960.
227. **Klun, J. A., Tipton, C. L., and Brindley, T. A.,** 2,4-Dihydroxy-7-methoxy-1, 4-benzoxazin-3-one (DIMBOA), an active agent in the resistance of maize to the European corn borer, *J. Econ. Entomol.,* 60, 1529, 1967.
228. **Allen, E. H. and Laird, S. K.,** Improved preparation of 6-methoxybenzoxazaline, *J. Org. Chem.,* 36, 2004, 1971.
229. **Richey, J. D., Caskey, A. L., Miller, J. N.,** 6-Methoxy-2-benz-oxazoline synthesis in 1,3-butane diol, *Agric. Biol. Chem.,* 40, 2413, 1971.
230. **Hutchins, J. E. C. and Fife, T. H.,** Fast intramolecular nucleophilic attack by phenoxide ion on carbamate ester groups, *J. Am. Chem. Soc.,* 95, 2282, 1973.
231. **Birk, Y. and Applebaum, S. W.,** Effect of soybean trypsin inhibitors on the development and midgut proteolytic activity of *Tribolium castaneum* larvae, *Enzymologia,* 22, 318, 1960.
232. **Steffens, R., Fox, F. R., and Kassell, B.,** Effect of trypsin inhibitors on growth and metamorphosis of corn borer larvae *Ostrinia nubilalis, J. Agric. Food Chem.,* 26, 170, 1978.
233. **Niemala, P., Aro, E. M., and Haukioja, E.,** Birch leaves as a resource for herbivores. Damage-induced increase in leaf phenols with trypsin-inhibiting effects, *Rep. Kevo Subarctic Res. Stn.,* 15, 37, 1979.
234. **Ishii, S., Hirano, C., Iwata, Y., Nakasawa, M., and Miyagawa, H.,** Isolation of benzoic and salicylic acids from the rice plant as growth inhibiting factors for the rice stem borer *(Chilo suppressalis* Walker) and some rice plant fungus pathogens, *Jpn. J. Appl. Entomol. Zool.,* 61, 281, 1962.
235. **Reese, J. C. and Beck, S. D.,** Effects of plant allelochemics on the black cutworm, *Agrotis ipsilon:* effects of *p*-benzoquinone, hydroquinone, and duroquinone on larval growth, development and utilization of food, *Ann. Entomol. Soc. Am.,* 69, 59, 1976a.
236. **Reese, J. C. and Beck, S. D.,** Effects of allelochemics on the black cutworm, *Agrotis ipsilon:* effects of resorcinol, phloroglucinol, and gallic acid on larval growth, development, and utilization of food, *Ann. Entomol. Am.,* 69, 999, 1976c.
237. **Reese, J. C.,** Chronic effects of plant allelochemics on insect nutritional physiology, *Entomol. Exp. Appl.,* 24, 625, 1978.
238. **Rafaeli-Bernstein, A. and Mordue, W.,** The effects of phlorozin, phloretin and ouabain on the reabsorption of glucose by the Malpighian tubules of *Locusta migratoria nigratorioides, J. Insect Physiol.,* 25, 241, 1979.
239. **Faith, W. L., Keyes, D. B., Clark, R. L.,** *Industrial Chemicals,* 3rd ed., John Wiley & Sons, New York, 1965, 625.
240. **Leistner, E.,** Biosynthesis of Plant Quinones, in *The Biochemistry of Plants, A Comprehensive Treatise,* Vol. 7, Stumpf, P. K. and Conn, E. E., Eds., Academic Press, New York, 1981.
241. **von Bolhuis, F. and Kiers, C. Th.,** Refinement of the crystal structure of *p*-benzoquinone at −160°C, *Acta Cryst. Sect. B,* 34, 1015, 1978.
242. **Heertjes, P. M.,** Phloroglucinol from picryl chloride, *Rec. Trav. Chim.,* 78, 452, 1959.
243. **McKillop, A., Howarth, B. D., and Kobylecki, R. J.,** A simple and inexpensive preparation of phloroglucinol and phloroglucinol trimethyl ether, *Syn. Commun.,* 4, 35, 1974.
244. **Shipchandler, M. T., Peters, C. A., and Hurd, C. D.,** Synthesis of gallic acid and pyrogallol, *J. Chem. Soc. Perkin Trans. I,* 1400, 1975.
245. **Haslam, E., Haworth, R. D., and Knowles, P. F.,** Gallotannins. IV. The biosynthesis of gallic acid, *J. Chem. Soc.,* p. 1854, 1961.
246. **Kato, Y. and Yasue, M.,** Pyrogallol and gallic acid derivatives, *Nagoya Shiritsu Daigaku Yakugakubu Kiyo,* 10, 59, 1962 (*Chem. Abstr.,* 59, 11322c).

247. **Cannizzaro, S.,** On benzoic acid and its corresponding alcohol, *Liebig's Ann. Chem.,* 88, 129, 1853.

248. **Vogel, A. D.,** *Practical Organic Chemistry,* 3rd ed., Longmans, London, 1959, 711.

249. **Lane, C. F., Myatt, H. L., Daniels, J., and Hopps, H. B.,** Organic synthesis using borane-methyl sulfide. II. Reduction of aromatic carboxylic acids in the presence of trimethyl borate, *J. Org. Chem.,* 39, 3052, 1974.

250. **Rinderknecht, H. and Niemann, C.,** Esterification of acylated alpha-amino acids, *J. Am. Chem. Soc.,* 70, 2605, 1948.

251. **Tiemann, F. and Haarmann, W.,** Ueber des coniferin und seine umwandlung in des aromatioche princop der vanille, *Chem. Ber.,* 7, 608, 1874.

252. **Mauthner, F.,** Untersuchungen uber die δ-resorcylsaure. II., *J. Prakt. Chem.,* 124, 319, 1930.

253. **Chatterjee, A., Sanguly, D., and Sen, R.,** New synthesis of 4-phenyl coumarins: dalbergin and nordalbergin, *Tetrahedron Lett.,* 32, 2407, 1976.

254. **Senhofer, Von C. and Sarlay, L.,** Uber directe einfuhyrun von carboxylguppin in phenole und aromatische sauren, *Monatsch. Chem.,* 2, 448, 1881.

255. **Gatenbach, S. and Linnroth, I.,** The biosynthesis of gentisic acid, *Acta Chem. Scand.,* 16, 2298, 1962.

256. **Lowenthal, J. and Pepper, J. M.,** An improved preparation of gentisic acid, *J. Am. Chem. Soc.,* 72, 3292, 1950.

257. **Pearl, I. A.,** Vanillic acid. I. Silver oxide method. II. Caustic fusion method, *Org. Syn. C.V. IV,* 972, 1950.

258. **Zemplen, G. and Bognar, R.,** Synthesis of naturally-occurring phlorrhizins, *Chem. Ber.,* 75B, 1040, 1942.

259. **Fischer, E. and Nouri, O.,** Synthese des phloretins und darstellung der nitrile von phenol-carbonsauren, *Chem. Ber.,* 50, 611, 1917.

260. **Neelakantan, S., Padmasami, R., and Seshadri, T. R.,** Synthesis of atranorin, *Tetrahedron Lett.,* p. 287, 1962.

261. **Ritter, J. J. and Ginsburg, D.,** The action of t-butyl hypochlorite on α-pinene, *J. Am. Chem. Soc.,* 72, 2381, 1950.

262. **Strubell, W. and Baumgartel, H.,** Weitere ergebrisse uber die arvacrol synthese. IV. Mitteilung uber *p*-cymol und seine derivate, *Arch. Pharm.,* 291, 66, 1958.

263. **Dreyer, D. L., Binder, R. G., Chan, B. G., Waiss, A. C., Jr., Hartwig, E. E., and Beland, G. L.,** Pinitol, a larval growth inhibitor for *Heliothis zea* in soybeans, *Experientia,* 35, 1182, 1979.

264. **Smissman, E. E., Beck, S. D., and Boots, M. R.,** Growth inhibition of insects and a fungus by indole-3-acetonitrile, *Science,* 133, 1961.

265. **Blau, P., Feeny, P., Contardo, L., and Robson, D. S.,** Allylgluco-sinolate and herbivorous caterpillars: a contrast in toxicity and tolerance, *Science,* 200, 1296, 1978.

266. **Erickson, J. M. and Feeny, P.,** Sinigrin: a chemical barrier to the black swallowtail butterfly, *Papilio polyxenes, Ecology,* 55, 103, 1974.

267. **Hoffman, H., Wagner, I., and Hoffman-Ostehof, O.,** Biosynthesis of cyclitols. XXIV. A soluble enzyme from *Vinca rosea* methylating myo-inositol to L-bornesitol, *Z. Physiol. Chem.,* 350, 1465, 1969.

268. **Jacobson, M.,** Constituents of *Heliopais* species. IV. The total synthesis of *trans*-affinin, *J. Am. Chem. Soc.,* 77, 2461, 1955.

269. **Benn, H. M. and Ettlinger, M. G.,** The synthesis of sinigrin, *Chem. Comm.,* 445, 1965.

270. **Waser, J. and Watson, W. H.,** Crystal structure of sinigrin, *Nature (London),* 198, 1297, 1963.

271. **Billek, J. and Schmook, F. P.,** The biosynthesis of gentisic acid, *Monatsch Chem.,* 98, 1651, 1967.

ACUTE INSECT TOXICANTS FROM PLANTS

Caleb W. Holyoke and John C. Reese

INTRODUCTION

Mankind has been using plant products to control insects for many centuries. While some plant products have probably acted as repellents, others have been immediately toxic to the insect. This review focuses on the latter category.

The use of plant parts and extractives as insecticides is covered in numerous texts,[80,110,112,126] general reviews,[1,5,91] and chemically specialized reviews (to be mentioned later). Rosenthal and Janzen[103] approach the subject from a broad ecological perspective of plant-herbivore interactions while Schoonhoven's[122] recent review more specifically addresses plant substances which are either toxic to or reduce growth of insects. Jacobson[26,27] and more recently Datta[3] have compiled reports of plants or materials derived from plants which are toxic to insects. Jacobson[2] and Crosby[87] have both addressed the general question of finding insecticidal materials from plants. (Volume 2 of this series[7] provides thorough coverage of isolation and identification techniques for naturally occurring pesticides.)

Recent research on natural products that affect insects appears to have moved away from toxicity studies to behavioral studies. This shift is made apparent by the contents of recent symposia volumes edited by Whitehead and Bowers[79] and Takahashi et al.[291] In both of these volumes the vast majority of new results address behavioral changes (antifeedant or repellent) induced by allelochemicals and there are few recent efforts to identify toxicants.

Although the word "toxicant" has been used to describe a wide variety of chemicals, many of these compounds may not function as acute toxicants. For example, toxicants may reduce assimilation, ingestion, or efficiency of conversion of food. However, total insect mass produced from a given amount of diet is not necessarily a measure of acute toxicity.[4] Even mortality at 6 days post-treatment is difficult to identify as acute toxicity.[5] Starvation effects from a strong antifeedant could easily be a contributing factor. The authors have tried, therefore, in compiling the following tables, to look for reports in which the activity exceeded simple starvation effects. Antifeedants are discussed elsewhere in this volume. Compounds which produce toxic effects by acting on insect hormone systems (e.g., phytoectysteroids, precocenes, etc.) are also covered elsewhere in this volume. Therefore, this review excludes those compounds which are antifeedants or hormone agonists/antagonists.

The coverage has been limited to well-characterized, pure compounds, rather than including biologically active extracts of unknown composition. Application methods differ widely from one report to the next (topical application, diet incorporation, injection, etc.), thus, reporting concentrations needed for activity is difficult; however, as much information as possible has been included for the reader. Finally, a large number of different species have been used, including species that would never attack the toxic plant.[6]

In addition, there is often difficulty in finding reliable reports of pure compounds being tested against a range of insect pests. Rather, one tends to find a wealth of data for crude plant extracts with little, if any, data on the pure isolated components.[87] This can occasionally lead to mistaken identification of active components. Crosby[87] shows structures for the putative active components in *Mammea* extracts which were later shown by Crombie et al.[263] to be due to somewhat different structures. That sort of problem also arises when a major constituent of a natural insecticide is known and presumed to be the active component. For example, Iwalu et al.[28] reported the insecticidal activity of danettia oil as due to two insect pests. They noted that the principal component of danettia oil was β-phenylnitroethane and concluded: "The active principle of danettia oil is β-phenylnitroethane." This conclusion was shown to be erroneous by Harvey et al.[29], leaving the question of the active component an interesting mystery. This review has been restricted to those materials which have been tested as pure, well-characterized compounds. There are polymeric natural products which are toxic to insects, but are not well characterized. Interested readers are directed to chapters on the subject in Rosenthal and Jansen.[103]

ALKALOIDS (Tables 1a, b)

Pelletier[290] points out that there are several definitions of an alkaloid. In this review only alkaline nitrogen-containing compounds and their quatemary salts are considered alkaloids, thus the nonbasic isobutyl- amides and maytansanoids are considered elsewhere.

Historically, the most important class of alkaloids for insect control has been the nicotinoids. Reviews by Schmeltz,[124] and Yamamoto[136] cover the chemistry, structure-activity relationships, and mode of action of nicotinoids. These compounds have been used in the form of tobacco extracts as insecticides for nearly 300 years[124] and have consequently developed an enormous literature. Schmeltz[124] has pointed out that the nicotinoids are most effective against "minute insects with soft bodies such as aphids". Their biosynthesis has been recently reviewed by Enzell et al.[86] and Leete.[106]

There are a number of other chemically and biogenetically distinct classes of alkaloids which are toxic to insects. Crosby[87] has discussed the *Sabadilla* alkaloids and *Ryania* alkaloids derived from the corresponding plant insecticides. The *Sabadilla* alkaloids of commerce are derived from the seeds of the lily *Schoenocaulon officinale* A. Gray. These alkaloids are also found in a number of *Veratrum* species and are known generally as "Veratrum alkaloids". Preparations containing these compounds appeared to be photo- or oxidatively unstable.[87] This handicap limited their use despite the wide range of insects controlled. The *Veratrum* alkaloids also have a wide range of biological activities in mammals, some of which are discussed by Crosby.[87]

Rayanodine, the toxic principle of *Ryania*, appears to have a fairly wide spectrum of activity against insect pests and has enjoyed some success. Crosby[87] notes its use as a stomach poison for control of lepidopterous larvae with use mainly against European corn borer *(Ostrinia nubilalis)*. The data for the commonly used ground plant or extractive[111] supports the range of activity.

The alkaloid physostigmine is the only known natural carbamate ester and is a very active anticholinesterase agent. Despite this, it has negligible insecticidal activity to several species by either contact or ingestion;[295] however, by injection it proved toxic to locusts.[296]

The remaining alkaloids follow no distinct pattern. Given the wide range of biological activities of alkaloids, it is probable that there are more insect toxicants from this class awaiting discovery. Biosynthetic pathways for alkaloids and the *N*-heteromatic nitrogenous compounds are discussed by Torsell.[132]

AMINO ACIDS (Tables 2a, b)

The nonprotein amino acids reviewed by Rosenthal and Bell[116] and Fowden[293] typically show a lower order of acute toxicity than the other classes of compounds in this review. However, plants producing them will often respond by producing high concentrations of these toxins in vulnerable tissue, the seeds of *Dioclea megacarpa* containing up to 13% dry weight of canavanine.[116] These insect toxins appear to be primarily of interest from an ecological point of view and do not appear to have potential as insect control chemicals by man. Fowden[293] has reviewed their biosynthesis and distribution. Torsell[132] has reviewed amino acid synthesis in general terms.

UNSATURATED ISOBUTYLAMIDES (Tables 3a, b)

These pepper-derived toxicants offer protection to the plants which produce them at relatively low concentrations. As is pointed out in most early references, they are also potent sialogogues for mammals and leave a burning taste. Those properties confer broad protection against most herbivores, save humans who prize the "bite" which comes with pepper. Jacobson[101] has reviewed these compounds but there has since been consistent effort in this area. More recently, Miyakado et al.[117] have reviewed the chemistry of *Piperaceae* amides and their own biological results.

Biogenetically, these compounds appear to share the fatty acid pathway, although there are no direct studies on their biosynthesis. Torsell[132] briefly summarized the biosynthesis of unsaturated fatty acids.

It should be noted that although the authors show piperine, the only nonisobutylamide in the tables as being active,[179] Miyakado et al.[117] and Ressler et al.[212] report that they are unable to substantiate this activity.

This class of compounds offers perhaps the most appealing natural product model for development of new insecticides since the pyrethroids. The problems for these compounds to overcome seem to be similar to the hurdles for the natural pyrethroids — photostability and susceptibility to metabolism in target species.

PYRETHROIDS (Tables 4a, b)

The pyrethroids are the most broadly active class of naturally occurring insecticides, so it is not surprising that they were the first, and to date only, natural product which has spawned a new class of synthetic insecticides. The pyrethroids have been reviewed in some detail in Cassida's book *Pyrethrum, the Natural Insecticide,*[83] by Matsui and Yamamoto,[109] and by several authors in Whitehead and Bowers' *Natural Products for Innovative Pest Management.*[79] The chemistry of the natural pyrethroids was summarized by Elliott and Janes[257] and the biosynthesis by Epstein and Poulter.[89] Recent reviews have focused almost exclusively on the synthetic pyrethroids.

The biological data for pure constituents of the pyrethrum extract are much more sparse than for the other principal natural insecticides, nicotine and rotenone. This is probably because of the ease of separation and purification of both nicotine and rotenone relative to the pyrethroids. Despite our inclusion of bicomponent pyrethroid mixtures the data tabulated in this chapter fail to show the range of effectiveness of "pyrethroids" because the screening was done usually with the article of commerce, crude "pyrethrum". Those interested in more detail should look first at Negherbon's[111] compilation.

STEROIDS AND OTHER TRITERPENES (Tables 5a, b)

There are relatively few nonecdysteroid triterpenes reported to be insect toxicants. Despite the lack of the characteristic steroid skeleton, the quassinoids are biogenetically derived from steroid precursors,[298] so the quassinoids are considered in this section. The alert reader will note the absence of data for quassin and neoquassin, the principal constituents of "Quassia". This void is caused by an apparent lack of biological data for pure quassin and neoquassinas, which was noted by Crosby.[87] There are some data available on the related quassinoid cytotoxins, bruceine B and glaucarubinone. Steroid biosynthesis is discussed by both Manitto[294] and Torsell.[132]

OTHER TERPENOIDS (Tables 6a, b)

This section includes the monoterpenes from pines which are repellent to many species and toxic to some. Also included are the norditerpene lactones from *Podocarpus* species which are perhaps better known as cytotoxins. The third interesting group is the terpenes from *Gossypium* species which are toxic to cotton pests. The chemistry, biosynthesis, and biological activity of the latter group is reviewed by Stipanovic et al.[131]

Biosynthesis of some classes of terpenes is reviewed from an ecological perspective by Mabry and Gill[107] and in more detail by Manitto[294] and Torsell.[132]

COUMARINS (Tables 7a, b)

Included in this category are coumarin itself, a representative linear furanocoumarin of some notoriety, and the interesting prenylated coumarins of the mammein/surangin group. A group of the mammeins was isolated, characterized, and reported to be the insecticidal principles of *Mammea* extractives by several groups;[87] however, Crombie et al.[263] isolated and chemically characterized the actual insecticidal components of *Mammea*. These coumarins are very highly substituted and synthesis will be an interesting problem.

The biosynthesis and distribution of these phenylpropanoids have been reviewed by Brown,[81] Manitto,[294] and Torsell.[132] Murray et al.[73] have recently published a comprehensive overview of naturally occurring coumarins, including chemistry, biochemistry, and distribution.

LIGNINS (Table 8a, b)

There are relatively few of these phenylpropanoid dimers which have shown acute toxicity to insects. Their biosynthesis has been reviewed by Manitto.[294]

FLAVONOIDS (Tables 9a, b)

This group includes rotenone and its relatives. Rotenone, first used as a fish poison, has long been an article of commerce for plant protection. Fukami and Nakajima[93] have reviewed the history and structure-activity relationships of rotenoids. Negherbon[111] gathered an immense range of data for pure constituents of Derris root or cubé. Hahlbrook[96] has reviewed biosynthesis of the flavonoids and Ingham[100] the isoflavonoids. Harborne[59] exhaustively reviewed the earlier work on flavonoid biosynthesis and distribution, and more recently, Harborne and Mabry[55] have examined advances in flavonoid research. Manitto[294] and Torsell[132] also discuss biosynthesis of flavonoids.

As is clear from Table 9a, the rotenoids are active against a wide range of insect pests. Unfortunately, this promise has not resulted in new synthetic insect control agents.

OTHER PHENYL PROPANOIDS (Tables 10a, b)

The compounds in this section do not show an extraordinary level or range of toxicity and the methylenedioxyphenyl compounds here are best known as synergists. Their biosynthesis is reviewed by Manitto[294] and Torsell.[132]

FATTY ACIDS (Tables 11a, b)

Like the phenyl propanoids in the preceding section, these compounds are not of exceptional insecticidal potency. The biosynthesis is well established.[132,294]

THIOCYANATES AND OTHER SULFUR COMPOUNDS (Tables 12a, b)

These compounds appear to have protective value for the cruciferous crops in which they are found.[8,9] The naturally occurring isothyocyanates also had their parallels in synthetic isothiocyanate insecticides and Negherbon[111] discusses these compounds. The biosynthesis and distribution of isothiocyanates and related compounds is reviewed by Larsen.[105]

MISCELLANEOUS COMPOUNDS (Tables 13a, b)

This catch-all holds the fascinating group of maytansanoids reported by Powell et al.[233,286] These compounds are related to a large group of antitumor antibiotics, more of which might be of value as insect control agents.

CONCLUSION

There are many examples of insect control chemicals in nature. The major classes have been covered well in prior reviews; however, one particularly interesting correlation is evident from this compilation, the apparent correlation between cytotoxicity and insect toxicity. Examples of this are found in the norditerpene lactones, quassinoids, neriifolin, and maytansanoids.

This chapter is a comprehensive compilation of acute insect poisons from higher plants. Dr. Holyoke would particularly appreciate learning of errors or omissions in this compilation and of appropriate additions in the future.

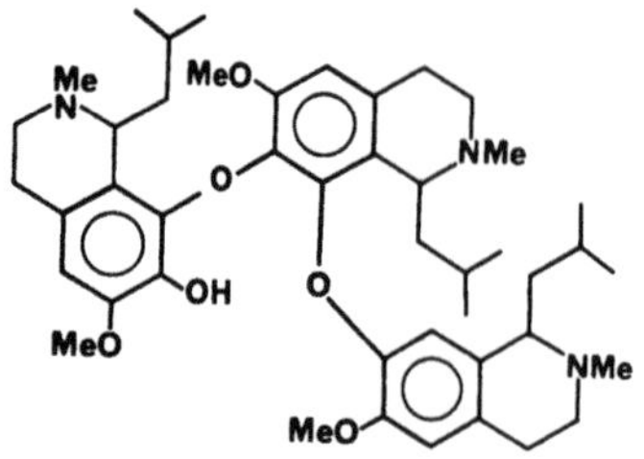

1. Pilocereine

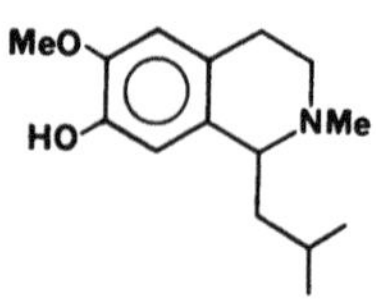

2. Lophocereine

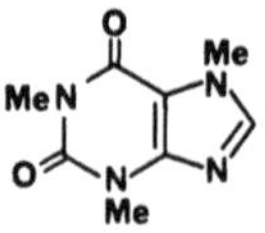

3. Caffeine

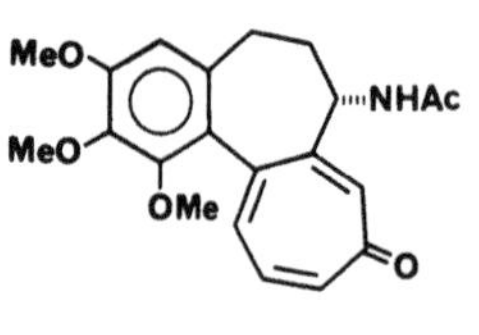

4. Colchicine

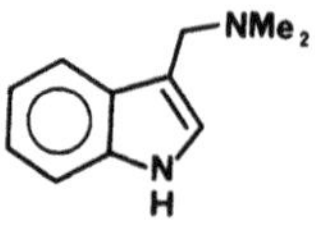

5. Gramine

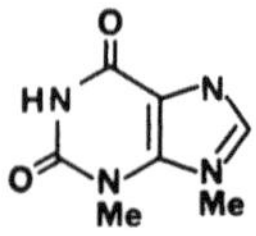

6. Theobromine

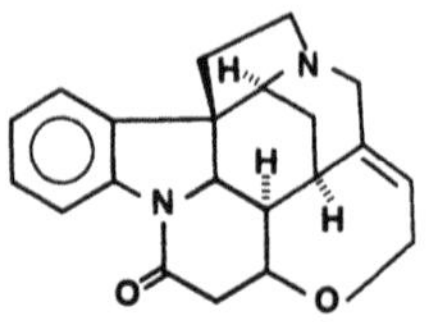

7. Strychnine

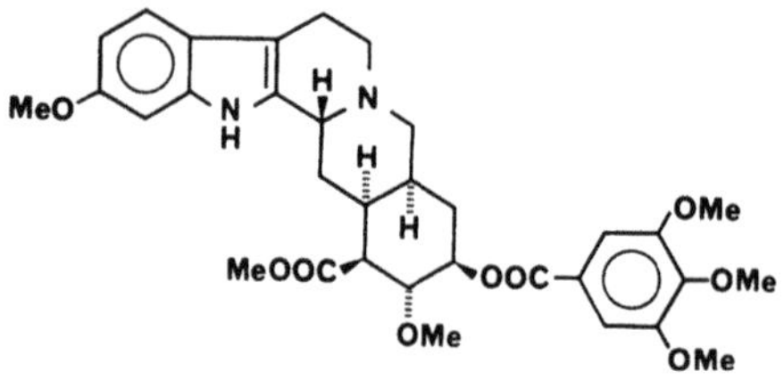

8. Reserpine

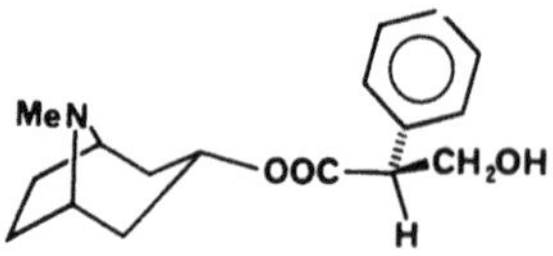

9. Atropine

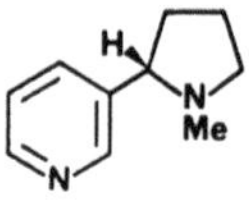

10. Nicotine

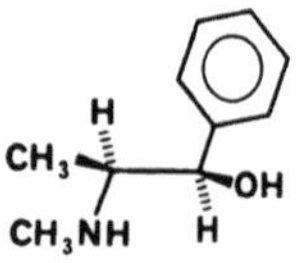

11. L-Ephedrine

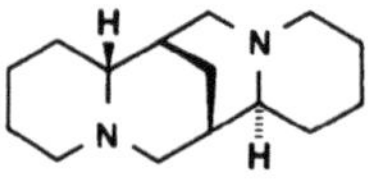

12. Sparteine

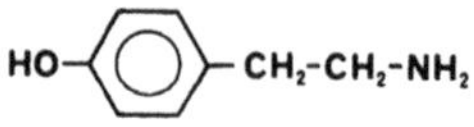

13. Tyramine

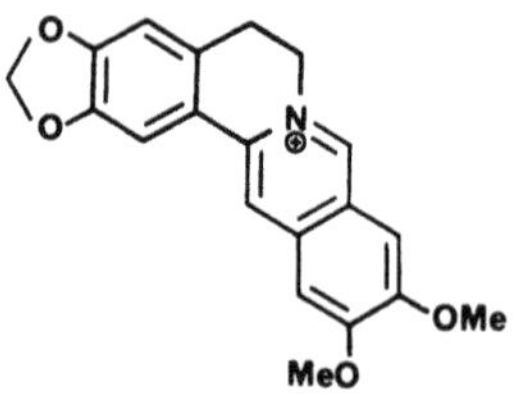

14. Berberine

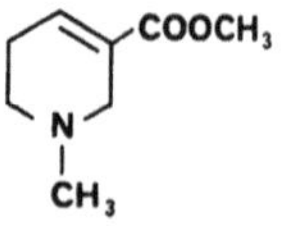

15. Arecoline

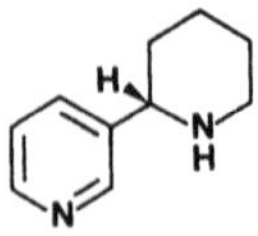

16. Anabasine

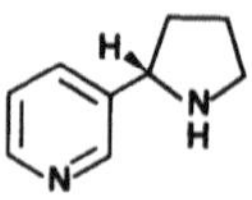

17. Nornicotine

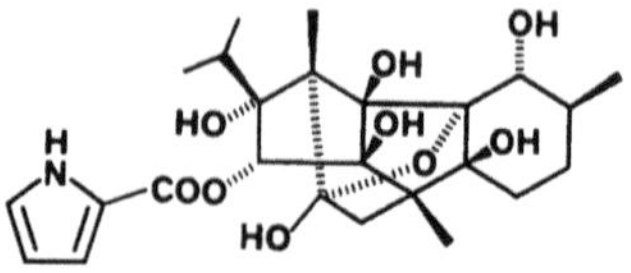

18. Ryanodine

Structures of alkaloids (Table 1).

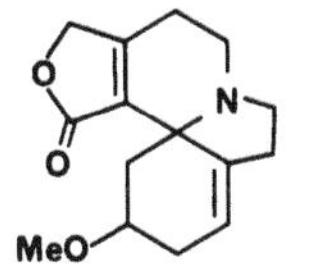

19. Cocculolidine

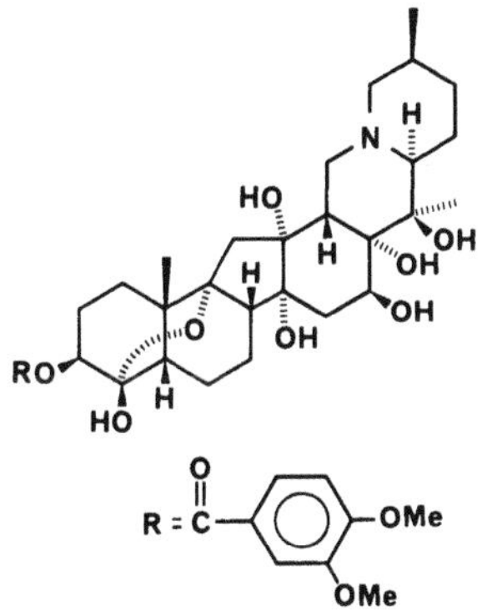

$R = \overset{O}{\overset{\|}{C}}$ —⟨ ⟩—OMe
         OMe

20. Veratridine

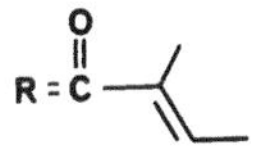

21. Cevadine

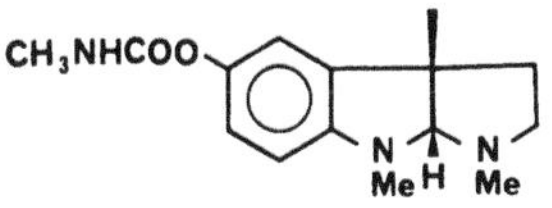

22. Physostigmine

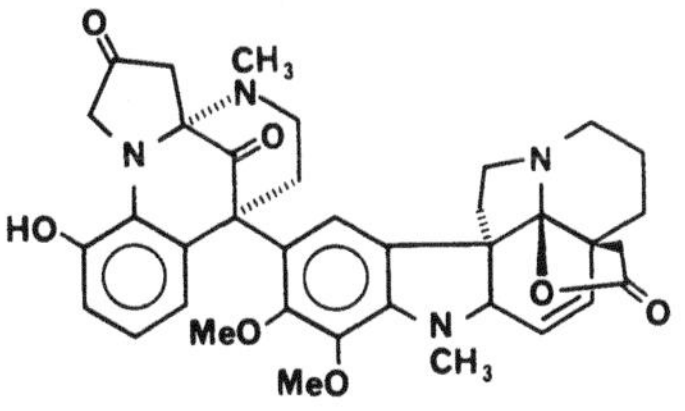

23. Haplophytine

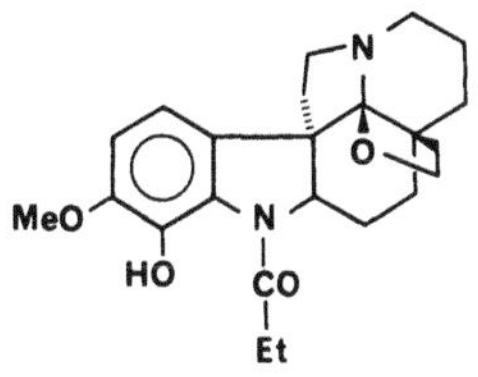

24. Cimicidine

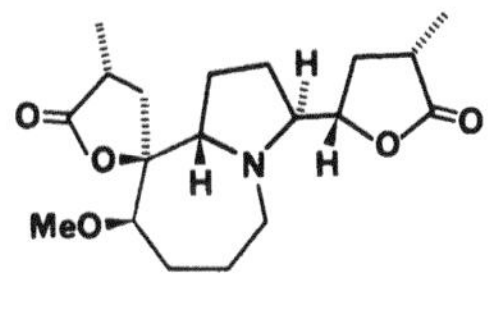

25. Stemospironine

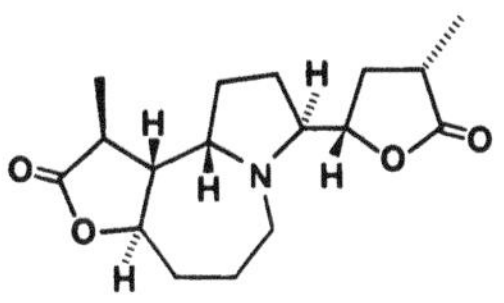

26. Stemonine

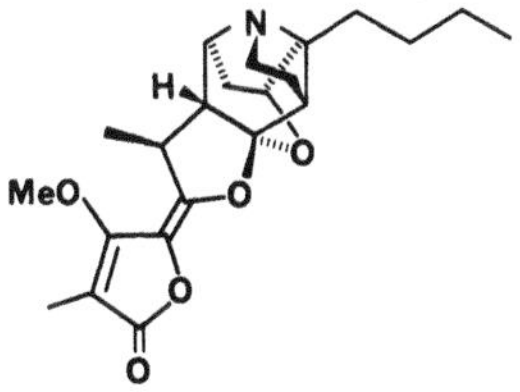

27. Stemofoline

Alkaloids continued.

Table 1a
ALKALOIDS AFFECTING INSECT MORTALITY

Insect	Plant	Compound (%) [No.]	Ref.
Drosophila sp. (8 species used; *D. pachea* only species not killed)	*Lophocereus schotti* (Engelm.) Britt. and Rose	Pilocereine (1) [1]	33
Drosophila hamatofila	*Lophocereus schotti* (Engelm.) Britt. and Rose	Lophocereine (1) [2]	33
Callosobruchus maculatus		Caffeine (1) [3]	4
		Colchicine (0.1) [4]	4
		Gramine (0.1) [5]	4
		Theobromine (1) [6]	4
		Strychnine (0.1) [7]	4
		Reserpine (0.1) [8]	4

Table 1a (continued)
ALKALOIDS AFFECTING INSECT MORTALITY

Insect	Plant	Compound (%) [No.]	Ref.
		Atropine (0.1) [9]	4
		Nicotine (0.1) [10] (as sulfate)	4
		L-Ephedrine (0.1) [11]	4
		Sparteine (0.1) [12] (as sulfate)	4
		Tyramine (1) [13]	4
Bombyx mori	*Phellondendri cortex*	Berberine [14]	21
	Areca semen	Arecoline [15]	21
	Melia cortex	Arecoline [15]	21
	Nicotiana spp.	Nicotine (<1) [10]	124
Bombyx mori L.	*Areca* spp. & *Melia* spp.	Arecoline [15]	21
"Cockroach"	*Nicotiana* spp.	Nicotine (1% injection) [10]	124
"Flies"	*Nicotiana* spp.	Nicotine (0.1) [10]	124
Apis mellifera	*Nicotiana* spp.	Nicotine (38 mg/g parenteral) [10] (0.94 mg/g topical)	111
Popillia japonica	*Nicotiana* spp.	Nicotine (85 mg/g parenteral) [10] (15 mg/g topical)	111
Oncopeltus fasciatus	*Nicotiana* spp.	Nicotine (0.4 mg/g parenteral) [10] (0.3 mg/g topical)	111
Galleria mellonella	*Nicotiana* spp.	Nicotine (3.8 mg/g parenteral) [10] (0.6 mg/g topical)	111
Anasa tristis	*Nicotiana* spp.	Nicotine (1.25 mg/g topical) [10] (0.35 mg/g injection)	111
Bombyx mori (larvae)	*Nicotiana* spp.	Nicotine (8 μg/g topical) [10] (7 μg/g injection)	111
Ceratomia catalpae (larvae)	*Nicotiana* spp.	Nicotine (0.2 mg/g topical) [10] (0.15 mg/g injection)	111
Oncopeltus fasciatus	*Nicotiana* spp.	Nicotine (0.45 mg/g topical) [10]	111
Periplaneta americana	*Nicotiana* spp.	Nicotine (4—5 mg/g oral) [10] (0.5—1.3 mg/g topical) (0.14—0.2 mg/g injection)	111
Popillia japonica	*Nicotiana* spp.	Nicotine (1 mg/g topical) [10] (0.9 mg/g injection)	111
Terebrio molitor	*Nicotiana* spp.	Nicotine (4.4 mg/g topical) [10]	111
Eriosoma americanum (adult)	*Nicotiana* spp.	Nicotine (<0.003 mg/ℓ fumigant) [10]	111
Aphis gossypii (adult)	*Nicotiana* spp.	Nicotine (0.004 mg/ℓ fumigant) [10]	111
Macrosiphoniella Sanborni (adult)	*Nicotiana* spp.	Nicotine (<0.009 mg/ℓ fumigant) [10]	111
Aphis rumicis (adult)	*Nicotiana* spp.	Nicotine (0.006 mg/ℓ fumigant) [10]	111
A. forbesii (adult)	*Nicotiana* spp.	Nicotine (0.006 mg/ℓ fumigant) [10]	111
Myzus solanifolii (adult)	*Nicotiana* spp.	Nicotine (0.007 mg/ℓ fumigant) [10]	111
Macrosiphium pisi (adult)	*Nicotiana* spp.	Nicotine (0.01 mg/ℓ fumigant) [10]	111
Myzus porosus (adult)	*Nicotiana* spp.	Nicotine (0.01 mg/ℓ fumigant) [10]	111
Brevicoryne brassicae (adult)	*Nicotiana* spp.	Nicotine (0.02 mg/ℓ fumigant) [10]	111
Myzis persicae (adult)	*Nicotiana* spp.	Nicotine (0.008 mg/ℓ fumigant) [10]	111
Trialeurodes packardi (adult)	*Nicotiana* spp.	Nicotine (0.01 mg/ℓ fumigant) [10]	111
T. vaporariorum (adult)	*Nicotiana* spp.	Nicotine (0.03 mg/ℓ fumigant) [10]	111
Empoasca fabae (adult)	*Nicotiana* spp.	Nicotine (0.02 mg/ℓ fumigant) [10]	111
Orthezia insignis (adult)	*Nicotiana* spp.	Nicotine (0.04 mg/ℓ fumigant) [10]	111
Phenococcus gossypii (adult)	*Nicotiana* spp.	Nicotine (0.2 mg/ℓ fumigant) [10]	111
Carpocapsa pomonella (adult)	*Nicotiana* spp.	Nicotine (<0.008 mg/ℓ fumigant) [10]	111
Bombyx mori (6th instar)	*Nicotiana* spp.	Nicotine (<0.001 mg/ℓ fumigant) [10]	111
Prodonia eridania (larvae)	*Nicotiana* spp.	Nicotine (0.05 mg/ℓ fumigant) [10]	111
Bregmatothrips iridis (adult)	*Nicotiana* spp.	Nicotine (0.01 mg/ℓ fumigant) [10]	111
Thrips tabaci (adult)	*Nicotiana* spp.	Nicotine (0.02 mg/ℓ fumigant) [10]	111
Taeniothrips simplex (adult)	*Nicotiana* spp.	Nicotine (0.02 mg/ℓ fumigant) [10]	111

Table 1a (continued)
ALKALOIDS AFFECTING INSECT MORTALITY

Insect	Plant	Compound (%) [No.]	Ref.
T. simplex (larvae)	*Nicotiana* spp.	Nicotine (0.03 mg/ℓ fumigant) [10]	111
Heliothrips femoralis (adult)	*Nicotiana* spp.	Nicotine (0.03 mg/ℓ fumigant) [10]	111
Thrips nigropilosus (adult)	*Nicotiana* spp.	Nicotine (0.07 mg/ℓ fumigant) [10]	111
T. nigropilosus (larvae)	*Nicotiana* spp.	Nicotine (<0.03 mg/ℓ fumigant) [10]	111
Epitrix parvula (adult)	*Nicotiana* spp.	Nicotine (0.15 mg/ℓ fumigant) [10]	111
Epicauta pennsylvanica (adult)	*Nicotiana* spp.	Nicotine (0.28 mg/ℓ fumigant) [10]	111
Epilachna varivestis (adult)	*Nicotiana* spp.	Nicotine (0.25 mg/ℓ fumigant) [10]	111
Aphidus phorodontis (adult)	*Nicotiana* spp.	Nicotine (0.008 mg/ℓ fumigant) [10]	111
Reticulotermas flavipes (adult)	*Nicotiana* spp.	Nicotine (0.01 mg/ℓ fumigant) [10]	111
Tetranychus bimaculates (adult)	*Nicotiana* spp.	Nicotine (0.15 mg/ℓ fumigant) [10]	111
Musca domestica (adult)	*Nicotiana* spp.	Nicotine (0.2 mg/ℓ fumigant) [10]	111
Periplaneta americana (adult)	*Nicotiana* spp.	Nicotine (<0.17 mg/ℓ fumigant) [10]	111
Apis wellifera (adult)	*Nicotiana* spp.	Nicotine (<0.26 mg/ℓ fumigant) [10]	111
Aphis rumicis	*Nicotiana* spp.	Nicotine (0.12 mg/g contact) [10]	111
Blattella germanica	*Nicotiana* spp.	Nicotine (3.2 mg/g contact) [10]	111
Bombyx mori (larvae)	*Nicotiana* spp.	Nicotine (1.4 mg/g injection) [10]	111
Calliphora erythrocephala (adult)	*Nicotiana* spp.	Nicotine (1.1 mg/g injection) [10]	111
Carpocapsa pumonella	*Nicotiana* spp.	Nicotine (0.85 mg/g injection) [10]	111
Lucilia sericata (adult)	*Nicotiana* spp.	Nicotine (0.4—0.7 mg/g injection) [10]	111
Lygus kalmii	*Nicotiana* spp.	Nicotine (6.8 mg/g contact) [10]	111
F. tormia regina	*Nicotiana* spp.	Nicotine (3.3 mg/g injction) [10]	111
Prodenia eridania	*Nicotiana* spp.	Nicotine (2 mg/g injection) [10]	111
Protoparce quinquemaculata	*Nicotiana* spp.	Nicotine (3.5 mg/g oral) [10]	111
Thermobia domestica	*Nicotiana* spp.	Nicotine (4—5 mg/g contact) [10]	111
Aphis rumicis	*Nicotiana* spp.	Anabasine (1.7 mg/g contact) [16]	111
	Nicotiana spp.	Nornicotine (1 mg/g contact) [17]	111
Blattella germanica (female)	*Ryania speciosa*	Ryanodine (25 μg/g topical) [18]	111
B. germanica (male)	*Ryania speciosa*	Ryanodine (5 μg/g topical) [18]	111
Oncopeeltus fasciatus	*Ryania speciosa*	Ryanodine (25 μg/g topical) [18]	111
Nephotettix bipunctatus cinticeps Uhler.	*Cocculus trilobus* DC.	Cocculolidine (0.025) [19]	19
Callosobruchus chinensis Linne.	*Cocculus trilobus* DC.	Cocculolidine (0.05) [19]	19
Niliparvata lugens Stal	*Nicotiana* spp.	Nicotine (as sulfate) 0.1) [10]	19
	Nicotiana spp.	Anabasine (as sulfate) 0.025) [16]	19
Nephotettix bipunctatus cinticeps Uhler.	*Nicotiana* spp.	Anabasine (as sulfate) 0.1) [16]	19
Oncopeltus fasciatus	*Schoenocaulon* spp.	Veratridine (1% dust) [20]	232
	Schoenocaulon spp.	Cevadine (0.014% dust) [21]	232
	Schoenocaulon spp.	Veratridine (0.25% sol'n) [20]	232
	Schoenocaulon spp.	Cevadine (0.07% sol'n) [21]	232
Melanoplus Femur-rubrum (DeGeer)	*Schoenocaulon* spp.	Cevadine (0.1% dust) [21]	232
Musca domestica	*Schoenocaulon* spp.	Cevadine (0.12% sol'n) [21]	254
	Schoenocaulon spp.	Veratridine (0.03% sol'n) [20]	254
Locusta migratoria	*Physostigma venenosum* (Balfour)	Physostigmine (injection) [22]	296
Blatella germanica	*Haplophyton cimicidum* A.D.C	Haplophytine (18 μg/g contact) [23]	6
	Haplophyton cimicidum A.D.C	Cimicidine (60 μg/g contact) [24]	6
Bombyx mori L.	*Stemona japonica* Miq.	Stemospironine (100 ppm/diet) [25]	304
	Stemona japonica Miq.	Stemonine (100 ppm/diet) [26]	304
	Stemona japonica Miq.	Stemofoline (50 ppb/diet) [27]	304

Table 1b

CHEMICAL DATA FOR ALKALOIDS AFFECTING INSECT MORTALITY

Name [No.]	Chem. Abstr. reg. no.	Formula	Physical constants[a]	Synthesis (Ref.)	Biosynthesis (Ref.)	Comments
Pilocercine [1]	2552-47-8	$C_{45}H_{65}N_3O_6$	mp 176.5—177	37	34, 35	
Lophocereine [2]	19485-63-3	$C_{15}H_{23}NO_2$	Picrate mp 183—185°		36	
Caffeine [3]	58-08-2	$C_8H_{10}N_4O_2$	mp 238°	137	190	Crystal structure[138] PMR[206] PMR[90] CMR[102]
Colchicine [4]	64-86-8	$C_{22}H_{25}NO_6$	mp 157° $[\alpha]_D^{17} - 121°$ (CHCl$_3$)	139	140, 191 (CMR)	Crystal structure[141]
Gramine [5]	87-52-5	$C_{11}H_{14}N_2$	mp 138—139° HCl mp 191° (d)	142	143	PMR[206]
Theobromine [6]	83-67-0	$C_7H_8N_4O_2$	mp 357°			PMR[90] CMR[102] PMR[206]
Strychnine [7]	57-24-9	$C_{21}H_{22}N_2O_2$	mp 268—290° bp$_5$ 270° $[\alpha]_D^{18} - 139.3°$ (CHCl$_3$)	148		Review[144] PMR, CMR[145] Crystal structure[192]
Reserpine [8]	50-55-5	$C_{33}H_{40}N_2O_9$	mp 264—265° $[\alpha]_D^{23} - 164°$ (Py) $[\alpha]_D^{23} - 118°$ (CHCl$_3$)	146 147		PMR[193] CMR[194] MS[195]
Atropine (Hyoscamine) [9]	(S)101-31-5 (RS)51-55-8 (Sulfate: 5908-99-6)	$C_{17}H_{23}NO_3$	mp 108—111° $[\alpha]_D^{15} - 22°$ (aq EtOH) Picrate mp 165° $^{1}/_{2}$ H$_2$SO$_4$ mp 190—194°	149 196	197 198	PMR[199] CMR[200] Crystal structure[201] MS[195]
Nicotine [10]	54-11-5	$C_{10}H_{14}N_2$	bp$_{745}$ 247° bp$_{17}$ 123—125° n_D^{20} 1.5282 $[\alpha]_D^{20} - 169°$	CMR data[150]	150 106	PMR[202] PMR[206]
L-Ephedrine [11]	299-42-3	$C_{10}H_{15}NO$	mp 40° bp 225°		203	PMR[206] CMR[204]

Compound	CAS No.	Molecular formula	Physical constants	Ref.	Ref.	References
Sparteine (Lupinidine) [12]	90-39-1	$C_{15}H_{26}N_2$	HCl mp 216—220° $[\alpha]_D^{25}$ − 6.5° (EtOH) bp_8 173° $[\alpha]_D^{21}$ − 16.4° (EtOH) n_D^{20} 1.5312	151	152	CMR[205]
Tyramine [13]	51-67-2	$C_8H_{11}NO$	mp 164—165° bp_{25} 205—207° bp_2 166° HCl salt mp 269°	162		Crystal structure[163]; MS, UV, IR[207]; Crystal structure[208]; PMR[209]
Berberine (Umbellatine) [14]	2086-83-1	$C_{20}H_{18}NO_4^+$	Hydroxide hydrate mp variable Iodide mp 250°		56	IR, PMR[155]; Review[157]
Arecoline [15]	63-75-2	$C_8H_{13}NO_2$	bp_{760} 209° bp_{17} 92—93° n_D^{20} 1.4302 HCl mp 158°	64		PMR[206]
Anabasine [16]	494-52-0	$C_{10}H_{14}N_2$	mp 25—30° bp 276° bp_2 104—105° $[\alpha]_D^{20}$ − 82.2° dipicrate: mp 205°	301	106	MS[302]
Nornicotine [17]	494-97-3	$C_9H_{12}N_2$	bp_{11} 130—131° bp_1 120° $[\alpha]_D^{22}$ − 88.8° $n_D^{18.5}$ 1.5378	150	106	PMR, MS[303]
Ryanodine [18]	15662-33-6	$C_{25}H_{35}NO_9$	mp 219—220° $[\alpha]_D^{25}$ + 26° (MeOH)	270		Review, PMR[134,271]; Crystal structure[128]; IR, PMR, UV, MS[169]
Cocculolidine [19]	13497-04-6	$C_{15}H_{19}NO_3$	mp 144—146° $[\alpha]_D^{25}$ + 237° $(CHCl_3)$[19] $[\alpha]_D^{25}$ + 273° $(CHCl_3)$[169]			
Veratridine [20]	71-62-5	$C_{36}H_{51}NO_{11}$	mp 180° $[\alpha]_D^{20}$ + 12.8° (EtOH)			Review[289]
Cevadine [21]	62-59-9	$C_{32}H_{49}NO_9$	mp 213—214.5° (d) $[\alpha]_D^{20}$ + 12.8° (EtOH)			Review[289]
Physostigmine (Eserine) [22]	57-47-6	$C_{15}H_{21}N_3O_2$	mp 86—87° or 105—106° (dimorphic) $[\alpha]_D$ − 82° $(CHCl_3)$	56		PMR[58]; CMR[70]; Crystal structure[74]; MS[92]

Table 1b (continued)
CHEMICAL DATA FOR ALKALOIDS AFFECTING INSECT MORTALITY

Name [No.]	Chem. Abstr. reg. no.	Formula	Physical constants[a]	Synthesis (Ref.)	Biosynthesis (Ref.)	Comments
Haplophytine [23]	16625-20-0	$C_{37}H_{40}N_4O_7$	mp 300—302° (d) $[\alpha]_D^{25} + 109°$ (CHCl$_3$)			UV, IR, PMR, MS, crystal structure[104] IR, UV, PMR, CMR[113] Crystal structure[135]
Cimicidine [24]	7096-81-3	$C_{23}H_{28}N_2O_5$	mp 268—270° (d) $[\alpha]_D^{25} + 23°$ (CHCl$_3$)			UV, PMR, CMR, IR, MS[300]
Stemospironine [25]	66267-46-7	$C_{19}H_{29}NO_5$	$[\alpha]_D^{27} - 8.2°$ (CHCl$_3$) HBr: mp 283—284° (d)			PMR, CMR, MS, UV, IR, crystal structure[304]
Stemonine [26]	27498-90-4	$C_{17}H_{25}NO_4$	$[\alpha]_D - 114°$ mp 151° HBr 1/2 H$_2$O mp 275°			Crystal structure, IR, UV, PMR[305] UV, IR, PMR, MS[309]
Stemofoline [27]	29881-57-0	$C_{22}H_{29}NO_5$	mp 87—89° $[\alpha]_D + 273°$ (MeOH)			Crystal structure, IR, UV, PMR, MS[306]

[a] °C.

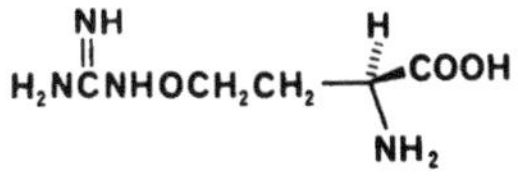

1. L-Canavanine

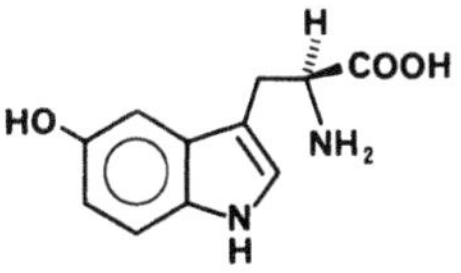

2. 5-Hydroxytryptophane

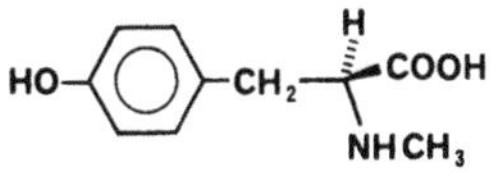

3. *N*-Methyltyrosine

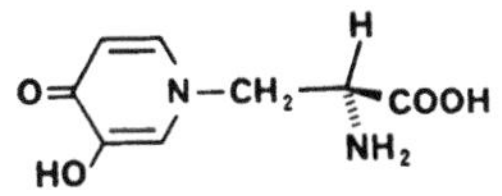

4. L-Mimosine

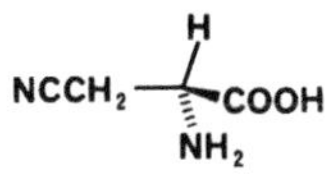

5. β-Cyano-L-alanine

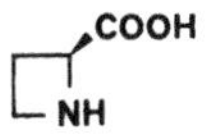

6. Azetidine-2-carboxylic acid

Structures of amino acids (Table 2).

Table 2a
AMINO ACIDS AFFECTING INSECT MORTALITY

Insect	Compound (conc.) [No.]	Ref.
Bombyx mori L.[a]	L-Canavanine [1]	21
Musca autumnalis DeGeer	L-Canavanine (800 ppm) [1]	22
Haematobia irritans (L.)	L-Canavanine (800 ppm) [1]	22
Stomoxys calcitrans (L.)	L-Canavanine (800 ppm) [1]	22
Prodenia eridania (Cramer)	5-Hydroxytryptophan (0.5%) [2]	23
Callosobruchus maculatus	*N*-Methyltyrosine (1%) [3]	4
	5-Hydroxytryptophan (1%) [2]	4
	L-Mimosine (0.1%) [4]	4
	β-Cyano-L-alanine (0.1%) [5]	4
	Azetidine-2-carboxylic acid (0.1%) [6]	4
Locusta migratoria	β-Cyano-L-alanine (1%) [5]	161
		123
Spodoptera litteralis (Boisd.)	Azetidine-2-carboxylic acid [6]	97

[a]   Plant — *Astragalus radix*.

Table 2b
CHEMICAL DATA FOR AMINO ACIDS AFFECTING INSECT MORTALITY

Name [No.]	Chem. Abstr. No.	Formula	Physical constants[a]	Synthesis (Ref.)	Comments
L-Canavanine [1] Free base	543-38-4	$C_5H_{17}N_4O_3$	mp 184° $[\alpha]_D^{20} + 7.9°$ (H_2O)	24	Distribution in plant kingdom[25] Review[210]
Sulfate	2219-31-0	$C_5H_{17}N_4O_3$ H_2SO_4	mp 172° (dec) $[\alpha]_D^{17} + 19.4°$ (H_2O)		Crystal structure[211]
L-5-Hydroxy-tryptophan [2]	4350-09-8	$C_{11}H_{12}N_2O_3$	mp 273° (dec) $[\alpha]_D^{20} - 32.5°$ (H_2O) $[\alpha]_D^{20} + 16.0°$ (4 N HCl)	158 159	Distribution in plant kingdom[77] Crystal structure[160]
N-Methyl tyrosine (Surinamine, Ratankin, Angalin, etc.) [3]	537-49-5	$C_{10}H_{13}NO_3$	mp 257° $[\alpha]_D^{23} + 16°$ (1 M HCl) $[\alpha]_D^{23} + 31.45°$ (1 M NaOH)	219	Occurrence, PMR, CMR[220]
L-Mimosine (Leucenol) [4]	500-44-7	$C_8H_{10}N_2O_4$	mp 235° $[\alpha]_D^{22} - 20°$ HCl salt mp 175° (d)	164	Crystal structure[165]
β-Cyano-L-alanine[b] [5]	6232-19-5	$C_4H_6N_2O_2$	mp 218—218.5° $[\alpha]_D^{26} - 2.9°$ (H_2O)	212 115	
Azetidine-2-carboxylic acid[c] [6]	2517-04-6	$C_4H_7NO_2$	mp >270° (d) $[\alpha]_D^{20} - 99°$ ($CHCl_3$)	215	Crystal structure[217] Review[218]

a °C.
b See References 213 and 214 for biosynthesis.
c See References 166 and 216 for biosynthesis.

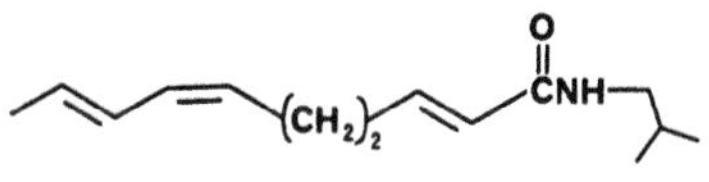

1. Spilanthol

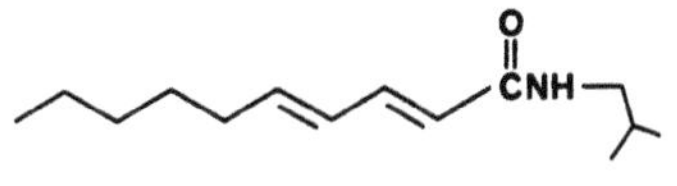

2. Pellitorine

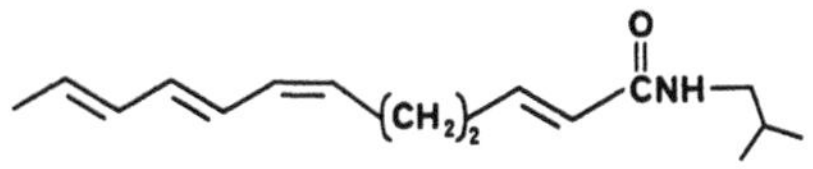

3. Neoherculin

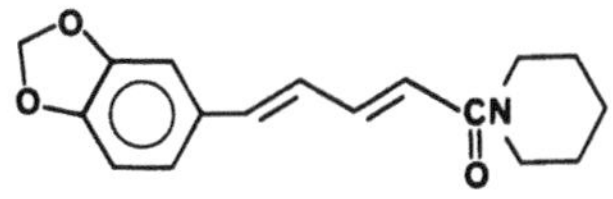

4. Piperine

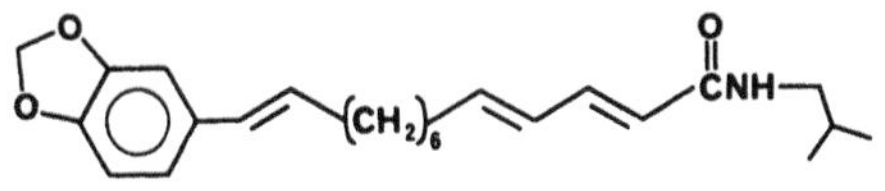

5. Guineensine

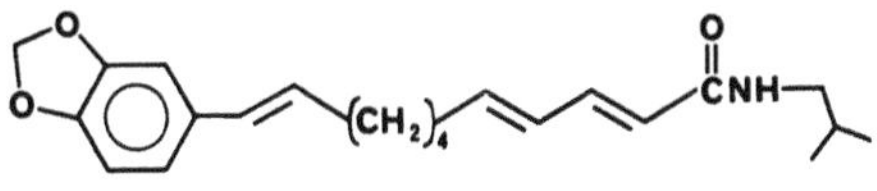

6. Pipercide

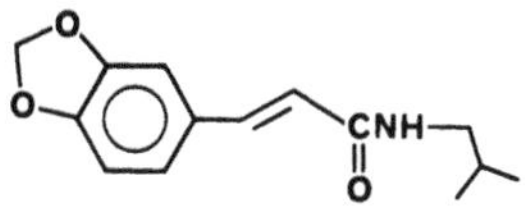

7. Fagaramide

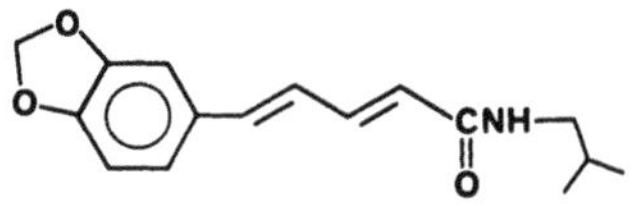

8. Piperlonguminine

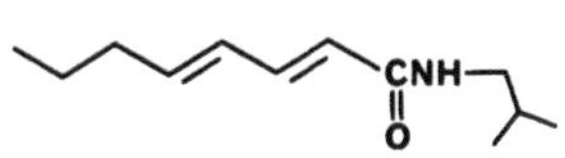

9. *N*-Isobutyl-2E,4E-octadienamide

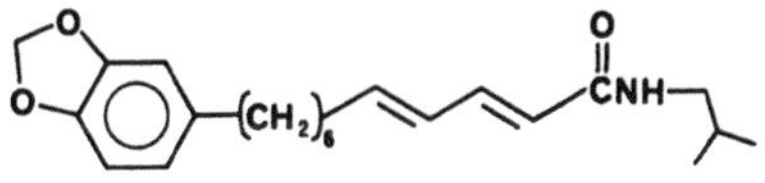

10. Dihydropipercide

Structures of unsaturated isobutyramides (Table 3).

Table 3a
UNSATURATED ISOBUTYRAMIDES AFFECTING INSECT MORTALITY

Insect	Plant	Compound (conc.) [No.]	Ref.
Bombyx mori L.	*Spilanthes acmella* var. *oleracea* Clarke	Spilanthol (200 ppm) [1] (affinin)	38
Musca domestica	*Piper* spp. *Fagara* spp.	Pellitorine [2]	170—172
Anopheles quadrimaculatus	*S. oleraceae* Jacq.	Spilanthol (0.02%) [1] (affinin)	101
	Heliopsis longipes (A. Gray) Blake	Spilanthol (0.02%) [1] (affinin)	173
Musca domestica	*H. longipes* (A. Gray) Blake	Spilanthol (0.02%) [1] (affinin)	173
	Heliopsis longipes (A. Gray) Blake	Spilanthol (0.2%) [1] (affinin)	60
Tenebrio molitor	*H. longipes* (A. Gray) Blake	Spilanthol (0.2%) [1] (affinin)	60
M. domestica	*Zanthoxylum* spp. (*Echinacea* spp.)	Neoherculin [3] (α-Sanshool, Echinacein)	175
	Zanthoxylum spp. (*Echinacea* spp.)	Neoherculin [3] (α-Sanshool, Echinacein)	176
Tenebrio molitor	*Zanthoxylum* spp. (*Echinacea* spp.)	Neoherculin [3] (α-Sanshool, Echinacein)	177
Aedes triseriatus	*Achillea millefolium*	Pellitorine (5 ppm) [2]	178
Sitophilus oryzae (larvae)	*Piper nigrum*	Piperine (0.125% diet) [4]	179
Callosobruchus maculatus	*P. nigrum*	Pellitorine (29—50 μg/insect) [2]	180
	P. nigrum	Guineensine (2—13 μg/insect) [5]	180
	P. nigrum	Pipercide (3—16 μg/insect) [6]	180
Culex pipiens	*Fagara macrophylla*	Fagaramide (15 ppm) [7]	181
	F. macrophylla	Piperlonguminine[a] (10 ppm) [8]	181
	F. macrophylla	Pellitorine (5 ppm) [2]	181
	F. macrophylla	N-Isobutyl-2E,4E-octadienamide (15 ppm) [9]	181
Pectinophera gossypiella	*F. macrophylla*	Pellitorine (25 ppm in diet) [2]	181
	F. macrophylla	N-Isobutyl-2E,4E-octadienamide (100 ppm in diet) [9]	181
Callosobruchus chinensis	*P. nigrum*	Pipercide (0.15—0.25 μg/insect) [6]	224
	P. nigrum	Pellitorine (2—7 μg/insect) [2]	224
Callosobruchus chinensis L.	*P. nigrum* L.	Pipercide (400 ppm-dip) [6]	292
	P. nigrum L.	Dihydropipercide (200 ppm-dip) [10]	292
	P. nigrum L.	Guineensine (300 ppm-dip) [5]	292
	P. nigrum L.	Pellitorine (5000 ppm-dip) [2]	292
Culex pipiens pallens	*P. nigrum* L.	Pipercide (0.3 ppm) [6]	117
	P. nigrum L.	Dihydropipercide (0.9 ppm) [10]	117
	P. nigrum L.	Guineensine (7 ppm) [5]	117
Musca domestica L.[b]	*P. nigrum* L.	Pipercide (400 ppm contact) [6]	117
	P. nigrum L.	Dihydropipercide (400 ppm contact) [10]	117
	P. nigrum L.	Guineensine (120 ppm contact) [5]	117

[a]   In Reference 181, this compound was incorrectly named piperlougumine.
[b]   Pyrethrin resistant strain.

Table 3b

CHEMICAL DATA FOR UNSATURATED ISOBUTYRAMIDES AFFECTING INSECT MORTALITY

Name [No.]	Chem. Abstr. reg. no.	Formula	Physical constants[a]	Synthesis (Ref.)	Comments
Spilanthol (Affinin) [1]	25394-57-4	$C_{14}H_{23}NO$	bp 0.2, 141°; mp 23°; n_D^{25} 1.5134	IR, UV, PMR[60]	PMR[183]
Pellitorine [2]	109-26-2	$C_{14}H_{25}NO$	mp 90°; mp 86—87°	108, 170, 172, 182	PMR, GC/MS, IR[178]; MS, PMR, IR, UV[180,226]
Neoherculin (Echinacein, α-Sanshool) [3]	18744-21-3	$C_{16}H_{25}NO$	mp 93.5—95°	UV, IR, PMR[174]	
Piperine [4]	94-62-2	$C_{17}H_{19}NO_3$	mp 129.5°	186, 307	PMR, CMR,[184] X-ray structure,[185] CMR[187]
Guineensine [5]	55038-30-7	$C_{24}H_{33}NO_3$	mp 114—116°	117, MS[292]	CMR,[187] MS, IR, UV, PMR[180,223,226]
Pipercide [6]	54794-74-0	$C_{22}H_{29}NO_3$	mp 120—122°	117, PMR[188]	MS, PMR, IR, UV,[180,224] CMR[187]
Fagaramide [7]	495-86-3	$C_{14}H_{17}NO_3$	mp 119.5°	PMR, IR, MS, UV[189]	PMR, CMR[14]
Piperlonguminine [8]	5950-12-9	$C_{16}H_{19}NO_3$	mp 166—168°		PMR, CMR,[184] IR, UV, PMR, MS[223]
N-Isobutyl-2E,4E-octadienamide [9]	23512-47-2	$C_{12}H_{21}NO$	mp 94°	UV[225]	PMR[181]
Dihydropipercide [10]	75022—26-3	$C_{22}H_{31}NO_3$		292	

[a] °C.

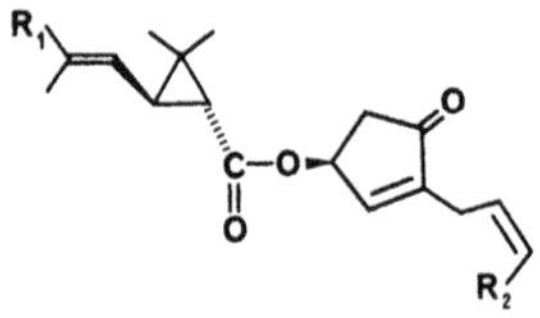

R_1	R_2	
CH_3	$CH=CH_2$	1. Pyrethrin I
$\overset{O}{\overset{\|}{C}}OCH_3$	$CH=CH_2$	2. Pyrethrin II
CH_3	CH_3	3. Cinerin I
$\overset{O}{\overset{\|}{C}}OCH_3$	CH_3	4. Cinerin II
$\overset{O}{\overset{\|}{C}}OCH_3$	C_2H_5	5. Jasmolin II

Structures of pyrethroids (Table 4).

Table 4a
PYRETHROIDS

Insect	Plant	Compound (conc.) [No.]	Ref.
Phaedon cochlearis	*Chrysanthemum cinerariaefolium*	Pyrethrin I [1]	133
	C. cinerariaefolium	Pyrethrin II [2]	133
	C. cinerariaefolium	Cinerin I [3]	133
	C. cinerariaefolium	Cinerin II [4]	133
Musca domestica	*C. cinerariaefolium*	Pyrethrin I [1]	84, 94, 99, 111, 118—121
	C. cinerariaefolium	Pyrethrin II [2]	84, 94, 99, 111, 118—121
	C. cinerariaefolium	Cinerin I [3]	84, 94, 99, 111, 118—121
	C. cinerariaefolium	Cinerin II [4]	84, 94, 99, 111, 118—121
Blattella germanica	*C. cinerariaefolium*	Pyrethrin I/cinerin I (10 mg/ℓ) [1/3]	111
	C. cinerariaefolium	Pyrethrin II/cinerin II (12.5 mg/ℓ) [2/4]	111
Aphis rumicis	*C. cinerariaefolium*	Pyrethrin I/cinerin I (10 mg/ℓ) [1/3]	111
	C. cinerariaefolium	Pyrethrin II/cinerin II (100 mg/ℓ) [2/4]	111
Triboleum castaneum	*C. cinerariaefolium*	Pyrethrin I/cinerin I [1/3]	111
	C. cinerariaefolium	Pyrethrin II/cinerin II [2/4]	111
Periplaneta americana	*C. cinerariaefolium*	Pyrethin I/cinerin I (1 mg/ℓ) [1/3]	111
	C. cinerariaefolium	Pyrethrin II/cinerin II (1.5 mg/ℓ) [2/4]	111
M. domestica	*C. cinerariaefolium*	Jasmolin II (1 μg/insect) [5]	95
Phaedon cochlearis	*C. cinerariaefolium*	Jasmolin II (0.01 μg/insect) [5]	95
Aedes aegypti	*C. cinerariaefolium*	Jasmolin II (0.075%) [5]	95
Tenebrio castaneum	*C. cinerariaefolium*	Jasmolin II (3.3 μg/insect) [5]	95
Daphnis magna	*C. cinerariaefolium*	Jasmolin II (0.75%) [5]	95

Table 4b
CHEMICAL DATA FOR PYRETHROIDS

Name [No.]	Chem. Abstr. reg. no.	Formula	Physical constants[a]	Synthesis (Ref.)	Biosynthesis (Ref.)	Comments
Pyrethrin I [1]	121-21-1	$C_{21}H_{28}O_3$	bp 0.005, 146—150° $[\alpha]_D^{20} - 14°$ (2,2,3-tri-methyl pentane)	256 Review[257]	221 Review[89]	PMR[222]
Pyrethrin II [2]	121-29-9	$C_{22}H_{28}O_5$	bp 0.007, 192—193° $[\alpha]_D^{19} + 14.7°$ (2,2,3-tri-methyl pentane)	Review[257]	221 Review[89]	PMR[222]
Cinerin I [3]	97-12-1	$C_{20}H_{28}O_3$	bp 0.005, 132°	255 Review[257]	Review[89]	PMR[222]
Cinerin II [4]	121-20-0	$C_{21}H_{28}O_5$	bp 0.001, 182—184° $[\alpha]_D^{16} + 16°$ (2,2,3-tri-methyl pentane)	Review[257]	Review[89]	PMR[222]
Jasmolin II [5]	1172-63-0	$C_{22}H_{30}O_5$		Review[257]	Review[89]	PMR[222]

[a] °C.

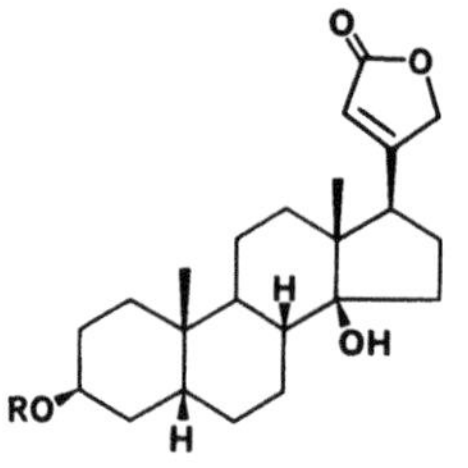

1. Neriifolin R = O-Thevetoside

2. Digitoxin R = O-D-digitoxyl-D-digitoxide

3. 2′ Acetylneriifolin R = 2′ Acetyl-O-thevoside

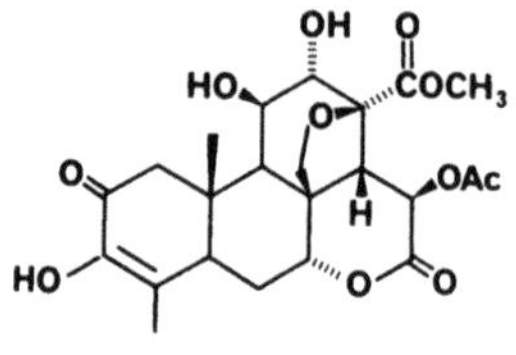

4. Bruceine B

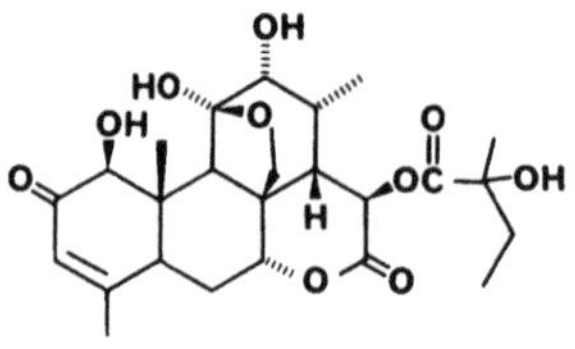

5. Glaucarubinone

Structures of steroids and other triterpenes (Table 5).

Table 5a
STEROIDS AND OTHER TRITERPENES AFFECTING INSECT MORTALITY

Insect	Plant	Compound [No.]	Ref.
Acalymma vittatum L. (larvae)	*Dacrydium intermedium* T. Kirk	Neriifolin (0.25% diet) [1]	227
Laspeyresia pomonella L. (larvae)	*D. intermedium* T. Kirk	Neriifolin (200 ppm diet) [1]	227
Bombyx mori L.	*Digitalis purpurea* L.	Digitoxin (100 ppm diet) [2]	38
Ostrinia nubilalis (Hubner)	*Thevetia thevetioides* (HBK) K. Schum	Neriifolin (30 ppm diet) [1]	39
	T. thevetioides (HBK) K. Schum	2′-Acetylneriifolin (192 ppm diet) [3]	39
Locusta migratoria migratorioides	*Quassia* spp.	Bruceine B (24 μg/g insect) [4]	297
	Quassia spp.	Glaucarubinone (12 μg/g insect) [5]	297

Table 5b
CHEMICAL DATA FOR STEROIDS AND OTHER TRITERPENES

Name [No.]	Chem. Abstr. reg. no.	Formula	Physical constants[a]	Synthesis (Ref.)	Biosynthesis (Ref.)	Comments
Neriifolin [1]	466-07-0	$C_{30}H_{96}O_8$	mp 218—25° $[\alpha]_D^{23} - 50.2°$ (MeOH)	41,42	43	
Digitoxin [2]	71-63-6	$C_{41}H_{64}O_{13}$	mp 256—257° $[\alpha]_D^{20} + 4.8$ (dioxane)	40		
2'-Acetyl neriifolin [3]	25633-33-4	$C_{32}H_{48}O_9$	mp 264—268° $[\alpha]_D^{15} - 76°$ (Py)	41,42	43	IR, PMR,[129]
Bruceine B [4]	25514-29-8	$C_{23}H_{28}O_{11}$				CMR,[153] related crystal structure,[167] Review[298]
Glaucarubinone [5]	1259-86-5	$C_{25}H_{34}O_{10}$	mp 225—228° $[\alpha]_D + 50°$ (MeOH)		299	CMR,[153] review[298]

a °C.

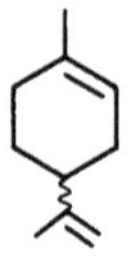

1. Limonene

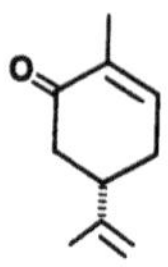

2. d-Carvone

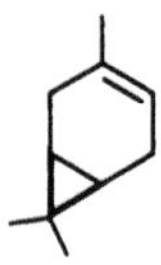

3. Δ-3-Carene

4. d-Limonene

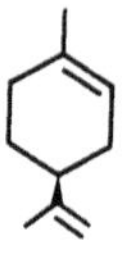

5. l-Limonene

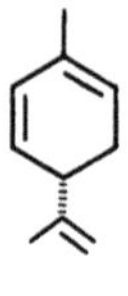

6. α-Phellandrene

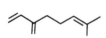

7. Myrcene

8. d-α-Pinene

9. l-α-Pinene

10. β-Pinene

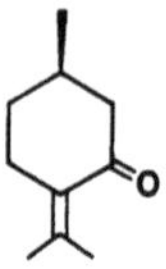

11. Pulegone

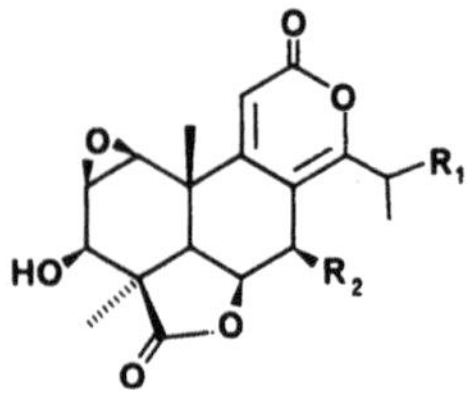

12. Nagilactone D R₁ = R₂ = H

13. Nagilactone C R₁ = CH₃, R₂ = OH

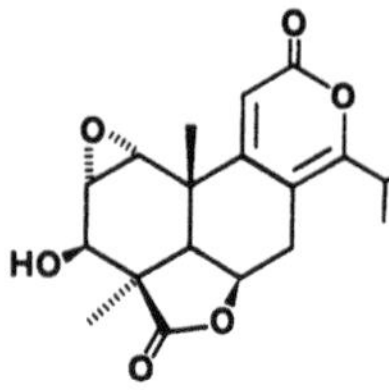

14. Hallactone A

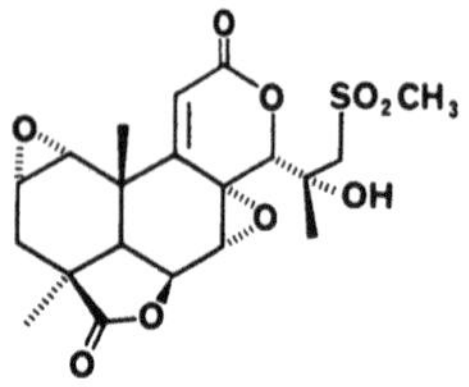

15. Hallactone B

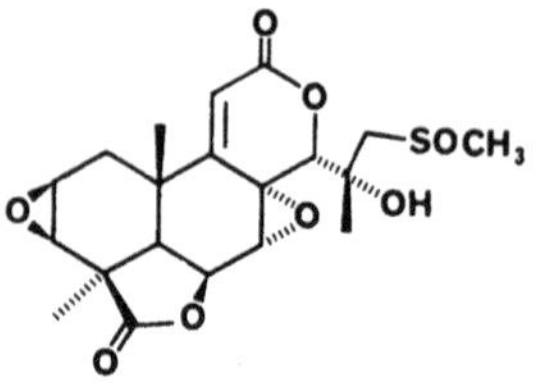

16. Podolactone C

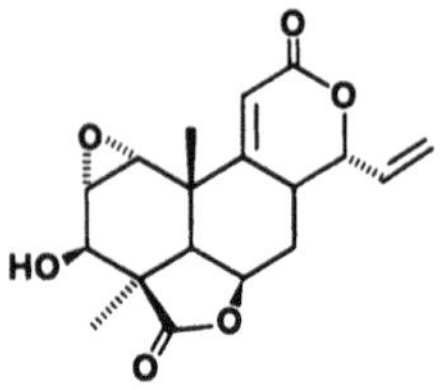

17. Podolactone E

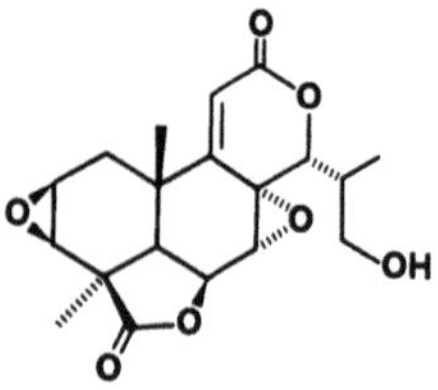

18. Sellowin A

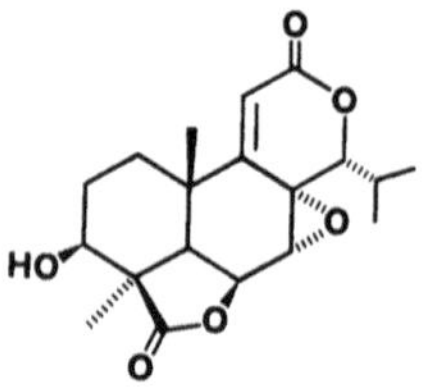

19. Nagilactone E

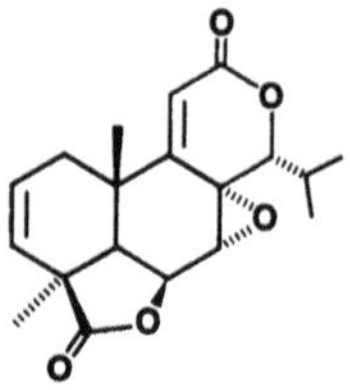

20. Podolide

Structures of terpenes (Table 6).

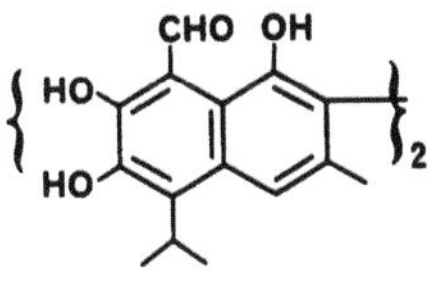

21. Gossypol

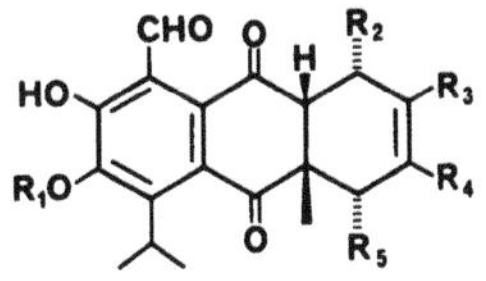

22. Heliocide H₁ R₁.₄.₅ = H; R₃ = CH₃; R₂ = CH₂-CH = C(CH₃)
23. Heliocide H₂ R₁.₂.₄.₅ = H; R₃ = (CH₂)₂ CH = C(CH₃)₂
24. Heliocide H₃ R₁.₂.₃.₅ = H; R₄ = (CH₂)₂ CH = C(CH₃)₂

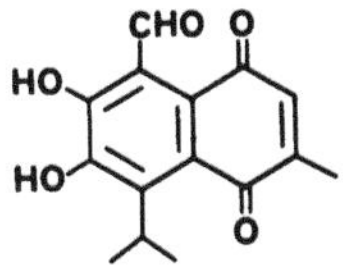

25. Hemigossypolone

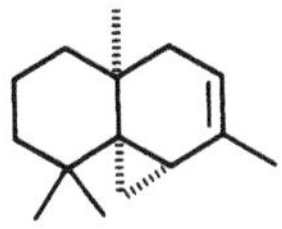

26. Thujopsene

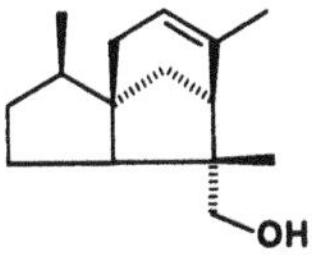

27. 8-Cedrene-13-ol

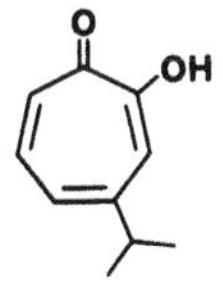

28. β-Thujaplicin

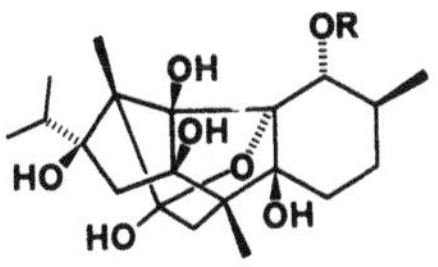

29. Cinnzeylanine R = OAc
30. Cinnzeylanol R = H

Terpenes continued.

Table 6a
OTHER TERPENES AFFECTING INSECT MORTALITY

Insect	Plant	Compound (conc.) [No.]	Ref.
Dendroctonus frontalis Zimmerman	*Pinus taeda* L.	Limonene [1]	32
Drosophila melanogaster Meig.	*Anethum graveolus* L.	d-Carvone (3624 ppm) [2]	51
Drosophila melanogaster	*A. graveolis*	d-Carvone [2]	51
Aedes aegypti	*A. graveolis*	d-Carvone [2]	51
Dendroctonus frontalis (Zimmerman)	*Pinus* spp.	Δ-3-Carene (0.36 μg/beetle) [3]	31
	Pinus spp.	d-Limonene (0.47 μg/beetle) [4]	31
	Pinus spp.	l-Limonene (0.55 μg/beetle) [5]	31
	Pinus spp.	α-Phellandrene (0.47 μg/beetle) [6]	31
	Pinus spp.	Myrcene (0.62 μg/beetle) [7]	31
	Pinus spp.	d-α-Pinene (0.64 μg/beetle) [8]	31
	Pinus spp.	l-α-Pinene (0.65 μg/beetle) [9]	31
	Pinus spp.	β-Pinene (0.95 μg/beetle) [10]	31
Dendroctonus brevicomis Le Conte	*P. ponderosa* Laws	β-Pinene [10]	127
	P. ponderosa Laws	Δ₃-Carene [3]	127
	P. ponderosa Laws	Limonene [1]	127
Spodoptera frugiparda (J. E. Smith)	*Mentha pulegium* L.	Pulegone [11]	308
M. domestica L.	*Pondocarpus* spp.	Nagilactone D (0.7 ppm) [12]	20
Lasperyesia pomonella (L.)	*Podocarpus* spp.	Nagilactone D (7.3 ppm) [12]	20
	Podocarpus spp.	Nagilactone C (6.3 ppm) [13]	18, 20
M. domestica L.	*Podocarpus halii* Kirk	Nagilactone C (12 ppm) [13]	72
	P. halii Kirk	Hallactone A (3.5 ppm) [14]	72
	P. neriifolius	Podolactone C (8.2 ppm) [16]	72
	P. neriifolius	Podolactone E (1.3 ppm) [17]	72
	Podocarpus spp.	Sellowin A (13 ppm) [18]	72
M. domestica	*P. halii* T. Kirk	Hallactone A [14]	229
	P. halii T. Kirk	Hallactone B (48 ppm diet) [15]	20
M. domestica L.	*Podocarpus* spp.	Nagilactone E (41 ppm diet) [19]	20
	Podocarpus spp.	Podolide (34 ppm diet) [20]	20
Epiphyas postvittana	*Podocarpus* spp.	Nagilactone C (76 ppm diet) [13]	20
	Podocarpus spp.	Nagilactone D (64 ppm diet) [12]	20
Aphis gossypii Glover	*Gossypium hirsutum* L.	Gossypol [21]	32
Anthonomous grandis thuberiae Pierce	*G. hirsutum* L.	Gossypol [21]	32
Estigmene acrea Drury	*G. hirsutum* L.	Gossypol [21]	32
Heliothis zea (Boddie)	*G. hirsutum* L.	Gossypol [21]	32
Pectinophera gossypiella (Saunders)	*Gossypium* spp.	Gossypol (0.1% diet) [21]	125
Heliothis virescens (F.)	*G. hirsutum* L.	Heliocide H₁ [22]	30
Heliothis spp.	*G. hirsutum*	Heliocide H₂ [23]	252
	G. hirsutum	Heliocide H₃ [24]	252
	G. hirsutum	Hemigossypolone [25]	251
Culex pipiens pallens	*Juniperus recurva*	Thujopsene (4.5 μg/mosquito) [26]	247
	J. recurva	8-Cedren-13-ol (6.6 μg/mosquito) [27]	247
Hylotropus bajulous	*J. recurva* & *Thuja plicta*	β-Thujaplicin [28]	76
Bombyx mori L.	*Cinnamonium zeylanicum* Nees	Cinnzeylanine (16 ppm) [29]	53
	C. zeylanicum Nees	Cinnzeylanol (16 ppm) [30]	53

Table 6b
CHEMICAL DATA FOR OTHER TERPENES

Name [No.]	Chem. Abstr. reg. no.	Formula	Physical constants[a]	Synthesis (Ref.)	Biosynthesis (Ref.)	Comments
dL-Limonene (dipentene) [1]	138-86-3	$C_{10}H_{16}$	bp$_{20}$ 178°; n$_D$ 1.4744	82,246		PMR, UV, IR[51]
D-Carvone [2]	2244-16-8	$C_{10}H_{14}O$	bp$_{735}$ 231°; bp$_{205-6}$ 91°; n$_D^{25}$ 1.4989	67		
Δ-3-Carene [3]	498-15-7	$C_{10}H_{16}$	bp 7.5 168—169°; $[\alpha]_D^{20}$ + 7.7°	236	237	
D-Limonene [4]	5989-27-5	$C_{10}H_{16}$	bp$_{20}$ 71°; $[\alpha]_D^{20}$ + 126.8°			
l-Limonene [5]	5989-54-8	$C_{10}H_{16}$	bp 177.6—177.8°; $[\alpha]_D$ − 122.6°			
α-Phellandrene [6]	4221-98-1	$C_{10}H_{16}$	bp 174°; $[\alpha]_D^{20}$ − 217°			
Myrcene [7]	123-35-3	$C_{10}H_{16}$	bp 116°; bp$_{85}$ 51—51.5°	242	243	MS[244]
D-α-Pinene [8]	7785-70-8	$C_{10}H_{16}$	mp 50°C; bp 155—156°; $[\alpha]_D$ + 51.1°	239		PMR[245] PMR[168], CMR[241]
l-α-Pinene [9]	7785-76-4	$C_{10}H_{16}$	bp 155—156°; $[\alpha]_D$ − 51.1°	239		PMR, CMR: as with D-α-pinene
β-Pinene [10]	127-91-3	$C_{10}H_{16}$	bp 163—164°; $[\alpha]_D$ − 22°	239		Related crystal structure[240]
Pulegone [11]	89-82-7	$C_{10}H_{16}O$	bp 224°; $[\alpha]_D^0$ + 23° (EtOH)	301	317	
Nagilactone D [12]	19891-53-3	$C_{18}H_{20}O_6$	mp 255—256° (dec); $[\alpha]_D$ + 90°			IR, PMR, UV[262]
Nagilactone C [13]	24338-53-2	$C_{19}H_{22}O_7$	mp 290° (dec); $[\alpha]_D$ +111°			IR, PMR, UV[262]
Hallactone A [14]	41787-72-8	$C_{19}H_{22}O_6$	mp 266—268° (dec)			IR, UV, PMR[229]
Hallactone B [15]	35470-59-8	$C_{20}H_{24}O_9S$	mp 325—330° (dec)			IR, UV, PMR,[229] IR, PMR, MS, UV

Table 6b (continued)
CHEMICAL DATA FOR OTHER TERPENES

Name [No.]	Chem. Abstr. reg. no.	Formula	Physical constants[a]	Synthesis (Ref.)	Biosynthesis (Ref.)	Comments
Podolactone C [16]	35467-31-3	$C_{20}H_{24}O_8$	mp 288—290°			IR, PMR, UV, MS,[281] crystal structure[311]
Podolactone E [17]	37070-59-0	$C_{18}H_{18}O_6$	mp 261—262°			IR, PMR, UV, MS,[288] crystal structure[287]
Sellowin A [18]	34198-79-3	$C_{19}H_{22}O_7$	mp 298° (dec) $[\alpha]_D$ +16° (Py)			Related crystal structure[321]
Nagilactone E [19]	36895-12-2	$C_{19}H_{24}O_6$	mp 295° $[\alpha]_D^{20}$ −14.1° (Py)			UV, IR, PMR[320]
Podolide [20]	55786-36-2	$C_{19}H_{22}O_5$	mp 296—298° $[\alpha]_D^{25}$ −12° (Py)			Crystal structure,[310,318]
Gossypol [21]	303-45-7	$C_{30}H_{30}O_8$	mp 184° (ether) mp 199° (CHCl$_3$) mp 214° (ligroin)	15	Distribution[250]	CMR, PMR[238] review[131]
Heliocide H$_1$ [22]	65024-84-2	$C_{25}H_{30}O_5$	mp 110—112°	CMR, PMR, IR, UV, MS[30]	Distribution[250]	CMR, PMR[238]
Heliocide H$_2$ [23]	63525-06-4	$C_{25}H_{30}O_5$	mp 123—126°	IR, PMR, CMR, UV, MS, X-ray structure[234]	Chemotaxonomy[250]	CMR, PMR,[238] crystal structure[131]
Heliocide H$_3$ [24]	64960-69-6	$C_{25}H_{30}O_5$	mp 128—134°	IR, PMR, CMR, UV, MS[235]		
Hemigossypolone [25]	35688-47-2	$C_{15}H_{14}O_5$	mp 166.5—169°	UV, PMR, CMR, MS[319]		
Thujopsene [26]	470-40-6	$C_{15}H_{24}$	$[\alpha]_D^6$ −94.3° bp$_{12}$ 121—122°	261		CMR[241]
8-Cedren-13-ol [27][b]		$C_{15}H_{24}O$	p-Nitro benzoate ester: mp 91—92° $[\alpha]_D^{25}$ −77.1°			IR, PMR, UV, MS[248]

β-Thujaplicin [28]	499-44-5	$C_{10}H_{12}O_2$	mp 52—52.5°	259
Cinnzeylanine [29]	62203-47-8	$C_{22}H_{34}O_8$	mp 265—267° (MeOH)	Crystal structure[260]
			$[\alpha]_D^{12}$ +45°	Identification,[71] crystal struc-
				ture,[253] IR, CMR, PMR, MS[53]
Cinnzeylanol [30]	62394-04-1	$C_{20}H_{32}O_7$	mp 278—279°	IR, PMR, CMR, MS[71]
			$[\alpha]_D^{26}$ + 60.8° (MeOH)	

a   °C.

b   We have been unable to find a CA registry number for the structure presented.

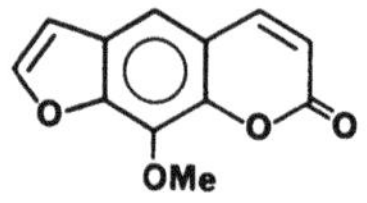

1. Xanthotoxin

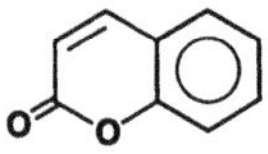

2. Coumarin

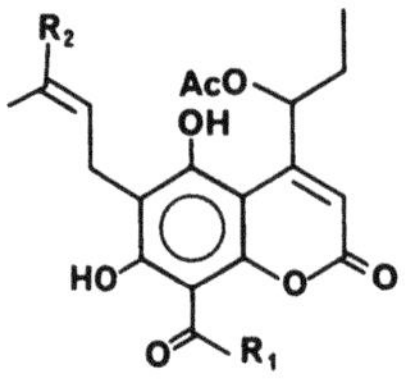

3. Mammea Xa R_1 = $CH_2CH(CH_3)_2$; R_2 = CH_3
   Mammea Xb R_1 = $CH(CH_3)CH_2CH_3$ R_2 = CH
4. Surangin B R_1 = $CH(CH_3)CH_2CH_3$ R_2 = CH_2CH_2CH = $C(CH_3)_2$

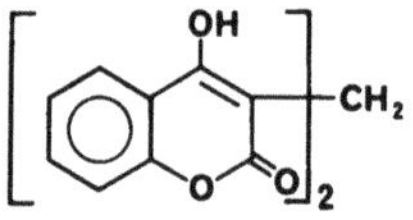

5. Dicoumarol

Structures of coumarins (Table 7).

Table 7a
COUMARINS AFFECTING INSECT MORTALITY

Insect	Plant	Compound (conc.) [No.]	Ref.
Spodoptera eridania (Cramer)	Families Rutaceae and Umbelliferae	Xanthotoxin (0.1%) [1] (8-methoxy psoralin)	54
Drosophilia melanogaster	*Eupatorium japonicum*	Coumarin [2] (ovicide 15 μg/cm²) (larvae 0.06% diet)	249
"Mustard beetle"	*Mammea americana L.*	Mammea Xa,b[a] (0.01%) [3]	263
Musca domestica	*M. americana L.*	Mammea Xa,b[a] (0.01%) [3]	263
"Mosquito larvae"	*M. longifola*	Surangin B [4]	263
M. domestica	*M. longifola*	Surangin B (0.01%) [4]	263
Callosobruchus maculata		Coumarin (0.1% diet) [4]	4
		Dicoumarol (1% diet) [5]	4

[a] Unnamed in this publication.

Table 7b
CHEMICAL DATA FOR COUMARINS

Name [no.]	Chem. Abstr. reg. no.	Formula	Physical constants[a]	Comments
Xanthotoxin (8-methoxy psoralen)[b] [1]	298-81-7	$C_{12}H_8O_4$	mp 148°	CMR,[230] crystal structure[231]
Coumarin [2]	91-64-5	$C_9H_6O_2$	mp 68.5—69.5°	
Mammea X a,b [3]	a. 98498-81-8[c]	$C_{24}H_{30}O_7$	mp 50—53° (mixture as isolated)	IR, PMR, MS, UV, ORD[263]
	26477-64-5			
	b. 26477-65-6			
Surangin B [4]	28319-38-2	$C_{29}H_{38}O_7$	mp 98—100° $[\alpha]_D^{24} - 30°$	IR, PMR, MS, UV[264]
Dicumarol [5]	66-76-2	$C_{19}H_{12}O_6$	mp 290—292°	

[a] °C.

[b] See Reference 228 for biosynthesis.

[c] There are 2 CA Reg's for this compound.

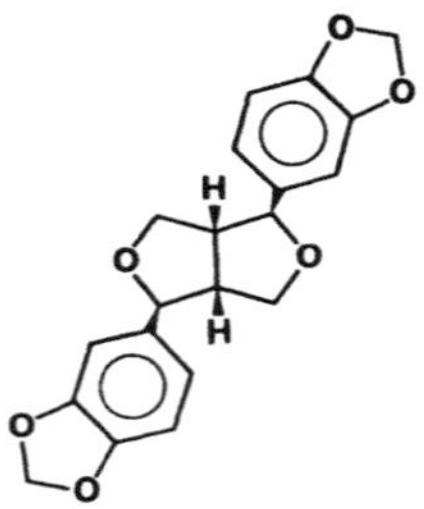

1. Sesamin

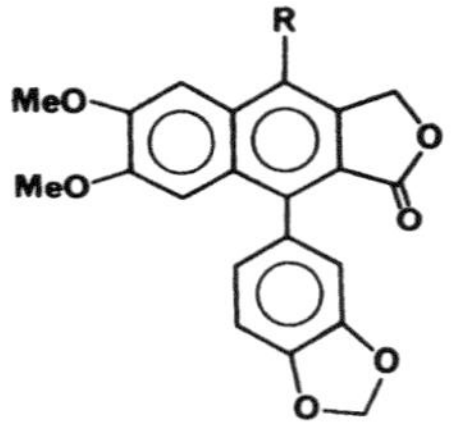

2. Justicidin A R = OCH₃
3. Justicidin B R = H

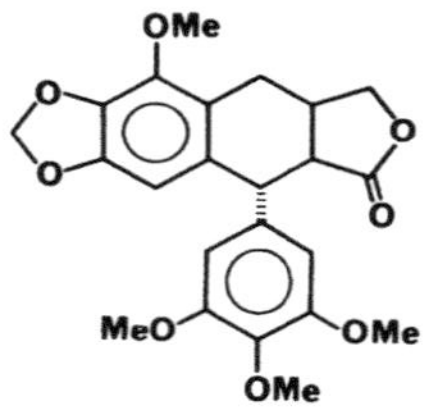

4. Peltatin methyl ether A

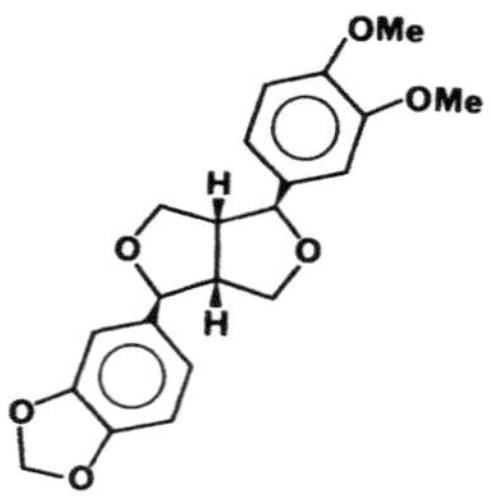

5. Kobusin

Structures of lignins (Table 8).

Table 8a
LIGNINS AFFECTING INSECT MORTALITY

Insect	Plant	Compound (conc.) (No.)	Ref.
Bombyx mori L.	*Magnolia kobus* DC	Sesamin (400 ppm) [1]	38
	Justica procumbens L.	Justicidin A (20 ppm) [2]	38
	J. procumbens L.	Justicidin B (20 ppm) [3]	38
Musca domestica L.	*Libocedrus biswillii* Hook. f.	Peltatin methyl ether A [4]	265
Bombyx mori L.	*Magnolia kobus* DC	Kobusin (200 ppm diet) [5]	38

Table 8b
CHEMICAL DATA FOR LIGNINS

Name [No.]	Chem. Abstr. reg. no.	Formula	Physical constants[a]	Synthesis (Ref.)	Comments
Sesamin [1]	13079-95-3	$C_{20}H_{18}O_6$	mp 122—124° $[\alpha]_D^{17} -68°$ (CHCl$_3$)	MS, IR[61]	Review,[98] PMR,[267,269] CMR[268]
Justicidin A [2]	25001-57-4	$C_{22}H_{18}O_7$	mp 263°	62,63	
Justicidin B [3]	17951-19-8	$C_{21}H_{16}O_6$	mp 240°	62,62	
Peltatin methyl ether A [4]	518-29-6	$C_{23}H_{24}O_8$	mp 162—163° $[\alpha]_D -119°$ (CHCl$_3$)		IR, PMR, UV[266]
Kobusin [5]	36150-23-9	$C_{21}H_{22}O_6$	$[\alpha]_D^{27} +63.9°$ (CHCl$_3$)		

[a] °C.

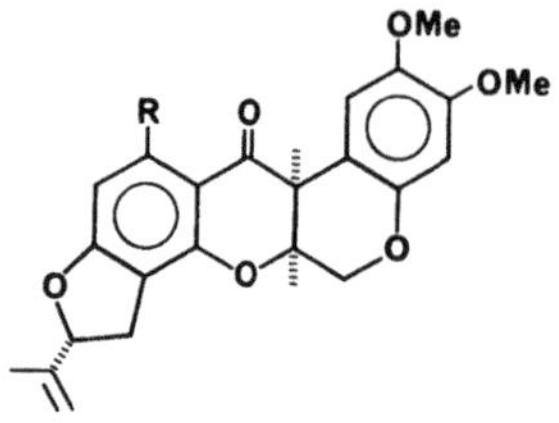

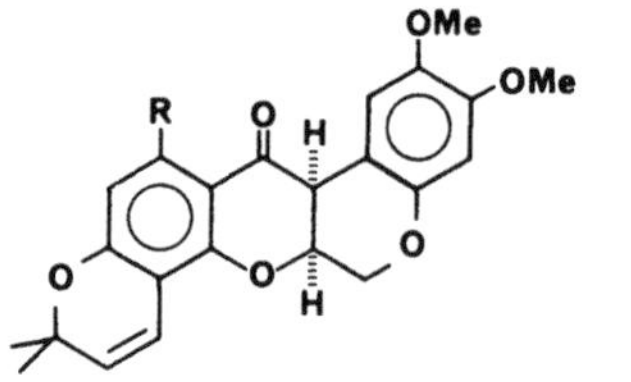

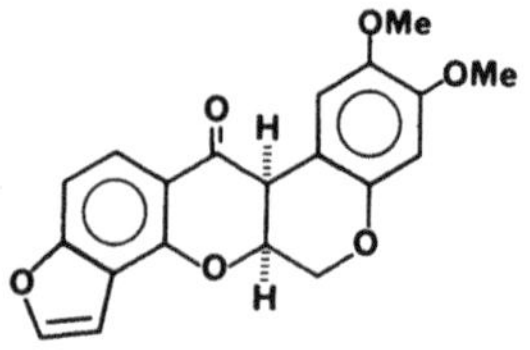

1. Rotenone
4. Sumatrol

2. Degulin
3. Toxicarol

5. Elliptone

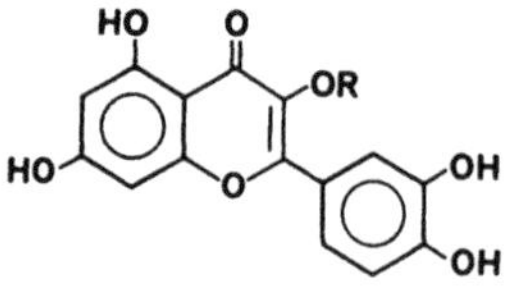

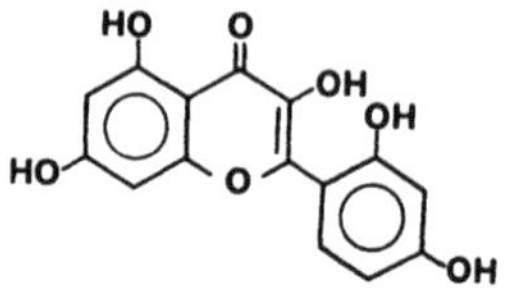

6. Isoquercitrin R = Glucose
7. Quercetin R = H
8. Quercitrin R = Rhamnose
9. Rutin R = Rutinose [Glc(6←1)-α-L-Rha]

10. Morin

Structures of rotenoids and flavanoids (Table 9).

Table 9a
ROTENOIDS AND OTHER FLAVONOIDS AFFECTING INSECT MORTALITY

Insect	Plant	Compound (conc.) [No.]	Ref.
Aphis rumicis	*Tephrosia* spp.	Rotenone (5 ppm topical) [1]	111
	Derris spp.		
	Lonchocarpus spp.		
	Millettia spp.		
	Mundulea spp.		
Apis mellifera		Rotenone (3 μg/g oral) [1]	111
Bombyx mori (4th instar)		Rotenone (3 μg/g oral) [1]	111
B. mori (5th instar)		Rotenone (7—10 μg/g inj.) [1]	111
Chironomous spp. (larvae)		Rotenone (6 ppm diet) [1]	111
Culex spp. (larvae)		Rotenone (5 ppm) [1]	111
Heliothis armigera		Rotenone (>490 ppm) [1]	111
Melanoplus femur-rubrum		Rotenone (4—7000 ppm) [1]	111
Musca domestica		Rotenone (0.3% contact) [1]	111
Periplaneta americana		Rotenone (6—15 μg/g inj.) [1]	111
Vanessa (=*Cynthia*) *cardui*		Rotenone (30 μg/g oral) [1]	111
Leptinotarsa decemlineata (5th instar)		Rotenone (0.2 μg/g oral) [1]	111
Bombyx mori: (4th instar)		Degulin (10—12 μg/g oral) [2]	111
Musca domestica		Degulin (0.6—2.8 μg/g contact) [2]	111
Aphis rumicis		Degulin (50 ppm topical) [2]	111
		Toxicarol (2000 ppm topical) [3]	111
		Sumatrol (650 ppm topical) [4]	111

Table 9a (continued)
ROTENOIDS AND OTHER FLAVONOIDS AFFECTING INSECT MORTALITY

Insect	Plant	Compound (conc.) [No.]	Ref.
Macrosiphoniella sanborni (adult)		Rotenone (10 mg/ℓ deposit) [1]	111
		Degulin (50 mg/ℓ deposit) [2]	111
		Elliptone (50 mg/ℓ deposit) [5]	111
		Toxicarol (60 mg/ℓ deposit) [3]	111
		Sumatrol (150 mg/ℓ deposit) [4]	111
Angeliastica alni		Rotenone (0.38 µg/cm$_2$ contact) [1]	111
Oryzaephilis surinamensis		Rotenone (62 mg/ℓ deposit) [1]	111
Aphis rumicis		Rotenone (3.3 ppm contact on nastertium) [1]	111
Brevicoryne brassicae		Rotenone (5 ppm contact on cabbage) [1]	111
Myzus persicae		Rotenone (5 ppm contact on cabbage) [1]	111
Trialeurodes vaporariorum (eggs)		Rotenone (50 ppm contact on beans) [1]	111
Malacosoma americana (larvae)		Rotenone (33 ppm contact on apple) [1]	111
Tetranychus telarius		Rotenone (500 ppm contact) [1]	111
Anuraphis rosae		Rotenone (20 ppm contact on apple) [1]	111
Aphis persicae-niger		Rotenone (25 ppm contact on peach) [1]	111
Aphis pomi		Rotenone (16 ppm contact on apple) [1]	111
Typhlociba comes (nymph)		Rotenone (10 ppm contact on grape) [1]	111
Thrips spp.		Rotenone (100 ppm contact on plantago) [1]	111
Trialeurodes vaporariorum (larvae)		Rotenone (10 ppm contact on beans) [1]	111
Thrips tabaci		Rotenone (50 ppm contact on beans) [1]	111
Doryphera decemlineata (adult)		Rotenone (500 ppm contact on potatoe) [1]	111
Doryphera decemlineata (larvae)		Rotenone (33 ppm contact on potato) [1]	111
Epilachna corrupta (adult)		Rotenone (100 ppm contact on bean) [1]	111
E. corrupta (larvae)		Rotenone (16 ppm contact on bean) [1]	111
Popillia japonica (adult)		Rotenone (75 ppm contact on *Polygonum*) [1]	111
Culex (larvae)		Rotenone (0.5 ppm) [1]	111
Heliothis zea (Boddie)	*Gossypium* spp.	Isoquercitrin (0.6% diet) [6]	125
	Gossypium spp.	Quercetin (0.6% diet) [7]	125
	Gossypium spp.	Quercitrin (0.6% diet) [8]	125
Heliothis virescens (F.)	*Gossypium* spp.	Isoquercitrin (0.2% diet) [6]	125
	Gossypium spp.	Quercetin (0.2% diet) [7]	125
	Gossypium spp.	Rutin (0.2% diet) [9]	125
	Gossypium spp.	Quercitrin (0.2% diet) [8]	125
Pectinophera Gossypiella (Saunders)	*Gossypium* spp.	Isoquercitrin (0.1% diet) [6]	125
	Gossypium spp.	Quercetin (0.1% diet) [7]	125
	Gossypium spp.	Rutin (0.1% diet) [9]	125
		Quercetin (0.1% diet) [8]	125
		Morin (0.2% diet) [10]	125

Table 9b

CHEMICAL DATA FOR ROTENOIDS AND OTHER FLAVONOIDS

Name [no.]	Chem. Abstr. reg. no.	Formula	Physical constants[a]	Synthesis (Ref.)	Comments
Rotenone[b] [1]	83-79-4	$C_{23}H_{22}O_6$	mp 163° (variable) $[\alpha]_D^{20} -231°$ (C_6H_6)	272	CMR,[274] PMR,[275] crystal structure,[276] review[93]
Degulin[2]	522-17-8	$C_{23}H_{22}O_6$	mp 165—171° $[\alpha]_D^{20} -23°$ (C_6H_6)	88	
Toxicarol [3]	82-09-7	$C_{23}H_{22}O_7$	mp 125—127° (variable) $[\alpha]_D^{20} -66°$ (C_6H_6)		CMR,[274] PMR[275]
Sumatrol [4]	82-10-0	$C_{23}H_{22}O_7$	mp 174.5—177.5° (variable) $[\alpha]_D^{20} -184°$ (C_6H_6)		CMR,[274] PMR[275]
Elliptone (Derride) [5]	478-10-4	$C_{20}H_{16}O_6$	mp 160° (variable) $[\alpha]_D^{20} -18°$ (C_6H_6)	277	CMR,[274] PMR[275]
Isoquercitrin [6]	21637-25-2	$C_{21}H_{20}O_{12}$	mp 234—236°	322	CMR[323,324]
Quercetin [7]	117-39-5	$C_{15}H_{10}O_7$	Dihydrate mp 313—314°		
Quercitrin [8]	522-12-3	$C_{21}H_{20}O_{11}$	mp 250—252° Dihydrate mp 182—185°		
Rutin [9]	153-18-4	$C_{27}H_{30}O_{16}$	mp 214—215° $[\alpha]_D^{23} +13.82°$ (EtOH)	322	CMR,[324] MS[325]
Morin [10]	480-16-0	$C_{15}H_{10}O_7$	Monohydrate mp 303—304°		

[a] °C.

[b] See Reference 273 for biosynthesis.

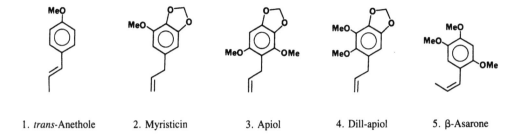

1. *trans*-Anethole   2. Myristicin   3. Apiol   4. Dill-apiol   5. β-Asarone

Structures of other phenylpropanoids (Table 10).

Table 10a
OTHER PHENYLPROPANOIDS AFFECTING INSECT MORTALITY

Insect	Plant	Compound (conc.) [No.]	Ref.
Musca domestica L.	*Pimpinella anisum* L.	*trans*-Anethole (75 µg/fly) [1]	50
Drosophila melanogaster Meig.	*Anethum graveolus* L.	Myristicin (0.34 mg/jar) [2]	51
Aedes aegypti L.	*A. graveolus* L.	Myristicin (10.8 mg/jar) [2]	51
D. melanogaster Meig.	*A. graveolus* L.	Apiol (0.29 mg/jar) [3]	51
A. aegypti L.	*A. graveolus* L.	Apiol (11.5 mg/jar) [3]	51
D. melanogaster Meig.	*A. graveolus* L.	Dill-apiol (0.27 mg/jar) [4]	51
A. aegypti	*A. graveolus* L.	Dill-apiol (11.0 mg/jar) [4]	51
D. melanogaster Meig.	*Pastinaca sativa* L.	Myristicin (200 ppm) [2]	52
Callosobruchus maculata		β-Asarone (1%) [5]	4

Table 10b

CHEMICAL DATA FOR OTHER PHENYLPROPANOIDS

Name [No.]	Chem. Abstr. reg. no.	Formula	Physical constants[a]	Synthesis (Ref.)	Comments
trans-Anethole [1]	104-46-1	$C_{10}H_{12}O$	mp 21.4° bp 202.3, 81—81.5° n_D^{25} 1.56145	66	
Myristicin [2]	607-91-0	$C_{11}H_{12}O_3$	bp760, 276—277° bp21, 157°	68	IR, PMR, MS, UV[51]
Apiol [3]	523-80-8	$C_{12}H_{14}O_4$	mp 29.5° bp 294°	69	IR, PMR, MS, UV[51]
Dillapiol [4]	484-31-1	$C_{12}H_{14}O_4$	mp 29.5° bp 285° bp4, 162° n_D^{35} 1.5278	68	IR, PMR, MS, UV[51]
β-Asarone [5]	5273-86-9	$C_{12}H_{16}O_3$	N_D^{20} 1.5641		

[a] °C.

CH₃(CH₂)ₙCOOH

1. Capric acid n = 8
2. Caprylic acid n = 6
3. Caproic acid n = 4

Structures of fatty acids (Table 11).

Table 11a
FATTY ACIDS AFFECTING INSECT MORTALITY

Insect	Compound (conc.) [No.]	Ref.
Tribolium confusum Jacqueline duVal	Capric acid [1]	44
Pseudosarcophaga affinis Auct. nec Fallen	Capric acid (0.00025 *M*) [1]	45
	Caprylic acid (0.054 *M*) [2]	45
	Caproic acid (0.155 *M*) [3]	45

Table 11b
CHEMICAL DATA FOR FATTY ACIDS

Name [No.]	Chem. Abstr. reg. no.	Formula	Physical constants[a]	Synthesis (Ref.)	Comments
Capric acid [1]	334-48-5	$C_{10}H_{20}O_2$	bp 270° n_D^{40} 1.4288		Review[46]
Caprylic acid [2]	124-07-2	$C_8H_{16}O_2$	bp 239.7° mp 16.7° $_D^{20}$ 1.4280	47	Review[46]
Caproic acid [3]	142-62-1	$C_6H_{12}O_2$	bp 205° mp 3.4° n_D^{20} 1.4163	48	Review[46]

[a] °C.

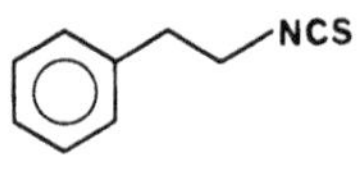

1. 2-Phenylethylisothiocyanate

2. 2-Propenylisothiocyanate

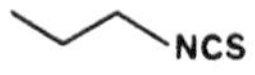

3. Propylisothiocyanate

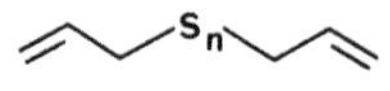

4. Diallyl disulfide n = 2
5. Diallyl trisulfide n = 3

Structures of sulfur compounds (Table 12).

Table 12a
ISOTHIOCYANATES AND OTHER SULFUR COMPOUNDS AFFECTING INSECT MORTALITY

Insect	Plant	Compound (conc.) [No.]	Ref.
Musca domestica L.	Many cruciferous crops	2-Phenylethylisothiocyanate (568 ppm) [1]	8
Drosophila melanogaster Meig.	Many cruciferous crops	2-Phenylethylisothiocyanate (568 ppm) [1]	8
	Brassica rapa L.	2-Phenylethylisothiocyanate (63 ppm) [1]	9
M. domestica L.	*B. rapa* L.	2-Phenylethylisothiocyanate [1]	9
Acrythosiphon pisum (Harris)	*B. rapa* L.	2-Phenylethylisothiocyanate [1]	9
Tetranychus atlanticus McGregor	*B. rapa* L.	2-Phenylethylisothiocyanate [1]	9
Epilachna varivestis Mulsant	*B. rapa* L.	2-Phenylethylisothiocyanate [1]	9
Anastrepha suspensa (Loew)		2-Propenylisothiocyanate (2.1 mg/ℓ) [2]	10
		Propylisothiocyanate (2.1 mg/ℓ) [3]	10
Culex pipiens quinquefasciatus Say	*Allium sativum* L.	Diallyl disulfide (5 ppm) [4]	11
	A. sativum L.	Diallyl trisulfide (5 ppm) [5]	11

Table 12b
CHEMICAL DATA FOR THIOCYANATES AND OTHER SULFUR COMPOUNDS

Name [No.]	Chem. Abstr. reg. no.	Formula	Physical constants[a]	Synthesis (Ref.)
2-Phenyl-ethyl-isothiocyantate [1]	2257-09-2	C_9H_9NS		
2-Propenyl-isothiocyanate (allyl isothiocyanate)[b] [2]	56-06-7	C_4H_5NS	fp $-80°$, bp $-151°$ bp12, $-44°$	12
Propyl isothiocyanate [3]	628-30-8	C_4H_7NS	bp 153°	13
Diallyl disulfide [4]	2179-57-9	$C_6H_{10}S_2$	bp16, 78—80°	14
Diallyl trisulfide [5]	2050-87-5	$C_6H_{10}S_3$		16,75

[a] °C.

[b] For comments on toxicology see Reference 17 (p. 531).

CH$_3$(CH$_2$)$_{10}$COCH$_3$

1. 2-Tridecanone

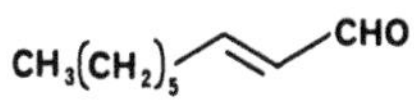

CH$_3$(CH$_2$)$_5$⌒CHO

2. *trans*-2-nonenal

H$_2$N—C(=NH)—NH—⌒—COOH

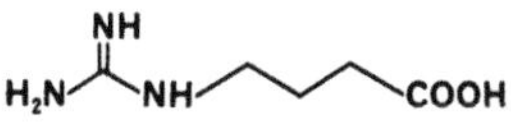

3. γ-Guandinobutyric acid

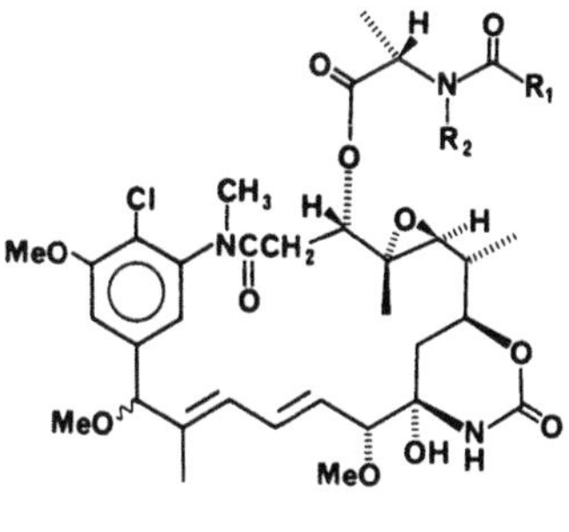

4. Trewiasine R$_1$ = CH(CH$_3$)$_2$, R$_2$ = CH$_3$

5. Dimethytrewiasine R$_1$ = CH(CH$_3$)$_2$, R$_2$ = H

7. Dehydrotrewiasine R$_1$ = C(CH$_3$) = CH$_2$, R$_2$ = CH$_3$

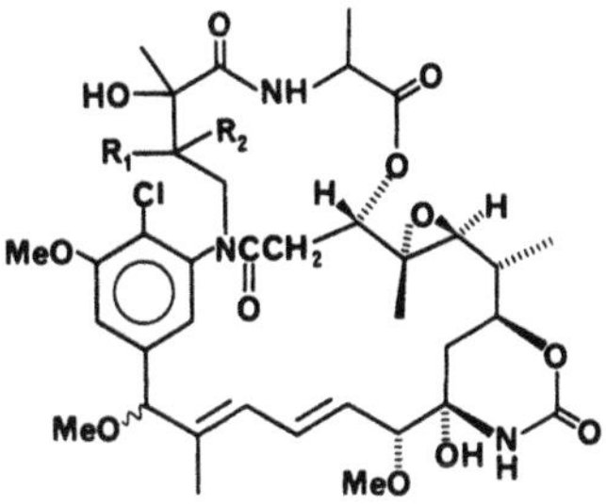

6. Treflorine R$_1$ = R$_2$ = H

8. *N*-Methyltrenudone R$_1$, R$_2$ together = O

9. Trenudine R$_1$ = H, R$_2$ = OH

O$_2$N—⌒—COOH

10. 3-Nitropropionic acid

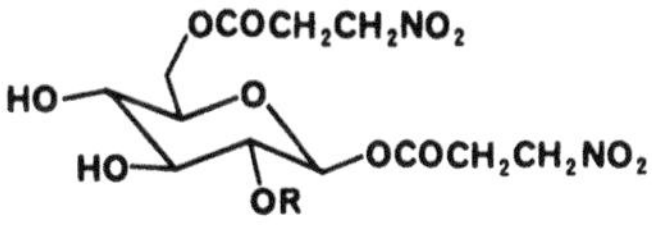

11. Karakin R = COCH$_2$CH$_2$NO$_2$
13. Cibarian R = H

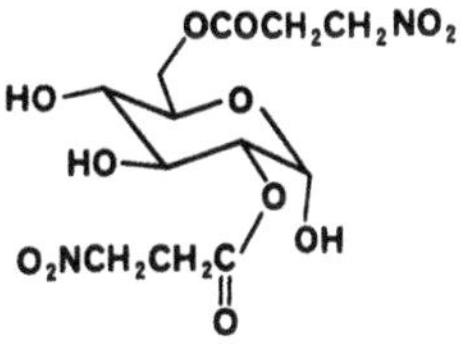

12. Coronarian

Structures of miscellaneous compounds (Table 13).

Table 13a

MISCELLANEOUS COMPOUNDS AFFECTING INSECT MORTALITY

Insect	Plant	Compound (conc.) [No.]	Ref.
Manduca sexta (L.)	*Lycopersicon hirsutum* f. *glabratum* C. H. Mull	2-Tridecanone (17.5 mg/cm^2) [1]	49
Heliothis zea (Boddie)	*L. hirsutum* f. *glabratum* C. H. Mull	2-Tridecanone (17.5 mg/cm^2) [1]	49
Aphis gossypii Glover	*L. hirsutum* f. *glabratum* C. H. Mull	2-Tridecanone (17.5 mg/cm^2) [1]	49
Psila rosae (F.)	*Daucus carota* L.	*trans*-2-Nonenal (2.17 mg/24 hr) [2]	57
Callosobruchus maculatus		γ-Guanidinobutyric acid [3]	4
Ostrinia nubilalis (Hübner)	*Trewia nudiflora*	Trewiasine (20 ppm diet) [4]	285
	T. nudiflora	Demethyl trewiasine (50 ppm diet) [5]	285
	T. nudiflora	Treflorine (12.5 ppm diet) [5]	285
	T. nudiflora	Dehydrotrewiasine (50 ppm diet) [7]	285
	T. nudiflora	*N*-Methyltrenudone (3 ppm diet) [8]	285
	T. nudiflora	Trenudine (3 ppm diet) [9]	285
Cydia pominella L.	*T. nudiflora*	Trewiasine (60 ppm diet) [4]	258
Acalymma vittatum F.	*T. nudiflora*	Trewiasine (≦3 ppm diet) [4]	258
Costelytra zealandica (White)	*Lotus pedunculatus* (Cav.)	3-Nitropropionic acid (0.1 mg/larva) [10]	81
	L. pedunculatus (Cav.)	Karakin (0.1 mg/larva) [11]	81
	L. pedunculatus (Cav.)	Coronarian (0.1 mg/larva) [12]	81
	L. pedunculatus (Cav.)	Cibarian (0.1 mg/larva) [13]	81
Trichoplusia ni (Hübner)	*Coronilla varia* L.	3-Nitropropionic acid (0.5% diet) [10]	312
Callosobruchus maculatus		3-Nitropropionic acid [10]	4

Table 13b

CHEMICAL DATA FOR MISCELLANEOUS COMPOUNDS AFFECTING INSECT MORTALITY

Name [No.]	Chem. Abstr. reg. no.	Formula	Physical constants[a]	Synthesis (Ref.)	Comments
2-Tridicanone [1]	593-08-8	$C_{13}H_{26}O$	mp 29° bp 260—265° bp$_{16}$, 160°	65	
trans-2-nonenal [2]	18829-56-6	$C_9H_{16}O$	bp$_{11}$, 88—90°	IR, UV,[154] PMR[78]	
γ-Guanidinobutyric acid[b] HCl salt [3]	463-00-3 13890-14-7	$C_5H_{11}N_3O_2$	mp 184°		Crystal structure[284]
Trewiasine [4]	78987-26-5	$C_{36}H_{49}ClN_2O_9$	mp 182—185° $[\alpha]_D^{23}$ −94° (CHCl$_3$)	284, 282, 283	IR, UV, PMR, CMR, MS[233,286]
Demethyltrewiasine [5]	78987-28-7	$C_{37}H_{47}ClN_2O_9$	mp 129—142° $[\alpha]_D^{23}$ −126° (CHCl$_3$)		IR, UV, PMR, CMR, MS[286]
Treflorine [6]	82390-93-0	$C_{35}H_{45}ClN_2O_{10}$	mp 205—208° (dec) $[\alpha]_D^{23}$ −138° (CHCl$_3$)		IR, UV, PMR, CMR, MS[233]
Dehydrotrewiasine [7]	78987-27-6	$C_{36}H_{47}ClN_2O_9$	mp 165—170° $[\alpha]_D^{23}$ −90° (CHCl$_3$)		IR, UV, PMR, CMR, MS[286]
N-Methyltrenudone [8]	82400-19-9	$C_{36}H_{45}ClN_2O_{11}$	mp 192—197° (dec) $[\alpha]_D^{23}$ −10° (CHCl$_3$)		IR, UV, PMR, CMR, MS[286]
Trenudine [9]	82390-94-1	$C_{35}H_{45}ClN_2O_{11}$	mp 200—205° (dec) $[\alpha]_D^{23}$ −114° (CHCl$_3$)		IR, UV, PMR, CMR, MS[233,286]
3-Nitropropionic acid[c] [10]	504-88-1	$C_3H_5NO_4$	mp 66—67°	278	Crystal structure[280]
Karakin [11]	1400-11-9	$C_{15}H_{21}N_3O_{15}$	mp 122—123°		IR, PMR, MS[314]
Coronarian [12]	63505-68-0	$C_{12}H_{18}N_2O_{12}$	mp 147.5—148°		Review, distribution IR, PMR, MS[315]
Cibarian [13]	39797-90-5	$C_{12}H_{18}N_2O_{12}$	mp 123.5—124°		IR, PMR[316]

a °C.
b For crystal structure see Reference 283.
c For biosynthesis see Reference 279.

REFERENCES

1. **Metcalf, R. L.,** Plant derivatives for insect control, in *Crop Resources,* Seigler, D. S., Ed., Academic Press, New York, 1977, 165.
2. **Jacobson, M.,** Isolation and identification of toxic agents from plants, in *Host Plant Resistance to Pests,* Hedin, P. A., Ed., ACS Symp. No. 62, American Chemical Society, Washington, D.C., 1977, 153.
3. **Datta, V. K.,** *Bibliography of Insecticide Materials of Vegetable Origin,* No. 133, Tropical Products Institute, London, 1979.
4. **Janzen, D. H., Juster, H. B., and Bell, E. A.,** Toxicity of secondary compounds to the seed-eating larvae of the bruchid beetle *Callosobruchus maculatus, Phytochemistry,* 16, 223, 1977.
5. **Jacobson, M. and Crosby, D. G., Eds.,** *Naturally Occurring Insecticides,* Marcel Dekker, New York, 1971.
6. **Rogers, E. F., Snyder, H. R., and Fischer, R. F.,** Plant insecticides. II. The alkaloids of *Haplophyton cimicidum, J. Am. Chem. Soc.,* 74, 1987, 1952.
7. **Mandava, N. B., Ed.,** *Handbook of Natural Pesticides: Methods,* Vol. 2, CRC Press, Boca Raton, Fla., 1985.
8. **Lichtenstein, E. P., Morgan, D. C., and Mueller, C. H.,** Naturally-occurring insecticides in cruciferous crops, *J. Agric. Food Chem.,* 12, 158, 1964.
9. **Lichtenstein, E. P., Strongf, F. W., and Morgan, D. C.,** Identification of 2-phenylethylisothiocyanate as an insecticide occurring naturally in the edible part of turnips, *J. Agric. Food Chem.,* 10, 30, 1962.
10. **Carroll, J. F., Morgan, N. O., and Weber, J. D.,** Bioassay of three isothiocyanates as fumigants against larvae of the Caribbean fruit fly and the apple maggot, *J. Econ. Entomol.,* 73, 321, 1980.
11. **Amonkar, S. V. and Banerji, A.,** Isolation and characterization of larvicidal principle of garlic, *Science,* 174, 1343, 1971.
12. **Mahuja, M. and Sherry, S. B.,** Optimized conditions for recovery of essential oil from rape and mustard seed cake, *Indian Chem. J.,* 11, 15, 1976.
13. **Katritzky, A. R., Gruntz, U., Mongelli, N., and Rezende, M. C.,** Conversion of primary aliphatic amines into thiocyanates and thiocarbonate esters, *J. Chem. Soc. Chem. Commun.,* 133, 1978.
14. **Nashimura, H. and Mizutani, J.,** Photochemistry and radiation chemistry of sulfur containing amino acids. A new reaction of the 1-propenylthiyl radicals, *J. Org. Chem.,* 40, 1567, 1975.
15. **Edwards, J. D.,** Synthesis of gossypol and gossypol derivatives, *J. Am. Oil Chem. Soc.,* 47, 441, 1970.
16. **Morel, G., Marchand, E., and Foucaud, A.,** A new use of the tetrahydrofuran/powdered potassium hydroxide system: convenient α-sulfenylation of ketones and preparation of dialkyl polysulfides, *Synthesis,* 918, 1980.
17. **Sax, N. I.,** *Dangerous Properties of Industrial Materials,* 5th ed., Van Nostrand-Reinhold, New York, 1979.
18. **Russell, G. B., Fenemore, P. G., and Singh, P.,** Insect control chemicals from plants. Nagilactone C, a toxic substance from the leaves of *Podocarpus nivalis* and *P. hallii, Aust. J. Biol. Sci.,* 25, 1025, 1972.
19. **Wada, K. and Munakata, K.,** An insecticidal alkaloid, cocculolidine from *Cocculus tribolus DC.* I. The isolation and insecticidal activity of cocculolidine, *Agric. Biol. Chem.,* 31, 336, 1967.
20. **Singh, P., Russell, G. B., Hayashi, Y., Gallagher, R. T., and Frederickson, S.,** The insecticidal activity of some norditerpene dilactones, *Entomol. Exp. Appl.,* 25, 121, 1979.
21. **Isogai, A., Chang, C., Murakoshi, S., Suzuki, A., and Tamura, S.,** Screening search for biologically active substances to insects in crude drug plants, *J. Agric. Chem. Soc. Jpn.,* 47, 443, 1973.
22. **Dahlman, D. L., Herald, F., and Knapp, F. W.,** L-Canavanine effects on growth and development of four species of Muscidae, *J. Econ. Entomol.,* 72, 678, 1980.
23. **Rehr, S. S., Bell, E. A., Janzen, D. H., Feeny, P. P.,** Insecticidal amino acids in legume seeds, *Biochem. Systematics,* 1, 63, 1973.
24. **Yamada, Y., Noda, H., and Okada, H.,** An improved synthesis of DL-canavanine, *Agric. Biol. Chem.,* 37, 2201, 1973.
25. **Turner, B. L. and Harborne, J. B.,** Distribution of canavanine in the plant kingdom, *Phytochemistry,* 6, 863, 1967.
26. **Jacobson, M.,** Insecticides From Plants: A Review of the Literature, 1941—1953, U.S. Department of Agriculture, Washington, D.C., 1958.
27. **Jacobson, M.,** Insecticides From Plants: A Review of the Literature, 1954—1971, U.S. Department of Agriculture, Washington, D.C., 1975.
28. **Iwalu, M. O. E., Osisiogu, I. V. W., and Agbakwuru, E. O. P.,** Dannettia oil, a potential new insecticide: tests with adults and nymphs of *Periplaneta americana* and *Zonocerus variegatus, J. Econ. Entomol.,* 74, 249, 1981.
29. **Hwang, Y.-S., Schaffrahn, R. H., Schultz, G. W., Reierson, D. A., Darwazeh, H. A., and Rust, M. K.,** Is β-phenylnithoethane the active principle of denettia oil?, *J. Econ. Entomol.,* 76, 427, 1983.

30. **Stipanovic, R. D., Bell, A. A., O'Brien, D. B., and Lukefahr, M. J.,** Heliocide H$_1$: a new insecticidal C$_{25}$ terpenoid from cotton *(Gossypium hirstumum), J. Agric. Food Chem.,* 26, 115, 1978.

31. **Coyne, J. F. and Lott, L. H.,** Toxicity of substances in pine oleoresin to southern pine beetles, *J. Ga. Entomol. Soc.,* 11, 301, 1976.

32. **Bottger, G. T., Sheehan, E. T., and Lukefahr, M. J.,** Relation of gossypol content of cotton plants to insect resistance, *J. Econ. Entomol.,* 57, 283, 1964.

33. **Kircher, H. W., Heed, W. B., Russell, J. S., and Grove, J.,** Senita cactus alkaloids: their significance to Sonoran Desert *Drosophila* ecology, *J. Insect Physiol.,* 13, 1869, 1967.

34. **Schutte, H. R. and Seelig, G.,** Biosynthesis of pilocerine and lophocerine, *Liebig's Ann. Chem.,* 730, 186, 1969.

35. **O'Donovan, D. G., Barry, E., and Horan, H.,** The biosynthesis of pilocereine. II., *J. Chem. Soc. (C),* 2398, 1971.

36. **O'Donovan, D. G. and Barry, E.,** Biosynthesis of lophocerine in *Lophocereus schottii.* II., *J. Chem. Soc. Perkin Trans. 1,* 2528, 1974.

37. **Bobbitt, J. M. and Chou, T. T.,** Synthesis of isoquinoline alkaloids. I. Lophocerine, *J. Org. Chem.,* 24, 1106, 1959.

38. **Murakoshi, S., Kamikado, T., Chang, C.-F., Sakurai, A., and Tamura, S.,** Effects of several components from the leaves of four species of plants on the growth of silkworm larvae, *Bombyx mori* L., *Jpn. J. Appl. Entomol. Zool.,* 20, 26, 1976.

39. **McLaughlin, J. L., Freeman, B., Powell, R. G., and Smith, C. R., Jr.,** Neriifolin and 2'-acetylneriifolin: insecticidal and cytotoxic agents from *Thevetia thevetiodes* seeds, *J. Econ. Entomol.,* 73, 398, 1980.

40. **Zorbach, W. W.,** Partial synthesis of Eudesmonoside, *J. Org. Chem.,* 27, 1766, 1962.

41. **Hartenstein, J. and Satzinger, G.,** Glycosidation with the help of Fetizon reagent, *Liebig's Ann. Chem.,* 1763, 1974.

42. **Yoshii, E., Koizumi, T., Ikeshima, H., Ozaki, K., and Hayashi, I.,** Studies on the synthesis of cardiotonic steroids. I. Efficient synthesis of cardenolides, *Chem. Pharm. Bull.,* 23, 2496, 1975.

43. **Aberhart, D. J., Lloyd-Jones, J. G., and Caspi, E.,** Biosynthesis of cardenolides in *Digitalis laneta, Phytochemistry,* 12, 1065, 1973.

44. **House, H. L. and Graham, A. R.,** Capric acid blended into foodstuffs for control of an insect pest, *Tribolium confusum, Can. Entomol.,* 99, 994, 1967.

45. **House, H. L.,** The nutritional status and larvicidal activities of C6- to C-14-saturated fatty acids in *Pseudosarcophaga affinis* (Diptera: Sarcophagidae), *Can. Entomol.,* 99, 384, 1967.

46. **Markley, K. S.,** *Fatty Acids,* Part 1, 2nd ed., Markley, K. S., Ed., Wiley Interscience, New York, 1960.

47. **Langenbeck, W. and Richter, M.,** Studies on the mechanism of the Maurer oxydation, *Chem. Ber.,* 89, 202, 1956.

48. **Vliet, E. B., Marvel, C. S., and Hsueh, C. M.,** 3-Methyl-1-pentanoic acid (valeric acid, β-methyl-), *Org. Syn. C. V.* II, 416, 1943.

49. **Williams, W. G., Kennedy, G. G., Yamamoto, R. T., Thacker, J. D., and Bordner, J.,** 2-Tridecanone: a naturally-occurring insecticide from the wild tomato *Lycopersicon hirsutum* f. *glabraturm, Science,* 207, 888, 1980.

50. **Marcus, C. and Lichtenstein, E. P.,** Biologically active components of anise: toxicity and interactions with insecticides in insects, *J. Agric. Food Chem.,* 27, 1217, 1979.

51. **Lichtenstein, E. P., Liang, T. T., Schulz, K. R., Schnoes, H. K., and Garter, G. T.,** Insecticidal and synergistic components isolated from dill plants, *J. Agric. Food Chem.,* 22, 658, 1974.

52. **Lichtenstein, E. P. and Casida, J. E.,** Myristicin, an insecticide and synergist occurring naturally in the edible parts of parsnips, *J. Agric. Food Chem.,* 11, 410, 1963.

53. **Isogai, A., Murakoshi, S., Suzuki, A., and Tamura, S.,** Chemistry and biological activities of cinnzeylanine and cinnzeylanol, new insecticidal substances from *Cinnamonum zeylanicum* Nes., *Agric. Biol. Chem.,* 41, 1779, 1977.

54. **Berenbaum, M.,** Toxicity of furanocoumarin to armyworms: a case of biosynthetic escape from insect herbivores, *Science,* 201, 532, 1978.

55. **Harborne, J. B. and Mabry, T. J.,** Eds., *The Flavanoids: Advances in Research,* Chapman & Hall, New York, 1982.

56. **Harley-Mason, J. and Jackson, A. H.,** Hydroxytryptamines. II. A new synthesis of physostigmine, *J. Chem. Soc.,* 3651, 1954.

57. **Guerin, F. M. and Ryan, M. F.,** Insecticidal effect of trans-2-nonenal, a constituent of carrot root, *Experientia,* 36, 1387, 1980.

58. **Newkome, G. R. and Bhacca, N. S.,** The absolute configuration of Physostigmine (Eserine). Application of the Nuclear Overhauser effect, *J. Chem. Soc. Chem. Commun.,* 385, 1969.

59. **Harborne, J. B.,** *Comparative Biochemistry of the Flavanoids,* Academic Press, New York, 1967.

60. **Crombie, L., Krasinski, A. H. A., Manzoor-I-Khuda, M.,** Amides of vegetable origin. X. The stereochemistry and synthesis of affinin, *J. Chem. Soc.,* 4970, 1963.

61. **Beroza, M. and Schechter, M. S.,** The synthesis of dl-sesamin and dl-asarinin, *J. Am. Chem. Soc.,* 78, 1242, 1956.
62. **Ohta, K., Marumo, S., Chen, Y.-L.,** Studies on the piscicidal components of *Justicia hayatai* var. *decumbens, Agric. Biol. Chem.,* 35, 431, 1971.
63. **Block, E. and Stevenson, R.,** Lignan lactones. Synthesis of (≡)-collinusin and justicidin B, *J. Org. Chem.,* 36, 3453, 1971.
64. **Mannich, C.,** A synthesis of the arecadine aldehydes and arecolines, *Chem. Ber.,* 75B, 1480, 1942.
65. **Hajek, M., Silhavy, P., Malek, J.,** Free radical addition reactions initiated by metal oxides. I. Antimarkovnikov addition of acetone to olefins initiated by argentic oxide, *Tetrahedron Lett.,* 3193, 1974.
66. **Mueller, E. and Roscheisen, G.,** Variation on the Wurtz synthesis. I. Catalyzed reactions of benzyl and allyl halides with alkali metals, *Chem. Ber.,* 90, 542, 1957.
67. **Suga, T.,** The synthesis of carvone from α-terpinyl acetate, *Bull. Chem. Soc. Jpn.,* 31, 569, 1958.
68. **Fujita, H. and Yamashita, M.,** The methylenation of several alkylbenzene-1,2-diol derivatives in aprotic dipolar solvents, *Bull. Chem. Soc. Jpn.,* 46, 3353, 1973.
69. **Baker, W. and Savage, R. I.,** Derivatives of 1,2,3,4-tetrahydroxybenzene. V. The synthesis of parsley apiole and derivatives, *J. Chem. Soc.,* 1062, 1938.
70. **Crooks, P. A., Robinson, B., and Meth-Cohn, O.,** The ^{13}C-nuclear magnetic resonance spectra of physostigmine and related compounds, *Phytochemistry,* 15, 1092, 1976.
71. **Yagi, A., Tokubuchi, N., Nohura, T., Nonaka, G., Nioshioka, I., and Koda, A.,** The constituents of cinnamomi cortex. I. Structures of cinncassiol A and its glucoside, *Chem. Pharm. Bull.,* 28, 1432, 1980.
72. **Singh, P., Fenemore, P. G., and Russell, G. B.,** Insect control chemicals from plants. II. Effects of five natural norditerpene dilactones on the development of the housefly, *Aust. J. Biol. Sci.,* 26, 911, 1973.
73. **Murray, R. D. H., Mendez, J., Brown, S. A.,** *The Natural Coumarins: Occurrence, Chemistry and Biochemistry.* Wiley-Interscience, New York, 1982.
74. **Pauling, P. and Petcher, T. J.,** Crystal and molecular structure of Eserine (Physostigmine), *J. Chem. Soc. Perkin Trans 2,* 1342, 1973.
75. **Banerji, A. and Kalena, G. P.,** A new synthesis of organic trisulfides, *Tetrahedron Lett.,* 21, 3003, 1980.
76. **Becker, G.,** Experiments on the influence of coniferous heartwood extractives and analogous compounds on the egg larvae of the longhorn beetle, *Holzforschung,* 17, 19, 1965.
77. **Bell, E. A. and Janzen, D. H.,** Medical and ecological consequences of L-Dopa and 5-HTP in seeds, *Nature (London),* 229, 136, 1971.
78. **Bestman, H. J., Vostrowsky, O., Paulus, H., Billmann, W., and Stransky, W.,** Pheromones. XI. A synthesis method for conjugated (E,Z)-dienes, *Tetrahedron Lett.,* p. 121, 1977.
79. **Whitehead, D. L. and Bowers, W. S., Eds.,** *Natural Products for Innovative Pest Management,* Pergamon Press, Oxford, 1983.
80. **Brown, A. W. A.,** *Insect Control by Chemicals,* McGraw Hill, New York, 1951.
81. **Brown, S. A.,** Coumarins, in *The Biochemistry of Plants, A Comprehensive Treatise,* Vol. 7, Conn, E. E., and Stumpf, P. K., Eds., Academic Press, New York, 1981, 269.
82. **Candlin, J. P. and Janes, W. H.,** The catalytic dimerization of dienes by nitrosocarbonyl transition-metal compounds, *J. Chem. Soc. (C),* p. 1856, 1968.
83. **Casida, J. E.,** *Pyrethrum, the Natural Insecticide,* Academic Press, New York, 1973.
84. **Chang. S. C. and Kearns, C. W.,** Effects of sesamex on toxicities of individual pyrethrins, *J. Econ. Entomol.,* 55, 919, 1962.
85. **Stumpf, P. K. and Conn, E. E., Eds.** *The Biochemistry of Plants, A Comprehensive Treatise,* Vol. 7, Academic Press, New York, 1981.
86. **Enzell, C. R., Wahlberg, I., and Aasen, A. J.,** Isoprenoids and alkaloids of tobacco, *Prog. Org. Chem. Nat. Prod.,* 34, 1, 1977.
87. **Crosby, D. G.,** Minor insecticides of plant origin, in *Naturally Occurring Insecticides,* Jacobson, M. and Crosby, D. G., Eds., Marcel Dekker, New York, 1971, 177.
88. **Omokawa, H. and Yamashita, K.,** Synthesis of (±) Degulin, *Agric. Biol. Chem.,* 38, 1731, 1974.
89. **Epstein, W. W. and Poulter, C. D.,** A survey of some irregular monoterpenes and their biogenetic analogies to presqualene alcohol, *Phytochemistry,* 12, 737, 1973.
90. **Twanmoh, L.-M., Wood, H. B., Jr., and Driscoll, J. S.,** NMR spectral characteristics of N-H protons in purine derivatives, *J. Heterocycl. Chem.,* 10, 187, 1973.
91. **Feinstein, L. and Jacobson, M.,** Insecticides occurring in higher plants, *Prog. Org. Chem. Nat. Prod.,* 10, 423, 1953.
92. **Bose, A. J., Fujiwara, H., Pramanik, B. N., Lazlo, E., and Spillert, C. R.,** Some aspects of chemical ionization mass spectroscopy using ammonia as reagent gas: a valuable technique for biomedical and natural products studies, *Anal. Biochem.,* 89, 284, 1978.
93. **Fukami, H. and Nakajima, M.,** Rotenone and the rotenoids, in *Naturally Occurring Insecticides,* Jacobson, M. and Crosby, D. G., Eds., Marcel Dekker, New York, 1971, 71.

94. **Gersdorff, W. A.,** Toxicity to houseflies of the pyrethrins, cinnerins and derivatives in relation to chemical structure, *J. Econ. Entomol.*, 40, 878, 1947.

95. **Godin, P. J., Stevenson, J. H., and Sawicki, R. M.,** The insecticidal activity of Jasmolin II and its isolation from pyrethrum *(Chrysanthemum cinerarifolium)*, *J. Econ. Entomol.*, 58, 548, 1965.

96. **Hahlbrock, K.,** Flavonoids, in *The Biochemistry of Plants, A Comprehensive Treatise*, Vol. 7, Stumpf, P. K. and Conn, E. E., Eds., Academic Press, New York, 1981, 425.

97. **Hassid, E., Applebaum, S. W., and Birk, Y.,** Azetidine-2-carboxylic acid: a naturally occurring inhibitor of *Spodoptera litturalis* (Boisd.) (Lepidoptera: Noctuidae), *Phytoparasiticae*, 4, 173, 1976.

98. **Budowski, P.,** Recent research on sesamin, sesamolin and related compounds, *J. Am. Oil Chem. Soc.*, 41, 280, 1964.

99. **Incho, H. H. and Greenberg, H. W.,** Synergistic effect of piperonyl butoxide with the active principles of pyrethrum and allethrolone esters of chrysanthemum acids, *J. Econ. Entomol.*, 45, 794, 1952.

100. **Ingham, J. L.,** Naturally occurring isoflavanoids (1855—1981), *Prog. Org. Chem. Nat. Prod.*, 43, 1, 1983.

101. **Jacobson, M.,** The unsaturated isobutylamides, in *Naturally Occurring Insecticides*, Jacobson, M. and Crosby, D. G., Eds., Marcel Dekker, New York, 1971, 137.

102. **Nicolau, C. and Hildenbrand, K.,** Carbon-13 nuclear magnetic resonance investigations of xanthine and its methylated derivatives, *Z. Naturforschung C*, 29, 475, 1974.

103. **Rosenthal, G. A. and Janzen, D. H., Eds.,** *Herbivores: Their Interaction with Secondary Plant Metabolites*, Academic Press, New York, 1979.

104. **Rae, I. D., Rosenberger, M. Szabo, A. G., Willis, C. R., Yates, P., Zacharias, D. E., Jeffrey, G. A., Douglass, B., Kirkpatrick, J. L., Weisbach, J. A.,** Haplophytine, *J. Am. Chem. Soc.*, 89, 3061, 1967.

105. **Larson, P. E.,** Glucosinolates, in *The Biochemistry of Plants, A Comprehensive Treatise*, Vol. 7, Stumpf, P. K. and Conn, E. E., Eds., Academic Press, New York, 1981, 502.

106. **Leete, E.,** The biosyntheses and metabolism of the tobacco alkaloids, in *Alkaloids, Chemical and Biological Perspectives*, Vol. 1, Pelletier, S. W., Ed., Wiley/Interscience, New York, 1983, 85.

107. **Mabry, T. J. and Gill, J. E.,** Sesquiterpene lactones and other terpenoids, in *Herbivores: Their Interactions with Secondary Plant Metabolites*, Rosenthal, G. A. and Janzen, D. H., Eds., Academic Press, New York, 1979, 501.

108. **Mandai, T., Gotoh, J., Otera, J., and Kawanda, M.,** A new synthetic method for pellitorine, *Chem. Lett.*, 313, 1980.

109. **Matsui, M. and Yamamoto, I.,** Pyrethroids, in *Naturally Occurring Insecticides*, Jacobson, M. and Crosby, D. G., Eds., Marcel Dekker, New York, 1971, 3.

110. **Metcalf, R. L.,** *Organic Insecticides, Their Chemistry and Mode of Action*, Wiley Interscience, New York, 1955.

111. **Negherbon, W. O.,** *Handbook of Toxicology*, Vol. 3, Negherbon, W. O., Ed., W. B. Saunders, Philadelphia, 1959.

112. **O'Brien, R. D.,** *Insecticides, Action and Metabolism*, Academic Press, New York, 1967.

113. **Yates, P., McLachlan, F. N., Rae, I. D., Rosenberger, M., Szabo, A. G., Willis, C. R., Cava, M. P., Behforouz, M., Lakshmikantham, M. V., and Zieger, W.,** Haplophytine. A novel type of indole alkaloid, *J. Am. Chem. Soc.*, 95, 7842, 1973.

114. **Rehr, S. S., Janzen, D. H., Feeny, P. P.,** L-Dopa in legume seeds: a chemical barrier to insect attack, *Science*, 181, 81, 1973.

115. **Ressler, C. and Ratzkin, H.,** Synthesis of β-cyano-L-alanine and γ-cyano-α-L-aminobutyric acid. Dehydration products of L-asparagine and L-glutamine; new synthesis of amino acid nitriles, *J. Org. Chem.*, 26, 3356, 1961.

116. **Rosenthal, G. A. and Bell, E. A.,** Naturally occurring, toxic non-protein amino acids, in *Herbivores: Their Interaction with Plant Secondary Metabolites*, Rosenthal, G. A., and Janzen, D. H., Eds., Academic Press, New York, 1979, 353.

117. **Miyakado, M., Nakayama, I., Ohno, N., and Yoshioki, H.,** Structure, chemistry and actions of the Piperaceae amides: new insecticidal constituents isolated from the pepper plant, in *Natural Products for Innovative Pest Management*, Whitehead, D. L. and Bowers, W. S., Eds., Pergamon Press, Oxford, 1983, 369.

118. **Sawicki, R. M.,** Insecticidal activities of pyrethrum extract and its four insecticidal constituents against houseflies. II. Synergistic activity of piperonyl butoxide with the four constituents, *J. Sci. Food Agric.*, 13, 260, 1962.

119. **Sawicki, R. M. and Elliot, M.,** Insecticidal activity of pyrethrum extract and its four insecticidal constituents against houseflies. VI. Relative toxicities of pyrethrin I and pyrethrin II against four strains of houseflies, *J. Sci. Food Agric.*, 16, 85, 1965.

120. **Sawicki, R. M., Elliot, M., Gower, J. C., Snary, M., Thain, E. M.,** Insecticidal activities of pyrethrum extract and its four insecticidal constituents against houseflies. I. Preparation and relative toxicity of the pure constituents; statistical analysis of mixtures of the components, *J. Sci. Food Agric.,* 13, 172, 1962.

121. **Sawicki, R. M. and Thain, E. M.,** Insecticidal activities of pyrethrum extract and its four insecticidal constituents against houseflies. IV. Knockdown activities of the four constituents, *J. Sci. Food Agric.,* 13, 292, 1962.

122. **Schoonhoven, L. M.,** Secondary plant substances and insects, *Rec. Adv. Phytochem.,* 5, 197, 1972.

123. **Schleisinger, H. M., Applebaum, S. W., and Birk, Y.,** Comparative uptake, tissue permeability and turnover of α,γ-diaminobutyric acid and β-cyano-L-alanine in insects, *J. Insect. Physiol.,* 23, 1311, 1977.

124. **Schmeltz, I.,** Nicotine and other tobacco alkaloids, in *Naturally Occurring Insecticides,* Jacobson, M. and Crosby, D. G., Eds., Marcel Dekker, New York, 1971, 99.

125. **Shaver, T. N. and Lukefahr, M. J.,** Effects of flavenoid pigments and gossypol on growth and development of the bollworm, tobacco budworm, and the pink bollworm, *J. Econ Entomol.,* 62, 643, 1969.

126. **Shepard, H. H.,** *The Chemistry and Action of Insecticides,* McGraw Hill, New York, 1951.

127. **Smith, R. H.,** Effects of monoterpene vapors on the western pine beetle, *J. Econ. Entomol.,* 58, 509, 1965.

128. **Srivistava, S. N. and Przybylska, M.,** The molecular structure of ryanodo-p-bromobenzyl ether, *Can. J. Chem.,* 46, 795, 1968.

129. **Polonsky, J., Baskevitch, Z., Gaudemer, A., and Das, B. C.,** Constituents of *Brucea amarissima,* structures of the Bruceines A, B and C, *Experientia,* 23, 424, 1967.

130. **Stedman, R. L.,** The chemical composition of tobacco smoke, *Chem. Rev.,* 68, 153, 1968.

131. **Stipanovic, R. D., Bell, A. A., Lukefahr, M. J.,** Natural Insecticides from Cotton, Am. Chem. Soc. Symp. Ser., 62, 197, 1977.

132. **Torsell, K. B. G.,** *Natural Product Chemistry. A Mechanistic and Biosynthetic Approach to Secondary Metabolism,* John Wiley & Sons, New York, 1983.

133. **Ward, J.,** Separation of the "pyrethrins" by chromatography, *Chem. Ind.,* p. 586, 1953.

134. **Weisner, K.,** The structure, stereochemistry and absolute configuration of anhydroryanodine, *Pure Appl. Chem.,* 7, 285, 1963.

135. **Cheng, P.-T., Nyburg, S. C., McLachlan, F. N., and Yates, P.,** X-ray analysis of the structure and correlation with the spectra of Haplophytine, *Can. J. Chem.,* 54, 726, 1976.

136. **Yamamoto, I.,** Nicotinoids as insecticides, *Adv. Pest Control Res.,* 6, 231, 1965.

137. **Bredereck, H. and Gotsmann, U.,** Purine synthesis. XVI. The synthesis of 5-alkyl- or 5-Arylsulfonyl amino-4-amino uracils, 4-amino-5-alkylamino uracils and 4-amino-5-[pyridino-methylene amino] uracil chlorides, *Chem. Ber.,* 95, 1902, 1962.

138. **Sutor, D. J.,** The structures of the pyrimidines and purines. VII. The crystal structure of caffeine, *Acta Cryst.,* 11, 453, 1958.

139. **Evans, D. A., Tanis, S. P., and Hart, D. J.,** A convergent total synthesis of (±) colchicine and (±) desacetamidoisocolchicine, *J. Am. Chem. Soc.,* 103, 5813, 1981.

140. **Leete, E.,** The biosynthesis of the tropolone ring of colchicine, *Tetrahedron Lett.,* p. 333, 1965.

141. **Lessinger, L. and Margulis, T. N.,** The crystal structure of colchicine. A new application of magic integers to multiple solution direct methods, *Acta Cryst.,* B34, 578, 1978.

142. **Kuhn, H. and Stein, O.,** On the condensation of indoles with formaldehyde and secondary amines. A new gramine synthesis, *Chem. Ber.,* 70, 567, 1937.

143. **Gower, B. G. and Leete, E.,** Biosynthesis of gramine: the immediate precursors of the alkaloid, *J. Am. Chem. Soc.,* 85, 3683, 1963.

144. **Smith, G. F.,** *Strychnos* alkaloids, *Alkaloids,* VIII, 591, 1965.

145. **Wenkert, E., Cheung, H. T. A., Gottlieb, H. E., Koch, M. C., Rabaron, A., and Plat, M. M.,** Carbon-13 nuclear magnetic resonance spectroscopy of naturally occurring substances. 56. *Strychnos* alkaloids, *J. Org. Chem.,* 43, 1099, 1978.

146. **Woodward, R. B., Bader, F. E., Bickel, H., Frey, A. J., and Kierstead, R. W.,** The total synthesis of reserpine, *J. Am. Chem. Soc.,* 78, 2023, 1956.

147. **Wender, P. A., Schaus, J. M., and White, A. W.,** General methodology for *cis-* hydroisoquinoline synthesis. Synthesis of reserpine, *J. Am. Chem. Soc.,* 102, 6157, 1980.

148. **Woodward, R. B., Cava, M. P., Ollis, W. D., Hunger, A., Daeniker, H. V., and Schenker, K.,** The total synthesis of strychnine, *Tetrahedron,* 19, 247, 1963.

149. **Schwenker, G., Prenntzell, W., Gassner, U., and Gerber, R.,** Notice of a new, generally useful tropane ester synthesis, *Chem. Ber.,* 99, 2407, 1966.

150. **Nakane, M. and Hutchinson, C. R.,** Biosynthetic studies of secondary plant metabolites with $^{13}CO_2$. *Nicotiana* alkaloids. II. New synthesis of nicotine and nornicotine. Quantitative carbon-13 NMR spectroscopic analysis of [2′,3′-N-CH$_3$-^{13}C] nicotine, *J. Org. Chem.,* 43, 3922, 1978.

151. **Bohlmann, F., Muller, H.-J., and Schumann, D.,** *Lupin* alkaloids. XL. Preparation and reactions of cyclic enamines and immonium salts, *Chem. Ber.,* 106, 3026, 1973.

152. **van Tamelen, E. E. and Foltz, R. L.**, The biogenetic type synthesis of dl-sparteine, *J. Am. Chem. Soc.*, 82, 2400, 1960.

153. **Polonsky, J., Baskevitch, Z., Gottlieb, H. E., Hagaman, E., Wenkert, E.**, Carbon-13 nuclear magnetic resonance spectral analysis of Quassinoid bitter principles, *J. Org. Chem.*, 40, 2499, 1975.

154. **Forss, D. A., Dunstone, E. A., Horwood, J. F., and Stark, W.**, The characterization of some unsaturated aldehydes in microgram quantities, *Aust. J. Chem.*, 15, 163, 1962.

155. **Kametani, T., Noguchi, I., Saito, K., and Kaneda, S.**, Studies on the synthesis of heterocyclic compounds. CCCII. Alternative total synthesis of (±) nandinine, (±) canadine and berberine iodide, *J. Chem. Soc. (C)*, p. 2036, 1969.

156. **Gear, J. R. and Spenser, I. D.**, The biosynthesis of hydrastine and berberine, *Can. J. Chem.*, 41, 783, 1963.

157. **Hahn, F. E. and Ciak, J.**, Berberine, in *Antibiotics*, Vol. 3, Corcoran, J. W. and Hahn, F. E., Eds., Springer-Verlag, Basel, 1975, 577.

158. **Shaw, K. N. F. and Morris, A. G.**, 5-Hydroxy-DL-tryptophan, *Biochem. Prep.*, 9, 12, 1962.

159. **Morris, A. J. and Armstrong, M. D.**, Preparation of 5-hydroxy-L-and D-tryptophan, *J. Org. Chem.*, 22, 306, 1957.

160. **Wakahara, A., Kido, M., Fujiwara, T., and Tomita, K.-I.**, The crystal and molecular structure of 5-hydroxy-DL-tryptophan, *Tetrahedron Lett.*, p. 3003, 1970.

161. **Schlesinger, H. M., Applebaum, S. W., and Birk, Y.**, Effect of β-cyano-L-alanine on the water balance of *Locusta migratoria*, *J. Insect Physiol.*, 22, 1421, 1976.

162. **Buck, J. S.**, Reduction of hydroxy-mandelonitriles. A new synthesis of tyramine, *J. Am. Chem. Soc.*, 55, 3388, 1933.

163. **Podder, A., Dattagupta, J. K., Saha, N. N., and Saenger, W.**, Crystal and molecular structure of a sympathomimetric amine, tyramine hydrochloride, *Acta Cryst.*, B35, 649, 1979.

164. **Adams, R. and Johnson, J. L.**, Leucenol. VI. A total synthesis, *J. Am. Chem. Soc.*, 71, 705, 1949.

165. **Mostad, A., Romming, C., and Rosenquist, E.**, The structure of L-mimosine, an L-DOPA analogue, *Acta Chem. Scand.*, 27, 164, 1973.

166. **Leete, E.**, The biosynthesis of azetidine-2-carboxylic acid, *J. Am. Chem. Soc.*, 86, 3162, 1964.

167. **Polonsky, J., Varenne, J., Prange, T. H., and Pascard, C.**, Antileukaemic quassinoids: structure (X-ray analysis) of bruceine C and revised structure of bruceantinol, *Tetrahedron Lett.*, 21, 1853, 1980.

168. **Bates, R. B. and Thalacker, V. P.**, Nuclear magnetic resonance parameters in bicyclo [3.1.1] heptanes. α-Pinene, myrtanal and verbenone, *J. Org. Chem.*, 33, 1730, 1968.

169. **Wada, K., Marumo, S., and Munakata, K.**, An insecticidal alkaloid, cocculolidine from *Cocculus trilobus* DC, *Tetrahedron Lett.*, p. 5179, 1966.

170. **Jacobson, M.**, Pellitorine isomers. II. The synthesis of N-isobutyl-*trans*-2-*trans*-4-decadienamide, *J. Am. Chem. Soc.*, 75, 2584, 1953.

171. **Crombie, L.**, The structure of an insecticidal isobutyl amide from pellitory root, *Chem. Ind.*, p. 1034, 1952.

172. **Crombie, L.**, Amides of vegetable origin. V. Stereochemistry of conjugated dienes, *J. Chem. Soc.*, p. 1007, 1955.

173. **Jacobson, M.**, The structure of spilanthol, *Chem. Ind.*, p. 50, 1957.

174. **Sonnett, P. E.**, Synthesis of the *N*-isobutyl amide of all-*trans*-2,6,8,10-dodecatetraenoic acid, *J. Org. Chem.*, 34, 1147, 1969.

175. **Crombie, L.**, Amides of vegetable origin. III. The structure and stereochemistry of neoherculin, *J. Chem. Soc.*, 995, 1955.

176. **Crombie, L.**, Amides of vegetable origin. I. Stereoisomeric *N*-isobutyl undeca-1:7-diene-1-carboxyamides and the structure of herculin, *J. Chem. Soc.*, 2997, 1952.

177. **Crombie, L. and Taylor, J. L.**, Amides of vegetable origin. VIII. The constitution and configuration of the sanshools, *J. Chem. Soc.*, 2760, 1957.

178. **LaLonde, R. T., Wong, C. F., Hofstead, S. J., Morris, C. D., and Gardner, L. C.**, *N*- (2-methyl propyl)-(E,E)-2,4-decadienamide. A mosquito larvacide from *Achillae millefoleum* L., *J. Chem. Ecol.*, 6, 35, 1980.

179. **Su, H. C. F.**, Insecticidal properties of black pepper to rice weevils and cowpea weevils, *J. Econ. Entomol.*, 70, 18, 1977.

180. **Su, H. C. F. and Horvat, R.**, Isolation, identification and insecticidal properties of *Piper nigrum* amides, *J. Agric. Food. Chem.*, 29, 115, 1981.

181. **Kubo, I., Matsumoto, T., Klocke, J. A., and Kamikawa, T.**, Molluscicidal and insecticidal activities of isobutyl amides isolated from *Fagara macrophylla*, *Experientia*, 40, 340, 1984.

182. **Nokami, J., Nishiuichi, K., Wakabayashi, S., and Okawara, R.**, Pyrolysis of sulfoxide. III. Conversion of 2-(phenylsulfinyl) enoate to 2,4-dienoate and synthesis of pellitorine, *Tetrahedron Lett.*, 21, 4455, 1980.

183. **Correa, J., Rocquet, S., and Diaz, E.**, Multiple NMR analysis of the affinin, *Org. Magnetic Reson.*, 3, 1, 1971.

184. **Wenkert, E., Cochran, D. W., Hagaman, E. W., Lewis, R. B., and Schell, F. M.,** Carbon-13 nuclear magnetic resonance spectroscopy with the aid of a paramagnetic shift reagent, *J. Am. Chem. Soc.,* 93, 6271, 1971.

185. **Grynpas, M. and Lindley, P. F.,** The crystal and molecular structure of l-piperoyl piperidine, *Acta Cryst.,* B31, 2663, 1975.

186. **Schulze, A. and Oediger, H.,** New approaches of aldol type reactions demonstrated by simple synthesis of piperine, *Liebig's Ann. Chem.,* p. 1725, 1981.

187. **Tabuneng, W., Bando, H., and Amiya, T.,** Studies on the constituents of the crude drug "Piperis Longa Fructus." On the alkaloids of fruits of *Piper longum* L., *Chem. Pharm. Bull.,* 31, 3562, 1983.

188. **Miyakado, M. and Yoshioka, M.,** The *Piperaceae* amides. II. Synthesis of pipericide, a new insecticidal amide from *Piper nigrum* L., *Agric. Biol. Chem.,* 43, 2413, 1979.

189. **Chatterjee, A. and Dutta, C. P.,** Alkaloids of *Piper longum* L. I. Structure and synthesis of piperlongumine and piperlonguminine, *Tetrahedron,* 23, 1769, 1967.

190. **Suzuki, T. and Takehashi, E.,** Caffeine biosynthesis in *Camilia sinensis, Phytochemistry,* 15, 1235, 1976.

191. **Battersby, A. R., Sheldrake, P. W., and Milner, J. R.,** Biosynthesis of colchicine: incorporation of a carbon-13 labelled precursor in a higher plant, *Tetrahedron Lett.,* p. 3315, 1974.

192. **Peerdeman, A. F.,** The absolute configuration of natural strychnine, *Acta Cryst.,* 9, 824, 1956.

193. **Rosen, W. G. and Shooerly, J. N.,** Rauwolfia alkaloids. XLI. Methyl neoreserpate, an isomer of methyl reserpate. Part 3. Conformations and NMR spectra, *J. Am. Chem. Soc.,* 83, 4816, 1961.

194. **Levin, R. H., Lallemand, J.-Y., and Roberts, J. D.,** Nuclear magnetic resonance spectroscopy. Applications of pulse and fourier transform carbon-13 nuclear magnetic resonance techniques to structure elucidation. Rauwolfia alkaloids, *J. Org. Chem.,* 38, 1983, 1973.

195. **Becker, O., Fuerstenua, N., Knippelberg, W., and Krueger, F. R.,** Mass spectra of substances ionized by fission fragment induced desorption, *Org. Mass Spectrom.,* 12, 461, 1977.

196. **Takeuchi, Y., Kogi, K., Shiori, T., and Yamada, S.-I.,** Amino acids and peptides. II. A one-step synthesis of atropine and other related alkaloids from dl-phenylalanine-3α-tropanyl ester, *Chem. Pharm. Bull.,* 19, 2603, 1971.

197. **Marion, L. and Thomas, A. F.,** A further note on the biogenesis of hyoscamine, *Can. J. Chem.,* 33, 1853, 1955.

198. **Barelle, F. E. and Gros, E. G.,** Biosynthesis of cuscohygrine and hyosciamine in *Atropa belladonna* from DL-α-N-methyl-[³H]ornithine and DL-δ-N-methyl-[³H]ornithine, *J. Chem. Soc. Chem. Commun.,* p. 721, 1969.

199. **Cannon, J. R., Joshi, K. R., Meehan, G. V., and Williams, J. R.,** The tropane alkaloids from three western australian *Anthocerus* species, *Aust. J. Chem.,* 22, 221, 1969.

200. **Sternberg, V. I., Narain, N. K., and Singh, S. P.,** Carbon-13 magnetic resonance spectra of the tropane alkaloids: cocaine and atropine, *J. Het. Chem.,* 14, 225, 1977.

201. **Kussäther, E. and Haase, J.,** The crystal and molecular structure of l-Hyoscamine hydrobromide, *Acta Cryst.,* B28, 2896, 1972.

202. **Whidby, J. F. and Seeman, J. I.,** The configuration of nicotine. A nuclear magnetic resonance study, *J. Org. Chem.,* 41, 1585, 1976.

203. **Yamasaki, K., Tamaki, T., Uzawa, S., Sankawa, U., and Shibata, S.,** Participation of the C_6-C_1 unit in the biosynthesis of ephedrine in *Ephedra, Phytochemistry,* 12, 2877, 1973.

204. **Baudet, M. and Gelbcke, M.,** Carbon-13 magnetic resonance spectra of β-amino alcohols derived from ephedrine and their oxazolidines, *Anal. Lett. B,* 12, 641, 1979.

205. **Bohlmann, K. and Zeisberg, R.,** ¹³C NMR spectra of Lupine alkaloids, *Chem. Ber.,* 108, 1043, 1975.

206. **Ma, J. C. N. and Warnhoff, E. W.,** On the use of nuclear magnetic resonance for the detection, estimation and characterization of N-methyl groups, *Can. J. Chem.,* 43, 1849, 1965.

207. **Boulten, A. A., Pollitt, R. J., and Majer, J. R.,** Identity of a urinary "pink spot" in schizophrenia and Parkinson's disease, *Nature (London),* 215, 132, 1967.

208. **Tamura, K., Wakahara, A., Fujiwara, T., and Tomita, K.-I.,** The crystal and molecular structure of tyramine hydrochloride, *Bull. Chem. Soc. Jpn.,* 47, 2682, 1974.

209. **Lambert, F., Ellenberger, M., and Cohen, Y.,** NMR of catechol and catecholamines, *Org. Magnetic Reson.,* 7, 66, 1975.

210. **Rosenthal, G. A.,** The biological effects and mode of action of L-canavanine, a structural analogue of L-argenine, *Q. Rev. Biol.,* 52, 155, 1977.

211. **Boyar, A. and Marsh, R. E.,** l-Canavanine, a paradigm for the structures of substituted guanidines, *J. Am. Chem. Soc.,* 104, 1995, 1982.

212. **Ressler, C., Nagarajan, G. R., Kirisawa, M., and Kashelikar, D. V.,** Synthesis and properties of α-cyanoglycine, L-β-cyano-β-alanine and L-γ-cyano-γ-amino butyric acid, *J. Org. Chem.,* 36, 3960, 1971.

213. **Dunnill, P. M. and Fowden, L.,** Enzymatic formation of β-cyano-L-alanine from cyanide by *Escherichia coli* extracts, *Nature (London),* 208, 1206, 1965.

214. **Floss, H. G., Hadwiger, L., and Conn, E. E.,** Enzymatic formation of β-cyano alanine from cyanide, *Nature (London),* 208, 1207, 1965.

215. **Phillips, B. A. and Cromwell, N. H.,** Azetidine-2-carboxylic acid derivatives, *J. Heterocycl. Chem.,* 10, 795, 1973.

216. **Leete, E.,** Biosynthesis of azetidine-2-carboxylic acid from methionine in *Nicotiana tabacum, Phytochemistry,* 14, 1983, 1975.

217. **Berman, H. A., McGandy, E. L., Burgner, J. W., II, and Van Etten, R. L.,** The crystal and molecular structure of azetidine-2-carboxylic acid. A naturally occurring homolog of proline, *J. Am. Chem. Soc.,* 91, 6177, 1969.

218. **Cromwell, N. H. and Phillips, B.,** The azetidines, recent synthetic developments, *Chem. Rev.,* 79, 331, 1979.

219. **Ebata, M., Takahashi, Y., and Otsuka, H.,** The preparation and some properties of *N*-methylated-L-amino acids, *Bull. Chem. Soc. Jpn.,* 39, 2535, 1966.

220. **Mwauluka, K., Bell, E. A., Charlwood, B. V., and Briggs, J. M.,** *N* -Methyl tyrosine from seeds of *Combretum zeyheri, Phytochemistry,* 14, 1657, 1975.

221. **Crowley, M. P., Godin, P. J., Inglis, H. S., Snarey, M., and Thain, E. M.,** Biosynthesis of "pyrethrins". I. Incorporation of ^{14}C-labelled compounds into flowers of *Chrysanthemum cinerareaefolium* and biosynthesis of chrysanthemum monocarboxylic acid, *Biochem. Biophys. Acta,* 60, 312, 1962.

222. **Bramwell, A. F., Crombie, L., Hemesley, P., Pattenden, G., Elliott, M., and Janes, N. F.,** Nuclear magnetic resonance spectra of the natural pyrethrins and related compounds, *Tetrahedron,* 25, 1727, 1969.

223. **Okogun, J. I. and Ekong, D. E. V.,** Extracts from the fruits of *Piper guineense* Schum. and Thonn., *J. Chem. Soc. Perkin Trans. 1,* 2195, 1974.

224. **Miyakado, M., Nakayama, I., Yoshioki, H., and Nakatani, N.,** The *Piperaceae* amides. I. Pipericide, a new insecticidal amide from *Piper nigrum* L., *Agric. Biol. Chem.,* 43, 1609, 1979.

225. **Jacobson, M.,** Pellitorine isomers. III. The synthesis of *N*-isobutyl-*trans*-4-*trans*-6-decadienamide and the structure of spilanthol, *J. Am. Chem. Soc.,* 78, 5084, 1956.

226. **Nakatani, N. and Inatani, R.,** Isobutyl amides from pepper *(Piper nigrum L.), Agric. Biol. Chem.,* 45, 1473, 1981.

227. **Reed, D. K., Freedman, B., and Ladd, T. L.,** Insecticidal and antifeedant activity of neriifolin against codling moth, striped cucumber beetle and japanese beetle, *J. Econ. Entomol.,* 75, 1093, 1982.

228. **Austin, D. J. and Brown, S. A.,** Furanocoumarin biosynthesis in *Ruta graveolens* cell cultures, *Phytochemistry,* 12, 1657, 1973.

229. **Russell, G. B., Fenemore, P. G., and Singh, P.,** The structures of Hallactones A and B, insect toxins from *Podocarpus hallii, J. Chem. Soc. Chem. Commun.,* 166, 1973.

230. **Elgamal, M. H. A., Elewa, N. H., Elkhrisy, E. A. M., and Duddek, H.,** Carbon-13 chemical shifts and carbon-proton coupling constants of some furanocoumarins and furochromones, *Phytochemistry,* 18, 139, 1979.

231. **Stemple, N. R. and Watson, W. H.,** The crystal and molecular structure of xanthotoxin, $C_{12}H_8O_4$, *Acta Cryst.,* B28, 2485, 1972.

232. **Allen, T. C., Link, K. P., Ikawa, M., and Brunn, L. K.,** The relative effectiveness of the principal alkaloids of Sabadilla seed, *J. Econ. Entomol.,* 38, 293, 1945.

233. **Powell, D. R. G., Weisleder, D., Smith, C. R., Jr., Koslowski, J., and Rohwedder, W. K.,** Treflorine, trenudine and *N*-methyl-trenudone: novel maytansanoid tumor inhibitors containing two fused macrocyclic rings, *J. Am. Chem. Soc.,* 104, 4929, 1982.

234. **Stipanovic, R. D., Bell, A. A., O'Brien, D. H., and Lukefahr, M. J.,** Heliocide H$_2$: an insecticidal sesterterperiod from cotton (Gossypium), *Tetrahedron Lett.,* p. 567, 1977.

235. **Stipanovic, R. D., Bell, A. A., O'Brien, D. H., and Lukefahr, M. J.,** Heliocide H$_3$: an insecticidal terpenoid from *Gossypium hirsutum, Phytochemistry,* 17, 151, 1978.

236. **Cocker, W., Geraghty, N. W. A., and Greyson, D. H.,** The chemistry of terpenes. XXV. A synthesis of Car-2, Car-3 and Car-3(10)ene(β-carene), *J. Chem. Soc. Perkin Trans. 1,* 1370, 1978.

237. **Akhila, A. and Banthorpe, D. V.,** Studies on the biosynthesis of the carene skeleton, *Phytochemistry,* 19, 1691, 1980.

238. **O'Brien, D. H. and Stipanovic, R. D.,** Carbon-13 magnetic resonance of cotton terpenoids. Carbon-proton long range coupling, *J. Org. Chem.,* 43, 1105, 1978.

239. **Thomas, M. T. and Fallis, A. G.,** The total synthesis os of (±)α- and (±)β-pinene. A general route to bicyclic mono and sesquiterpenes, *J. Am. Chem. Soc.,* 98, 1227, 1976.

240. **Richards, G. F., Moran, R. A., Heitman, J. A., and Scott, W. E.,** The molecular geometry of β-pinene as deduced from the crystal and molecular structure of *cis*-pinocarvyl *p*-nitrobenzoate, *J. Org. Chem.,* 39, 86, 1974.

241. **Bohlmann, F., Zeisberg, R., and Klein, E.,** Naturally occurring terpene derivatives. L. Carbon-13 NMR spectra of monoterpenes, *Org. Magnetic Reson.,* 7, 426, 1975.

242. **Vig, O. P., Vig, A. K., and Kumar, S. D.,** New synthesis of β-myrcene and β-farnesene, *Indian J. Chem.,* 13, 1244, 1975.

243. **Waller, G. R., Frost, G. M., Burleson, D., Brannon, D., and Zalkow, L. H.,** Biosynthesis of monoterpenoids by *Santolina chamaecyparissus* L., *Phytochemistry,* 7, 213, 1968.

244. **Beckley, H. D. and Hey, H.,** Combination of field ionization and electron impact mass spectra for structure determinations, particularly of monoterpenes, *Org. Mass Spectrom.,* 1, 47, 1968.

245. **Harris, R. K. and Cunliffe, A. V.,** Nuclear magnetic resonance studies of 1,3-butadienes. IX. The proton spectra of isoprene and related compounds, *Org. Magnetic Reson.,* 9, 483, 1977.

246. **Miyaura, N., Tagami, H., Itoh, M., and Suzuki, A.,** Isoprenylation of olefins via the reaction of iodine with lithium trialkyl isopropenyl borates obtainable from trialkyl boranes, *Chem. Lett.,* 1411, 1974.

247. **Oda, J., Ando, N., Nakajima, Y., and Inouye, Y.,** Studies on insecticidal constituents of *Juniperus recurva.* Buch., *Agric. Biol. Chem.,* 41, 201, 1977.

248. **Baggaley, K. H., Erdtman, H., and Norin, T.,** Some new cedrane derivatives from *Juniperus foetidissima* Wild. Configuration of cedroic acid, *Tetrahedron,* 24, 3399, 1968.

249. **Nakajima, S. and Kawazu, K.,** Coumarin and euponin, two inhibitors for insect development from leaves of *Eupatorium japonicum, Agric. Biol. Chem.,* 44, 2893, 1980.

250. **Bell, A. A., Stipanovic, R. D., O'Brien, D. H., and Fryxell, P. A.,** Sesquiterpenoid aldehyde-quinones and derivatives in pigment glands of *Gossypium, Phytochemistry,* 17, 1297, 1978.

251. **Stipanovic, R. D., Bell, A. A., Lukefahr, M. J.,** in *Proceedings of the Beltwide Cotton Production Research Conference,* National Cotton Council, Memphis, Tenn., 1976, 91.

252. **Lukefahr, M. J., Stipanovic, R. D., Bell, A. A., and Gray, J. R.,** in *Proceeding of the Beltwide Cotton Production Research Conference,* National Cotton Council, Memphis, Tenn., 1977, 97.

253. **Isogai, A., Suzuki, A., Tamura, S., Ohashi, Y., and Sasada, Y.,** Cinnzeylanine, a new pentacyclic diterpene acetate from *Cinnamonum zeylanicum, Acta Crystallogr.,* B33, 623, 1977.

254. **Ikawa, M., Dicke, R. J., Allen, T. C., Link, K. P.,** The principal alkaloids of Sabadilla seed and their toxicity to *Musca domestica* L., *J. Biol. Chem.,* 159, 517, 1945.

255. **Crombie, L. and Harper, S. H.,** Experiments in the synthesis of pyrethrins. IV. Synthesis of cinerone, cinerolone and cinerin I., *J. Chem. Soc.,* p. 1152, 1950.

256. **Crombie, L., Harper, S. H., and Newman, F. C.,** Experiments on the synthesis of the pyrethrins. XI. Synthesis of *cis*-pyrethrolone and pyrethrin I: introduction of the *cis*-penta-2:4-dienyl system by selective hydrogenation, *J. Chem. Soc.,* p. 3963, 1956.

257. **Elliott, M. and Janes, N. F.,** Chemistry of the natural pyrethrins, in *Pyrethrum,* Casida, J. E., Ed., Academic Press, New York, 1973, 55.

258. **Reed, D. K., Kwolek, W. F., and Smith, C. R., Jr.,** Investigation of antifeedant and other insecticidal activities of trewiasine towards the striped cucumber beetle and codling moth, *J. Econ. Entomol.,* 76, 641, 1983.

259. **Franck-Neumann, M., Brion, F., and Martina, D.,** Friedel-Crafts acylation of tropone-irontricarbonyl. Synthesis of β-thujaplicin and β-dolabrin, *Tetrahedron Lett.,* p. 5033, 1978.

260. **Derry, J. E. and Hamor, T. E.,** Crystal and molecular structure of 4-isopropyl tropolone (β-thujaplicin), *J. Chem. Soc. Perkin Trans.,* 2, 694, 1972.

261. **Branca, S. J., Lock, R. L., and Smith, A. B., III,** Exploitation of the vinylogous Wolff rearrangement. An efficient total synthesis of (±) Mayurone, (±) Thujopsene and (±) Thujopsadiene, *J. Org. Chem.,* 42, 3165, 1977.

262. **Hayashi, Y., Takahashi, S., Ona, H., and Sakan, T.,** The structures of nagilactone A, B, C, and D, novel nor- and bisnor-diterpenoids, *Tetrahedron Lett.,* p. 2071, 1968.

263. **Crombie, L., Games, D. E., Haskins, N. J., and Reed, G. F.,** Extractives of *Mammea americana* L. V. The insecticidal compounds, *J. Chem. Soc. Perkin Trans.* 1, 2255, 1972.

264. **Joshi, B. S., Karnat, Y. N., Govindachari, T. R., and Ganguly, A. K.,** Isolation and structure of Surangin A and Surangin B, two new coumarins from *Mammea longifolia* (Wight) Planch & Triana, *Tetrahedron,* 25, 1453, 1969.

265. **Russell, G. B., Singh, P., and Fenemore, P. G.,** Insect-control chemicals from plants. III. Toxic lignans from *Libocedrus bidwillii, Aust. J. Biol. Sci.,* 29, 99, 1976.

266. **Bianchi, E., Sheth, K., and Cole, J. R.,** Antitumor agents from *Bursera faragoides* (Burseraceae) (β-peltatin-A methyl ether and 5'-desmethoxy-β-peltaltin methyl ether), *Tetrahedron Lett.,* p. 2759, 1969.

267. **Anjaneyulu, A. S. R., Rao, A. M., and Row, L. R.,** Novel hydroxy lignans from the heartwood of *Gmelina arborea, Tetrahedron,* 33, 133, 1977.

268. **Pelter, A., Ward, R. S., Watson, D. J., Murray-Rust, P., and Murray-Rust, J.,** On the question of distinguishing between 2,6-diaryl and 2,4-diaryl-3,7-dioxabicyclo-[3.3.0] octanes, *Tetrahedron Lett.,* p. 1509, 1978.

269. **Kamikado, T., Chang, C.-F., Murakoshi, S., Sakurai, A., and Tamura, S.,** Isolation and structure elucidation of growth inhibitors on silkworm larvae from *Magnolia kobus* DC, *Agric. Biol. Chem.,* 39, 833, 1975.

270. **Belanger, A. D., Berney, J. F., Borschberg, H. J., Brousseau, R., Doutheau, A., Durand, R., Katayama, H., Lapalme, R., Leturc, D. M., Liao, C.-C., MacLachlan, F. N., Maffrand, J.-P., Marazza, F., Martino, R., Moreau, C., Saint-Laurent, L., Saintonge, R., Soucy, P., Ruest, L., and Deslongchamps, P.,** The total synthesis of ryanodol, *Can. J. Chem.,* 57, 3348, 1979.

271. **Weisner, K.,** The structure of ryanodine, *Adv. Org. Chem.,* 8, 295, 1972.

272. **Crombie, L., Freeman, P. W., and Whiting, D. A.,** A new synthesis of rotenoids. Application to 9-demethyl mundeserone, mundeserone, rotenoic acid, idalpenol and rotenone, *J. Chem. Soc. Perkin Trans.,* 1, 1277, 1973.

273. **Crombie, L., Dewick, P. M., and Whiting, D. A.,** Biosynthesis of rotenoids. Chalcone, isoflavone and rotenoid stages in the formation of amorphigenin by *Amorpha fruticosa* seedlings, *J. Chem. Soc. Perkin Trans.,* 1, 1285, 1973.

274. **Crombie, L., Kilbee, G. W., and Whiting, D. A.,** Carbon-13 magnetic resonance spectra of natural rotenoids and their relatives, *J. Chem. Soc. Perkin Trans.,* 1, 1497, 1975.

275. **Carlson, D. G., Wiesleder, D., and Tallent, W. H.,** NMR investigation of rotenoids, *Tetrahedron,* 29, 2731, 1973.

276. **Begley, M. J., Crombie, L., and Whiting, D. A.,** Conformation and absolute configuration of rotenone: examination of 8′-bromorotenone by X-ray methods, *J. Chem. Soc. Chem. Commun.,* p. 850, 1975.

277. **Anzevano, P. B.,** Roteroid interconversion. Synthesis of elliptone from rotenone, *J. Heterocycl. Chem.,* 16, 1643, 1979.

278. **Hass, H. B., Feuer, H., and Pier, S. M.,** The preparation of 3-nitropropionic acid, *J. Am. Chem. Soc.,* 73, 1858, 1951.

279. **Birkinshaw, J. H. and Dryland, A. M. L.,** Biochemistry of microorganisms. CXVI. Biosynthesis of β-nitro-propionic acid by the mold, *Penicillum atrovenetum, Biochem. J.,* 93, 478, 1964.

280. **Sutor, D. J., Calvert, L. D., and Llewellyn, F. J.,** The crystal structure of β-nitropropionic acid, *Acta Crystallogr.,* 7, 767, 1954.

281. **Galbraith, M. N., Horn, D. H. S., and Sasse, J. M.,** Podolactones C and D, terpene sulfoxides from *Podocarpus neriifolius, J. Chem. Soc. Chem. Commun.,* p. 1362, 1971.

282. **Pant, R.,** Guanylation of amino compounds, *Z. Physiol. Chem.,* 335, 272, 1964.

283. **Robin, Y.,** Biochemical preparation of γ-guanidino butyric acid and significance in the metabolism of guanidine derivatives, *Bull. Soc. Chem. Biol.,* 35, 285, 1953.

284. **Tomita, K., Fujiwara, T., and Tomita, K.-I.,** Crystal and molecular structure of ω-amino acids, ω-amino sulfonic acids and their derivatives. The crystal and molecular structure of γ-guanidino butyric acid hydrochloride and hydrobromide, *Bull. Chem. Soc. Jpn.,* 45, 3628, 1972.

285. **Freedman, B., Reed, D. K., Powell, R. G., Madrigal, R. V., Smith, C. R., Jr.,** Biological activities of *Trewia nudiflora* extracts against certain economically important insect pests, *J. Chem. Ecol.,* 8, 409, 1982.

286. **Powell, R. G., Weisleder, D., and Smith, C. R., Jr.,** Novel matansanoid tumor inhibitors from *Trewia nudiflora:* trewiasine, dehydrotrewiasine and demethyl trewiasine, *J. Org. Chem.,* 46, 4398, 1981.

287. **Poppleton, B. J.,** Podolactone A *p*-bromobenzoate. Stereochemistry and absolute configuration, *Cryst. Struct. Commun.,* 4, 101, 1975.

288. **Galbraith, M. N., Horn, D. H. S., and Sasse, J. M.,** Plant growth inhibitory lactones from *Podocarpus neriifolius.* Structure of podolactone E, *Experientia,* 28, 253, 1972.

289. **Kupchan, S. M. and By, A. W.,** Steriod alkaloids: the Veratrum groups, in *The Alkaloids,* 10, Manske, R. H. F., Ed., Academic Press, New York, 1968, 193.

290. **Pelletier, S. W.,** The nature and definition of an alkaloid, in *Alkaloids: Chemical and Biological Perspectives,* Vol. 1, Pelletier, S. W., Ed., John Wiley & Sons, New York, 1983, 1.

291. **Takahashi, N., Yoshioka, H., Misato, T., Munakata, S., Eds.,** *Pesticide Chemistry: Human Welfare and the Environment,* Vol. 2, Pergammon, Oxford, 1983.

292. **Miyakado, M., Nakayama, I., and Yoshioka, H.,** Insecticidal joint action of pipercide and co-occurring compounds isolated from *Piper nigrum* L., *Agric. Biol. Chem.,* 44, 1701, 1980.

293. **Fowden, L.,** Nonprotein amino acids, in *The Biochemistry of Plants: A comprehensive Treatise,* Vol. 7, Stumpf, P. K. and Conn, E. E., Eds., Academic Press, New York, 1981.

294. **Manitto, P.,** *Biosynthesis of Natural Products,* Halstead Press, New York, 1981.

295. **Kuhr, R. J. and Dorough, H. W.,** *Carbamate Insecticides: Chemistry, Biochemistry and Toxicology,* CRC Press, Boca Raton, Fla., 1976.

296. **Hopf, H. S.,** Studies on the mode of action of insecticides. I. Injection experiments on the role of cholinesterase inhibition, *Ann. Appl. Bot.,* 39, 193, 1952.

297. **Odjo, A., Piart, J., Polonsky, J., and Roth, M.,** Study of the insecticidal effect of two quassinoids on the larvae of *Locusta migratoria migratorioides, C. R. Acad. Sci.,* 293, 241, 1982.

298. **Polonsky, J.,** Quassinoid bitter principles, *Prog. Chem. Org. Nat. Prod.,* 30, 101, 1973.

299. **Moron J., Merrien, M.-A., and Polonsky, J.,** Onthe biosynthesis of quassinoids of *Simaruba glauca* (simaraubacae), *Phytochemistry,* 10, 585, 1971.

300. **Cava, M. P., Lakshmikantham, M. V., Talapatra, S. K., Yates, P., Rae, I. D., Rosenberger, M., Szabo, A. G., Douglas, B., and Weisbach, J. A.,** Cimicine and cimicidine, lactonic alkaloids from *Haplophyton cimicidum, Can. J. Chem.,* 51, 3102, 1973.

301. **Corey, E. J., Ensley, H. E., Suggs, J. W.,** A convenient synthesis of (S)-(−)-pulegone from (−)-citronellol, *J. Org. Chem.,* 41, 380, 1976.

302. **Duffield, A. M., Budzikiewicz, H., Djerassi, C.,** Mass spectrometry in structural and stereochemical problems. LXXII. A study of the fragmentation processes of some tobacco alkaloids, *J. Am. Chem. Soc.,* 87, 2926, 1965.

303. **Seeman, J. I.,** A new pyrroline synthesis. The use of an *N*-vinyl moiety as an NH protecting group, *Synthesis,* 498, 1977.

304. **Sakata, K., Akoi, K., Chang, C.-F., Sakurai, A., Tamura, S., and Murakoshi, S.,** Stemospironine, a new insecticidal alkaloid of *Stemona japonica* Miq. Isolation, structural determination and activity, *Agric. Biol. Chem.,* 42, 457, 1978.

305. **Koyama, H. and Oda, K.,** Crystal and molecular structure of stemonine hydrobromide hemihydrate, *J. Chem. Soc. B,* 1330, 1970.

306. **Irie, H., Masaki, N., Ohno, K., Osaki, K., Tagi, T., and Uyeo, S.,** The crystal structure of a new alkaloid, stemofoline, from *Stemona japonica, J. Chem. Soc. Chem. Commun.,* 1066, 1970.

307. **Nakai, T., Setoi, H., Kageyama, Y.,** Stereocontrolled synthesis of conjugated dienamides via the ynamine-claisen rearrangement with (arylthio) ynamine, *Tetrahedron Lett.,* 22, 4097, 1981.

308. **Zalkow, U., Gordon, M. M., and Lanir, N.,** Antifeedants from rayless goldenrod and oil of pennyroyal: toxic effects for fall armyworm, *J. Econ. Entomol.,* 72, 812, 1979.

309. **Irie, H., Harada, H., Ohno, K., Mizutane, T., and Uyeo, S.,** Studies of the alkaloid protostemonine, *J. Chem. Soc. Chem. Commun.,* 268, 1970.

310. **Bryan, R. F. and Smith, P. M.,** X-ray determination of the structure of podolide an antileukemic nor-diterpene lactone, *J. Chem. Soc. Perkin Trans.,* 2, 1482, 1975.

311. **Cassady, J. M., Lightner, T. K., McCloud, T. M., Hembree, J. A., Byrn, S. R., and Chang, C.,** Revised structure of podolactone C, the antileukemic component of *Podocarpus milanjianus* Rendle, *J. Org. Chem.,* 49, 942, 1984.

312. **Byers, R. A., Gustine, D. L., and Moyer, B. G.,** Toxicity of β-nitroproionic acid to *Trichoplusia ni, Environ. Entomol.,* 6, 229, 1977.

313. **Hutchins, R. F. N., Sutherland, O. R. W., Granasunderam, G., Greenfield, W. J., Williams, E. M., and Wright, H. J.,** Toxicity of nitro compounds from *Lotus pedunculatus* to grass grub *(Costelytra zealandica)* (Coleoptera: Scarabaeidae), *J. Chem. Ecol.,* 10, 81, 1984.

314. **Harlow, M. C., Stermitz, F. R., and Thomas, R. D.,** Isolation of nitro compounds from *Astralagus* species, *Phytochemistry,* 14, 1421, 1975.

315. **Moyer, B. G., Pfeffer, P. E., Moniot, J. L., Shamma, M., and Gustine, D. L.,** Corollin, coronillin and coronarian: three new 3-nitro-propanoyl-D-glucopyranoses from *Coronilla varia, Phytochemistry,* 16, 375, 1977.

316. **Stermitz, F. R., Lowry, W. T., Ubben, E., and Sharifi, I.,** 1,6-di-3-Nitropropanoyl-β--glucopyranoside, from *Astralagus cibarius, Phytochemistry,* 11, 3525, 1972.

317. **Allen, K. G., Banthorpe, D. V., Charlwood, B. V., Ekundayo, O., and Mann, J.,** Metabolic pools associated with monoterpene biosynthesis in higher plants, *J. Org. Chem.,* 41, 380, 1976.

318. **Kupchan, S. M., Baxter, R. L., Ziegler, M. F., Smith, P. M., and Bryan, R. F.,** Podolide, a new antileukemic norditerpene lactone from *Podocarpus gracilior, Experientia,* 31, 137, 1975.

319. **Gray, R. J., Mabry, T. J., Bell, A. A., Stipanovic, R. D., and Lukefahr, M. J.,** para-Hemigossypolone: a sesquiterpenoid aldehyde quinone from *Gossypium hirsutum, J. Chem. Soc. Chem. Commun.,* p. 109, 1976.

320. **Hayashi, Y., Yokoi, J., Watanabe, Y., Sakan, T., Masuda, Y., and Yamamoto, R.,** Structures of nagilactones E and F, and biological activity of nagilactones as plant growth regulator, *Chem. Lett.,* p. 759, 1972.

321. **Arora, S. K., Bates, R. B., Chou, P. C. C., Sanchez, W. E., and Brown, K. S.,** Structures of norditerpene lactones from *Podocarpus* species, *J. Org. Chem.,* 41, 2458, 1976.

322. **Horhammer, L., Wagner, H., Arndt, H. G., Kraemer, H., and Farkas, L.,** Synthesis of naturally occurring polyhydroxy flavanoid glycosides, *Tetrahedron Lett.,* p. 567, 1966.

323. **Wagner, H., Chari, M. M., and Sonnenbichler, J.,** ^{13}C NMR spectra of naturally occurring flavanoids, *Tetrahedron Lett.,* p. 1799, 1976.

324. **Wenkert, E. and Gottlieb, H. E.,** Carbon-13 nuclear magnetic resonance spectroscopy of flavanoid and isoflavanoid compounds, *Phytochemistry,* 16, 1811, 1977.

325. **Schmid, R. D.,** Structure determination of flavanoid disaccharides by mass spectrometry, *Tetrahedron,* 28, 3259, 1972.

INSECT GROWTH-DISRUPTING AND FECUNDITY-REDUCING INGREDIENTS FROM THE NEEM AND CHINABERRY TREES

Heinrich Schmutterer

INTRODUCTION

The search for secondary plant metabolites which disturb the endocrine system of arthropods began with the accidental discovery of the so-called "paper factor" in paper derived from wood of *Abies balsaminea*.[1] During the following years various phytojuvenoids and phytoecdysoids were isolated from different species of plants. The chemical structures of some of them were identical with those of juvenile and molting hormones which naturally occur in insects and other arthropods. 20-Hydroxyecdysone, for instance, is found in numerous plants, particularly among conifers and ferns.[2,3] A further step forward was the discovery of antijuvenoids, the precocenes, in the herb *Ageratum houstonianum*.[4] The growth-disrupting (growth regulating) compound azadirachtin, isolated from seed kernels of *Azadirachta indica* (neem) and *Melia azedarach* (chinaberry) appears to be at present a very promising plant ingredient for pest control purposes due to its outstanding mode of action and its activity against a wide range of major insect pests without serious side effects on nontarget organisms.

Some hormonal and antihormonal active substances from plants seem to be desirable biorational, environmentally sound insecticides which have a mode of action selective to pest growth and development rather than direct toxic action. Their effects lead to mortality and reduced fecundity, resulting in lower pest populations, hopefully to below the economic threshold levels.

Numerous observations in Africa, the Americas, and Asia about antifeedant effects against insect pests, caused by extracts from various parts of neem and chinaberry trees stimulated subsequent scientific investigations.[5] In 1962, work in India demonstrated a strong antifeedant effect with aqueous neem seed kernel extracts (NSKEs) against the desert locust, *Schistocerca gregaria*.[6] The same had been found in 1937 by testing chinaberry leaf extracts in Algeria.[7] The antifeedant was isolated and shown to be the tetranortriterpenoid azadirachtin.[8]

Special attention of insect physiologists and endocrinologists was attracted after the discovery of strong growth-disrupting and fecundity-reducing properties of neem and chinaberry extracts and azadirachtin during the first years of the 1970s.

The interesting biological activities of neem and chinaberry ingredients, their efficacy in most developmental instars and stages of insects of various orders, the apparent lack of toxicity of neem to warm-blooded organisms, the availability of the raw material in a number of tropical and subtropical countries, and other important reasons have led to a worldwide interest during recent years. Consequently, it seemed to be appropriate to compile an overview on the present status of neem and chinaberry research with special emphasis on growth-disrupting and fecundity-reducing properties of ingredients from these plants.

SOURCES

The neem tree, Indian lilac, or margosa tree *(Azadirachta indica* A. Juss = *Antelaea azadirachta* L., *Melia azadirachta* L.), a member of the mahogany family Meliaceae, is the main source of the growth-regulating compound azadirachtin. This usually evergreen, sometimes deciduous, fast-growing plant reaches a height of about 20 m, sometimes up to 25 m (Figure 1).[9] It originates from south and southeast Asia where it is widely distributed in Bangladesh, Burma, India, Indonesia (east of Java), Malaysia (mainland), Thailand, and

FIGURE 1.   Neem tree in the coastal plain of Kenya.

Pakistan. During the last 100 to 150 years the tree has been introduced, mainly by Indian immigrants, into Africa and South America, and during the last 10 to 20 years also for the purpose of utilizing it in agroforestry systems. In Africa, *A. indica* is currently concentrated in a belt running across the continent from Somalia in the east up to Mauretania in the west, but its main concentration is observed in semiarid areas between the 10th and 14th degree of latitude north of the equator. In the Americas, neem trees are abundant in Surinam[10] and Haiti,[11] whereas in Brazil, Puerto Rico, Cuba, and Nicaragua at present (1984) only a limited number of neem trees exist which have been planted during the last 5 to 15 years. In the Philippines and in Papua New Guinea, neem was introduced between 1981 and 1983. Neem does also occur, mainly as a shade or avenue tree, in Saudi Arabia and in the Yemens. In other words, *A. indica* has a worldwide distribution and perhaps toward the end of this century it will be found in all tropical/subtropical countries suitable for it. Neem trees generally grow best in areas with a warm to very warm climate and an extended dry season. They are able to tolerate severe drought and do well on poor shallow, stony, or sandy soils where agricultural crops give low yield, or fail altogether. In addition, a soil improving capacity of neem has been recorded.[12]

Azadirachtin and related triterpenoids are mainly found in ripe seed kernels; lower concentrations may occur in the leaves and in other parts of the tree.[13] The triterpenoids 7-desacetyl-7-benzoyl-azaradione, 7-desacetyl-7-benzoyl-gedunin, *cis*-(β-epoxy)azadiradione, and 17β-hydroxyazadiradione, which also possess insect growth disrupting properties, were isolated from whole dried neem fruits.[14,15]

Depending on the origin, the age of the seed kernels, the extracting solvent and other reasons, a considerable variation exists concerning the contents of the active principle(s) and cosequently also of the biological activity[13,16,17] (Table 1). This may create difficulties regarding standardization of extracts.

A. indica produces its fruits on drooping panicles usually once a year; about 50 kg can be expected from a mature tree.[9] In many countries the fruiting period lasts from May to

Table 1
YIELD OF AZADIRACHTIN FROM NEEM
SEEDS FROM VARIOUS SOURCES[13]

Investigators	Source	Yield (g/kg^{-1})
Butterworth & Morgan, 1971	Commercial India	0.75
Zanno et al., 1975	Kenya	1.0
Uebel et al., 1979	Commercial India	0.2
Percy, 1972	Nigeria	2.0
Morgan (unpublished)	Ghana	3.5
Morgan (unpublished)	Old Indian seeds	<0.2

FIGURE 2.   Fruit of *Azadirachta indica* (Indonesia).

August, depending on climatic conditions, latitude, etc. In West Africa (Togo) two fruiting periods are observed, one lasting from February to March and another one from July to August. The ripe neem fruits are yellowish, more or less oval and smooth (Figure 2). Their length ranges from 1.4 to 2.0 cm, the width is about 1.0 cm. They fall on the ground. The whitish pulp has a sweetish taste. The seed kernels — usually one, sometimes two per fruit — are enclosed in a hard, white shell (Figure 3). They have a brown skin and a greenish content. Their odor is strong, and garlic-like. The oil content of the seeds ranges from 40 to 60%. Neem oil contains normally relatively low amounts of azadirachtin and also has a strong odor between garlic and peanuts.

FIGURE 3.   Dried fruit (right), seeds in hard shell (middle), and seed kernels (left) of *Azadirachta indica.*

FIGURE 4.   Flowers, young fruit, and leaves of *Azadirachta indica* (Indonesia).

The medium to dark green neem leaves are impairy pinnate and up to 32 cm long (Figure 4). The serrate, asymmetric leaflets number 7 to 17 and are up to 6 to 7 cm long. The Thai neem tree *(Azadirachta indica* var. *siamensis)* (Figure 5) has often entire leaflets (Figure 6). Old leaves are shed during the dry season, in many countries in February and March (South and Southeast Asia). The flowers are white and small (Figure 3).

Melia azedarach L., the chinaberry tree, pride of India, bead tree, or Persian lilac, is

FIGURE 5.   "Edible" neem tree, *Azadirachta indica* var. *siamensis* (Burma).

FIGURE 6.   Leaves of *Azadirachta indica* var. *siamensis* (Burma).

another source of azadirachtin and/or related compounds.[18] This plant, also a member of the Meliaceae, probably originates from western Asia but today it exists in many countries of Africa, the Americas, and Asia, as well as in Australia. It is often planted as an ornamental. Due to its lower temperature requirements it exists not only in tropical and subtropical countries but also in the northern regions of the Mediterranean basin. As an avenue tree it is common in Greece, Argentina, Australia, and in the northern parts of New Zealand. It reaches a height of about 15 m, but some trees may become 25 to 40 m tall, for instance on the Pacific islands (Tonga), and in the tropical rain forests of Queensland (Australia) and Papua New Guinea.[19]

The flowers (Figure 7) are lilac and scented, the fruits globular (length 1.4, width 1.5 cm), and contain embedded in a very hard shell, 3 to 4 elongate-oval (Figure 8) seed kernels (0.8 × 0.3 cm) with a blackish-brown to black skin and a white content. Azadirachtin has been extracted from the seed kernels, but may also occur in the leaves.[18]

The seed kernels of other *Melia* species, for instance *M. volkensii*, may also contain azadirachtin or related triterpenoids because growth-regulating effects in insects have been observed after application of extracts from whole fruits.[20] These plants are distributed in parts of Africa and Asia.

STRUCTURAL FORMULA AND ISOLATION

Structural Formula

The first isolation[8] and proposal for a structural formula of azadirachtin was made in England, and a molecular formula of $C_{35}H_{44}O_{16}$ was assigned to it.[21] As azadirachtin could not be obtained in crystalline form, a confirmation of the complete structure by X-ray

FIGURE 7.   Flowers and young fruit of *Melia azedarach*.

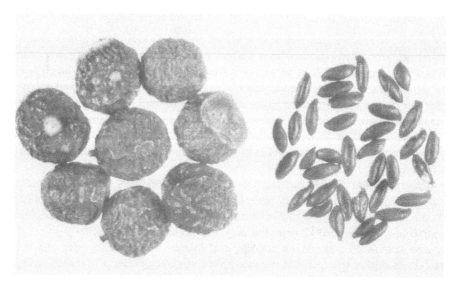

FIGURE 8.   Dried fruit (left) and seed kernels (right) of *Melia azedarach*.

chrystallography was not possible. A more correct elucidation of the structure was feasible by simultaneous use of partially relaxed Fourier transform and continuous-wave decoupling ^{13}C-NMR.[22] Some doubts remained concerning the configuration of one of the ring carbon atoms. Due to the complex structure of the compound there seems to be no economic way for its synthetic preparation at present or in the near future.

Recently, a final structural, formula was proposed (Figure 9).[117]

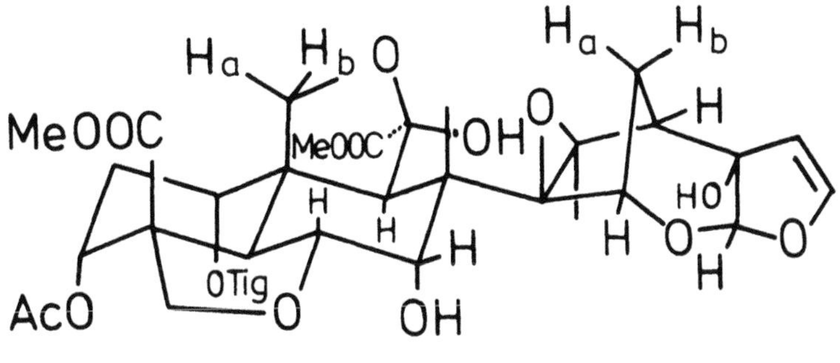

FIGURE 9.   Structure of azadirachtin.[117]

Isolation

Azadirachtin was isolated by grinding neem seed kernels in ethanol, partitioning the extract between methanol and light petroleum, chromatographing the methanol-soluble portion in Floridin earth, and subjecting the active fractions to preparative thin-layer chromatography (TLC) on silica gel.[23] The purity was assessed from the appearance of the NMR spectrum and its behavior in TLC on silica gel, at R_f 0.60 (ethyl acetate), 0.50 (ether-acetone, 4:1), 0.40 (chloroform-acetone, 7:3), and 0.15 (ether-acetone, 49:1).[21]

Another method for isolation is ethanolic extraction of ground neem seed kernels, followed by silica gel chromatography, and preparative TLC.[22] Some authors used open-column reversed-phase liquid chromatography on phase-bonded C-18 Hi-Flosil and high performance liquid chromatography (HPLC) on μBondapak C-18 columns for monitoring (217 nm) the purification (Figure 10). However, in this case the yield was rather poor (8.7 g of >90% pure azadirachtin from 48.2 kg of seed kernels of Indian origin), perhaps because of the old age of the plant material employed.[24]

A further possibility for azadirachtin isolation is extraction of seed kernels with methanol, extraction of the soluble material remaining after evaporation of methanol with hexane, fractionation of the polar material by open-column chromatography on silicic acid, elution of the fraction with strong growth-regulating activity in test insects with ethyl acetate, preparative TLC, and finally purification of the azadirachtin band on a RP-18 (Merck) HPLC column with 35% acetonitrile as solvent.[25]

BIOSSAY

There are numerous biotests to assess growth-disrupting and fecundity-reducing effects of neem and chinaberry ingredients but not all of them can be mentioned here. In most cases the test insects are fed for a varying period of time with treated leaves of their host plants and then kept on untreated leaves/plants in petri dishes or cages until they die or molt to the adult stage.[16,29] Artificial diets, into which certain amounts of the active principle(s) or extracts are incorporated, may also be employed for bioassays.[26,27,30,31]

After topical treatment by means of a microapplicator the larvae or pupae are placed in cages singly or in groups.[28,32,33,34] Control larvae are treated with the solvent (+ wetting agent) only. It is advisable to keep another completely untreated control to be able to find out whether any effects are caused by the solvent (+ wetting agent) control.

Another method is injection of azadirachtin by means of a micro-injector. In contrast to ordinary biotests with conventional synthetic pesticides, the bioassays with neem ingredients last relatively long, sometimes for weeks, due to the delayed effects of azadirachtin and related compounds.

Biotests for assessment of fecundity-reducing and egg-sterlizing effects are also carried

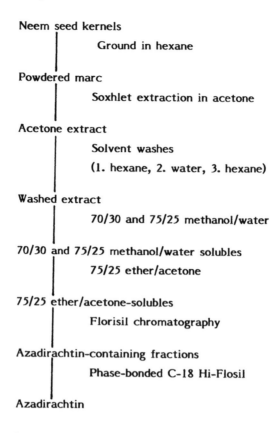

Neem seed kernels
| Ground in hexane

Powdered marc
| Soxhlet extraction in acetone

Acetone extract
| Solvent washes
| (1. hexane, 2. water, 3. hexane)

Washed extract
| 70/30 and 75/25 methanol/water

70/30 and 75/25 methanol/water solubles
| 75/25 ether/acetone

75/25 ether/acetone-solubles
| Florisil chromatography

Azadirachtin-containing fractions
| Phase-bonded C-18 Hi-Flosil

Azadirachtin

FIGURE 10.   Flow diagram of the isolation of azadirachtin from neem seed kernels.[24]

out in petri dishes, or better, cages of varying sizes.[35-38] The test insects are fed for a certain period with treated plants and afterwards kept in pairs or groups ($\delta\,\delta$ + $\,♀\,♀$) on untreated ones until the termination of the assay. The control insects are kept on plants treated with the solvent (+ adherent) and then also transferred to untreated plants. The eggs laid by the females are removed as soon as possible, counted, and put in petri dishes or small plastic boxes for further studies on possible egg-sterilizing effects.

The insect species suitable for bioassays with neem and chinaberry products are more or less the same as those used for biotests with synthetic pesticides, for instance, *Locusta migratoria migratorioides, Epilachna varivestis, Leptinotarsa decemlineata, Pieris brassicae, Plutella xylostella, Manduca sexta, Spodoptera littoralis, S. frugiperda,* and *Nilaparvata lugens.* *E. varivestis* and *P. brassicae* proved to be particularly sensitive to neem products.

INSECT GROWTH DISRUPTING EFFECTS

Effects on Embryonic Development

Very little information exists on the influence of neem ingredients on insect eggs. From the data available at present it may be concluded that application of high concentrations results in slight effects only, if any.

There was no effect of sprays containing 10 or 15 ppm/ℓ of azadirachtin or of 100 ppm/ℓ of a partially purified extract from whole dried neem fruit on *Epilachna varivestis* eggs.[39] In experiments (dipping method) with eggs of *Plutella xylostella*, concentrations as high as 5000 and 10,000 ppm/ℓ of ethanolic and aqueous extracts of dried NSKs gave some reduction of the rate of emergence; 500 and 1000 ppm/ℓ were slightly or not effective.[40] Spraying of eggs of *Dysdercus fasciatus* with a methanolic NSKE (conc. 0.01 and 0.02%) resulted in the same emergence rate of first instar nymphs as in untreated control eggs.[41]

Dipping of eggs of *Cnaphalocrocis medinalis* on leafcuts into neem oil (conc. 12, 25, and 50%) resulted in a low rate of emergence, but probably growth regulating ingredients in the oil were not solely responsible for this effect.[42] Possibly some of the eggs were choked when covered by a layer of oil.

As described in the chapter on neem effects on fecundity and egg fertility, some influence on egg fertility was observed after uptake of active neem principles per os by adults of *Leptinotarsa decemlineata* and after topical treatment of 5th instar nymphs of *Dysdercus fasciatus*, but in these cases the mode of action was different.

Effects on Postembryonic Development

Typical insect growth regulator (IGR) effects of neem ingredients were first observed in 1972 in the East African coffee bug, *Antestiopsis orbitalis bechuana*,[43,44] following a topical application of a crude methanolic extract from leaves and after employing azadirachtin by spraying food and test insects, such as *Pieris brassicae*, *Plutella xylostella*, *Heliothis virescens*, and *Dysdercus fasciatus* at the same time.[45] In both cases an influence of the neem ingredients on the hormonal system of the test insects was suggested. A few years earlier, growth retardant effects and development of malformed adults were found in *Spodoptera frugiperda* and *Heliothis zea* after incorporation of chloroform extracts of chinaberry leaves into an artificial diet and spraying of maize seedlings employed as foodplants for *H. zea*.[30]

During the last 12 years IGR effects of various neem products (aqueous, acetonic, methanolic, and ethanolic extracts of whole fruit, seed kernels and leaves, neem oil, extracts from neem oil, seed cake, enriched NSKEs, azadirachtin and related compounds from seed kernels) and of chinaberry products (methanolic and aqueous extracts of leaves, extracts of whole fruit, of seed kernels and of oil and seed cake) were described by numerous authors. An overview of the main results is given below, based on the various insect orders tested.

Orthoptera

Last instar nymphs of the American cockroach, *Periplaneta americana*, reacted after injection with azadirachtin (0.75 µg/g body weight) with a delay in molting from the 23rd to the 38th day.[46] Molting, if any, was partial, and the insects died during the process. The LD$_{50}$ value of azadirachtin in male and female last instar nymphs was 1.5 µg/g body weight.

In the African migratory locust, *Locusta migratoria migratorioides*, the effect of azadirachtin on metamorphosis was studied by injection of 4th and 5th instar nymphs within the first 2 days after molting. By this procedure, molt inhibition and mortality in a dose-dependent manner was induced.[47,48] Two micrograms azadirachtin per gram body weight prevented molts completely. The intermolt period of azadirachtin-injected nymphs varied from 8 to 60 days, that of control nymphs from 6 (4th instar) to 9 days (5th instar). Attempts of injected nymphs to shed the skin led to a rupture along the dorsal midline of the meso- and metathorax and over the pronotum and head capsule. The resorption of the exuvial fluid was incomplete or depressed. Nymphs which attempted ecdysis after azadirachtin injection possessed no or only a reduced number of sensory hairs in the hair field on the vertex, which means the integument was not correctly formed. The pigmentation and structure of the newly synthesized nymphal or adult cuticle also differed from that of the controls.

Injection of azadirachtin into locust nymphs altered feeding behavior too, as injected insects consumed less food than control specimens. Consequently, the feeding inhibition was not caused by a neuronal input from the palpi but probably by influencing neural control centers.[47]

In experiments with house crickets, *Acheta domesticus*, the test insects were fed a diet

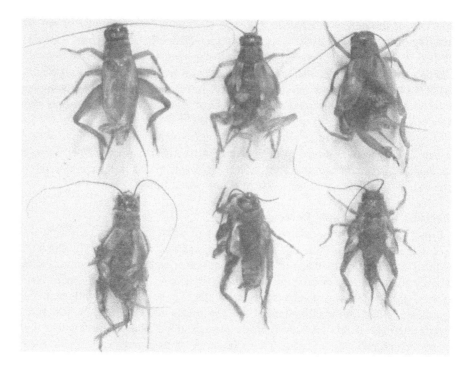

FIGURE 11.   House crickets that died during the nymphal adult molt after being treated topically with 25 μg azadirachtin.[49]

containing azadirachtin at different concentrations (10, 50, or 100 ppm).[49] There was a dose-dependant effect on the weight gain and duration of the nymphal development. Removal of the labial and maxillary palpi from the crickets did not prevent the nongustatory feeding-deterrent action of azadirachtin. Low doses of azadirachtin in the diet caused no difficulties in molting; crickets fed a diet with higher concentrations were observed with cast skins attached to their hindlegs, cerci, or ovipositors. After topical treatment of 7th instar nymphs (1 μℓ of an acetone solution containing 50 μg azadirachtin per insect) high mortality occurred before ecdysis. Treatment of 8th instar nymphs (conc. 25 μg/insect) disrupted the nymphal-adult molt and nymphs usually died after apolysis but before ecdysis (Figure 11).

Heteroptera

Topical treatment of last-instar nymphs of the East African coffee bug, *Antestiopsis orbitalis bechuana,* with a methanolic neem leaf extract (NLE) (1 μℓ per nymph) resulted in adults with shortened or rudimentary hemi-elytrons, bent hind wings, rudimentary scutellum, and shortened rudimentary pronotum (Figure 12).[43]

Results obtained with cotton stainer bugs *(Dysdercus* spp.) as test insects were somewhat contradictory. In trials with higher test doses of azadirachtin solution (0.2 mℓ sprayed on insects and food at the same time) all treated 5th instar nymphs of *Dysdercus fasciatus* died before ecdysis. In experiments with *D. cingulatus,* the application of acetone extracts from neem leaves and seed kernels gave surprising results, as topically treated 3rd instar nymphs molted into 6th instar supernumerary nymphs, skipping the 4th and 5th instars.[50] Treated 4th instars molted first into 5th instars and then into a supenumerary 6th instar. The 6th instar nymphs retained some of the nymphal characters, for instance 2-segmented tarsi and rudimentary wings; they were unable to copulate. Seed kernel extracts gave 88% mortality, leaf extracts 55%.

FIGURE 12. (Left) Normal adult of *Antestiopsis orbitalis bechuana;* other bugs deformed after topical application of a methanolic neem leaf extract.[44]

Table 2
EFFECT OF A CRUDE METHANOLIC NSKE AFTER SPRAYING IN DIFFERENT CONCENTRATIONS ON 5TH INSTAR *DYSDERCUS FASCIATUS* NYMPHS (n = 15)[41]

Conc. (%)	Dead 5th instars	0	I	II	III	IV	V	VI[a]	Dead adults	Mortality (%)
0.2	15	—	—	—	—	—	—	—	—	100[b]
0.1	5	—	—	—	—	1	2	7	10	100[b]
0.05	8	—	—	—	1	2	4	—	7	100[b]
0.02	9	1	—	4	1	—	—	—	6	100[b]
MeOH	4	9	—	2	—	—	—	—	—	26.67
Control	—	15	—	—	—	—	—	—	1	6.67

[a] 0, I, II, III, IV, V, VI = Degree of morphogenetic defects.
[b] = Significant at $p = 0.01$.

By topical application of acetone extracts of seed kernels to nymphs of various instars of *D. fasciatus* the above-mentioned deviating results could not be confirmed, as no supernumerary instars developed.[41] However, a few individuals resembled 6th instar nymphs when a part of their exuvia remained attached to the dorsal side of the thorax after the 5th instars had molted into adults. In this case the development of wings was prevented. Otherwise, depending on the concentration of a methanolic NSKE (0.02 to 0.2%) sprayed on the insects the treated nymphs died mostly either before ecdysis or during the molting process (Table 2). Some individuals were able to carry out the next molt, but died during the next one. The appearance of nymphs with asymmetric bodies after molting was quite common (Figure 13). Topical application of an acetone solution of azadirachtin (1 $\mu\ell$ per 3rd, 4th or 5th instar nymph) also did not result in supernumerary nymphal instars in *Dysdercus koenigii* but wingless adults were characteristic.[51]

No typical phase sensitive to growth-regulating compounds from neem could be found in 4th instar of *D. fasciatus* nymphs treated topically (conc. 1 μg per nymph) in different stages.[41] Treatment up to the 4th day after molting resulted in high mortality. This effect was, however, increasingly delayed the later the application took place. Nymphs treated shortly before molting reacted only slightly or not at all. According to the degree of morphogenetic defects in adults and nymphs (Figure 14) of *D. fasciatus* the following key was proposed for distinction:

FIGURE 13.   Nymph of *Dysdercus fasciatus* with asymmetric body after topical treatment with methanolic NSKE.[41]

FIGURE 14.   (Right) Normal adult of *Dysdercus fasciatus;* other bugs malformed after topical application of a methanolic NSKE on 5th instar nymphs.[41]

0	normal bugs
I	slightly damaged bugs; nymphs with partially deformed wing cases; adults with slightly to strongly deformed wings and often not foldable hindwings
II	more intensely damaged bugs; exuviae attached to legs
III	heavily malformed bugs; exuviae attached to last abdominal segments
IV	very heavily deformed bugs; exuviae attached to whole abdomen
V	no molt; exuviae ruptured in the middle of dorsal region of thorax, but insects enclosed in them

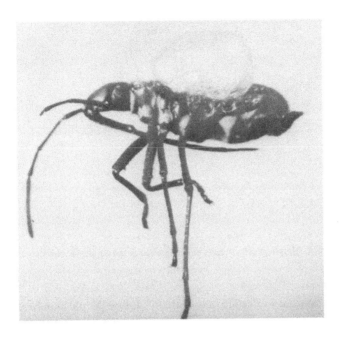

FIGURE 15.   Nymph of *Dysdercus fasciatus* showing a bubble filled with hemolymph after treatment with acetonic NSKE.[41]

Normally only bugs of grades I and II were viable but in rare cases so was grade III.

Acetone extracts were less effective than methanolic extracts and yielded different symptoms. Often the abdomen of tested nymphs became brownish, in other cases the whole body was brightened, and bubbles filled with hemolymph appeared on the dorsal side near the base of the wings (Figure 15).

Topical application of a partially purified extract from NSK (conc. 20 and 30 μg per insect) caused high mortality in treated young 5th instar nymphs of the beet leaf bug, *Piesma quadratum*. A few individuals reached the adult stage but showed malformed wings or could not cast the old skin completely.[32,52]

In 5th instar nymphs of the large milkweed bug, *Oncopeltus fasciatus*, incomplete adult ecdysis was caused by topical treatment of the ventral side of the abdomen with azadirachtin (10 μg per nymph).[33,53,54] The control insects, untreated or solvent treated began to molt on the 7th day after the last ecdysis, the treated molted on the 9th day. Death usually occurred within 24 hr after the beginning of the attempt to molt and dissection revealed adult cuticle.

Topical application of various azadirachtin-rich NSKEs (AZT-VR-K, MTB/H$_2$O-VR-K, etc.) dissolved in acetone and used at a dose of 1 to 10 μg per 3rd instar nymph of the rice bug, *Leptocorisa oratorius*, caused dramatic growth-disrupting effects in successive nymphal instars (see Figure 16). MIB/H$_2$O-VR-K was most effective with an LD$_{50}$ <1 μg. At 10 μg per insect none of the treated bugs survived. The adults obtained after application of lower concentrations were in most cases not viable.[55]

Homoptera

In contrast to *Orthoptera, Heteroptera, Lepidoptera,* and *Coleoptera,* no IGR effects could be observed in aphids after topical application of methanolic NSKE or tertiary methyl butyl ether (MTB) extracts. Via treated *Vicia faba* plants there was a systemic effect, especially when the extract was combined with lecithin and sesame oil.[56] Usually *Acyrthosiphon pisum* nymphs, which fed on treated plants died during their next molt. They tried

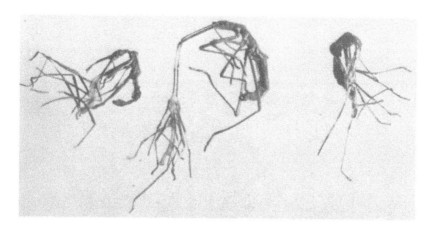

FIGURE 16.   Morphogenetic defects in *Leptocorisa oratorius* after application of an enriched NSKE.[55]

to cast their old skin but succeeded only partially. Either the old skin began to detach from the body of the insect, but could not be ruptered, or the skin ruptered above the head, but could not be cast completely.

Foliar application of neem oil and of enriched NSKEs led to disturbances during molts (Figure 17), extension of the nymphal period, and disturbances of the molting processes, resulting in a dose-dependent mortality of nymphal stages in the hoppers, *Nilaparvata lugens*, *Sogatella furcifera*, and *Nephotettix virescens* on rice.[55,57] Similar effects of neem products after application to the soil had to be attributed solely to systemic action of active principles.

Hymenoptera

Topical application of azadirachtin (0.2 to 5.0 μg per larva, dissolved in 1 μℓ methanol) to 3rd instar larvae of the honey bee, *Apis mellifera*, did not influence the weight gains in larvae treated with 0.05 to 0.25 μg each.[28,58] At the highest doses (0.25 and 0.5 μg per larva) abnormalities were observed 24 and 48 hr after treatment. The deviants were straighter, thinner, and longer than the control larvae, their skin was shrivelled and spotted, whereas both ends of the body were transparent. These features became more prominent about 1 day after appearance, and the larvae were dead a few hours later. High mortality occurred in the period between prepupa and pupa, especially before larval-pupal ecdysis.

IGR effects in honey bees were also observed in very small bee colonies after 2 sprayings of a concentration of 500 ppm/ℓ of an enriched NSKE (AZT-VR-K) on flowering plants, such as *Phacelia tanacetifolia*, in a field cage.[59]

Coleoptera

Growth-disrupting effects of various neem and chinaberry products (methanolic and aqueous extracts of neem and chinaberry leaves, partially purified fractions of neem leaf extracts, neem oil, extracts from neem oil, partially purified extracts of whole neem fruit, methanolic NSKE, purified fractions of NSKE, and azadirachtin) were first investigated in detail by employing the phytophagous coccinellid *Epilachna varivestis* as test insect. Histological studies were also carried out. The Mexican bean beetle, *E. varivestis*, is, therefore, one of the best known insects as far as neem effects are concerned. The results can be summarized as follows.

Spraying of crude methanolic and aqueous extracts (conc. 2 and 4%) of *A. indica* and *M. azedarach* leaves on bean leaves serving as food for various larval instars, prevented

A

B

FIGURE 17. Molt-disturbing effects of neem products in: (A) *Soga-tella furcifera* and (B) *Nephotettix virescens.*[55]

pupation of 4th instars completely.[32,39] The feeding activity of treated larvae during the first 2 to 3 days was more or less identical to that of control larvae but afterwards it was considerably reduced. In another experiment, employing 3rd and 4th instar larvae and topically applied doses of 150, 200, and 250 μg of a methanolic extract of neem leaves per insect, a few individuals developed into pupae and malformed adults; some other adults looked normal. Some specimens could be considered as some sort of "intermediates" between prepupae and pupae and between pupae and adults. The "prepupal/pupal inter-

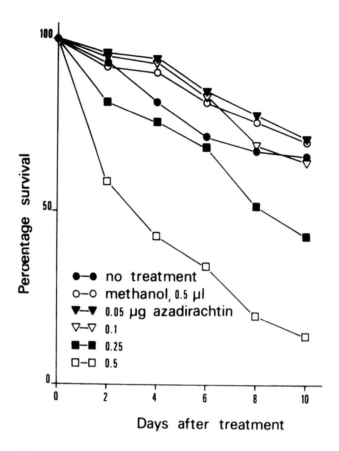

FIGURE 18.   Effect of azadirachtin, applied topically to 2-day-old larvae
in vitro, on development of *Apis mellifera*.[26]

mediates'' were partly covered by the larval skin, whereas the ''pupal/adult intermediates''
were characterized by specimens which developed within the pupal skin but could not shed
it. In some cases these pupal/adult intermediates were able to free their legs and mouthparts
partly so that feeding and walking was possible to a limited extent.* Sometimes these
creatures lived for several days. Some pupae exhibited swollen wing sheaths which often
burst during or shortly after the pupal molt leading to loss of hemolymph and consequently
death (Figure 20).

The various neem products tested were effective in all larval instars in a dose-dependent
manner. Higher concentrations led to mortality before the next ecdysis or during this process,
lower ones were effective during the next one or two molts.

Some pure fractions of a methanolic NSKE, including azadirachtin were applied as sprays
in different concentrations on bean leaves. The test insects, young 4th instar larvae, were
transferred to the leaves after spraying and 48 hr later to untreated food. No obvious
antifeedant effect was observed except in the azadirachtin treatment (pure fraction 18 to
79 ℓ). One fraction caused dark-brown to blackish coloration of the legs of the larvae; in
two others the appearance of 1 to 4 reddish-brown to black spots in the dorsal thoracic
region occurred (Figure 18). The application of the latter two pure fractions also resulted

---

*   Intermediates resulting from neem treatment are not identical with true intermediates obtained after application
     of JH, as their whole bodies or parts of their bodies are still covered by the larval skin. The cuticle underneath
     is more or less pupal or adult. Intermediates from JH treatments show, for instance, after a complete ecdysis
     a cuticle with larval and pupal features.

FIGURE 19.   Honeybee worker with crippled wings (left) and with part of pupal exuvia on abdomen (right).

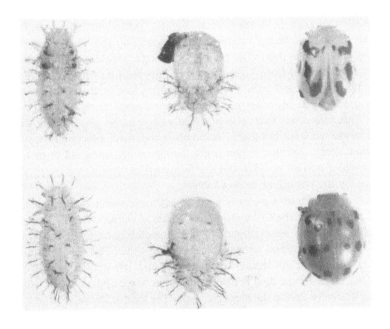

FIGURE 20.   Upper row from left to right: 4th instar *Epilachna varivestis* larva with black spots in the thoracic region after uptake of methanolic NSKE, pupa with deformed and inflated wing sheath, adult with deformed body, and atypical pigmentation. Lower row: normal larva, pupa, and adult.

in so-called "permanent larvae" (German: Dauerlarven) which had an intensive yellow color, lived up to 3 to 4 weeks, and finally died or more rarely molted into malformed pupae with inflated wing pads. The few resulting adults were all severely malformed and showed a strongly modified pigmentation on the elytra (Figures 20, 21).

FIGURE 21.   Adults of *Eilachna varivestis* with deformed and inflated forewings and unfoldable hindwings (left) as a result of uptake of methanolic NSKE by 4th instar larvae.

Experiments for assessment of a possible sensitive phase for neem effects in the 4th larval instar were carried out as feeding experiments with larvae of different ages (24, 24 to 48, 48 to 60, 66, 69, and 72 hr).[39] The concentration applied on foodplants was 100 ppm/ℓ of a partially purified extract from neem fruit. Treated individuals with an age up to 60 hr could not molt into pupae. In the variants with an age of 66 and 69 hr a few larvae reached the pupal and adult stages but most of the latter were severely malformed. The normal looking specimens derived from larvae which did not feed on treated leaves. In the 72 hr variant the majority of the test insects (12 of 20) molted into normal adults; four other adults were heavily malformed. Most normal adults derived from larvae which did not consume treated food. This means that there was practically no sensitive phase in the 4th larval instar although old larvae were not influenced and molted normally, a fact which can be explained by lack of food intake. In other experiments, larvae treated about 48 hr or less before the next molt showed a normal ecdysis, but the following ecdysis was affected if treated food was consumed by the preceding instar (Table 4).

Potato leaves sprayed with a partially purified extract of whole neem fruit, served as food for larvae of all instars of the Colorado potato beetle, *Leptinotarsa decemlineata*. The 1st, 2nd, and 3rd instar larvae were allowed to feed completely on treated leaves whereas 4th instar larvae were transferred to untreated leaves after 48 hr. The concentrations applied on the food of 1st, 2nd, and 4th instars were 10, 50, and 100 ppm/ℓ on that of 3rd instars. The 1st and 2nd instar larvae died before molting; the same applied to the 3rd instar.[36] The 4th instar died before or during the prepupal stage. The treated larvae showed a much smaller size than the untreated and exhibited a dark red color as well as a glassy appearance.

By topical application of a neem fruit extract in concentration of 1, 0.5, and 0.25% (1 $\mu\ell$ per insect) to 4th instar larvae of *L. decemlineata* the following results were obtained: 100% mortality occurred after employing 20, 10, and 5 μg per larva.[36] After application of 1 and 0.5 μg per insect most larvae reached the prepupal stage but could not molt into pupae. Under the influence of 0.1 μg per larva, 13 of 18 insects died as prepupae or pupae, but 7 specimens of the former could be considered as prepupal-pupal intermediates. The head capsule was more or less larval and the abdomen pupal whereas the wing-sheaths were inflated. Three pupae molted into adults with slightly crippled wings. The control insects developed into normal adults.

Table 3

GROWTH OF *NILAPARVATA LUGENS*, *SOGATELLA FURCIFERA*, AND *NEPHOTETTIX VIRESCENS*[a] ON RICE PLANTS SPRAYED WITH NEEM OIL AND NEEM SEED EXTRACTS CODED NE II, MTB, AZT-VR-K[155]

Neem seed derivative	Conc.	N. lugens			S. furcifera			N. virescens		
		Nymphs becoming adults (%)	Growth period (days)	Growth index[b]	Nymphs becoming adults (%)	Growth period (days)	Growth index[b]	Nymphs becoming adults (%)	Growth period (days)	Growth index[b]
Neem oil (%)	0 (control)	100	20	5.1	96	16	6.0	80	22	3.7
	0.5	62	20	3.1	52	18	2.9	54	24	2.3
	1.0	36	20	1.8	28	18	1.6	22	25	0.9
	3.0	8	23	0.3	12	18	0.7	2	25	0.1
	6.0	—	—	—	—	—	—	—	—	—
NE II (mg/kg)	0 (control)	94	19	4.4	82	18	5.4	94	19	4.9
	20	94	19	4.4	82	18	5.4	94	19	4.9
	50	40	19	3.2	60	18	2.2	50	20	2.5
	100	12	20	2.4	46	19	1.4	14	24	0.6
	200	8	20	1.6	22	19	0.4	4	25	0.2
MTB (mg/kg)	0 (control)	92	17	5.6	92	17	5.4	92	23	4.1
	10	84	17	5.1	80	18	4.6	62	23	2.7
	20	76	17	4.6	60	18	3.3	54	24	2.3
	50	26	17	1.5	52	19	2.8	26	24	1.1
	100	6	17	0.4	40	19	2.2	14	26	0.6
AZT-VR-K (mg/kg)	0 (control)	80	16	5.0	84	15	5.8	92	19	4.8
	10	62	16	3.9	56	15	3.9	50	22	2.3
	20	48	17	2.8	42	15	2.8	36	23	1.6
	50	16	17	0.9	18	16	1.2	16	25	0.6
	100	—	—	—	8	16	0.5	—	—	—

[a] First instar nymphs were caged on 30-day-old rice plants; average of 5 replications, 10 nymphs per replicate.

[b] Growth index calculated as the ratio of percentage of nymphs becoming adults, to to growth period.

[c] Dash indicates the nymphs died.

Table 4
EFFECT OF A PURIFIED EXTRACT FROM NEEM FRUIT ON 4TH INSTAR *EPILACHNA VARIVESTIS* LARVAE OF DIFFERENT AGE IN A FEEDING TEST ON BEAN LEAVES
(n = 20)[39]

Conc.	Age of larvae (hr)	Av. wt. at onset of trial (mg)	Mortality of larvae & prepupae	Mortality of pupae	Normal adults	Slightly deformed adults	Heavily deformed adults
100	24	22.2	20	—	—	—	—
100	24—48	33.5	20	—	—	—	—
100	48—60	39.5	20	—	—	—	—
100	66	44.8	16	—	1	—	3
100	69	50.1	12	2	3	—	3
100	72	52.1	3	1	12	—	4
Control (solvent)	24	21.7	1	—	19	—	—
Control (solvent)	66—69	46.4	2	1	16	1	—
Control (untreated)	72	52.4	—	—	20	—	—

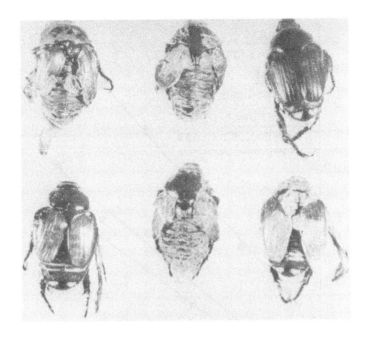

FIGURE 22.   Deformed pupae and adults of *Popillia japonica* after topical application of azadirachtin.[34]

Topical application of azadirachtin (3.5 $\mu\ell$ per insect) to larvae of the Japanese beetle, *Popillia japonica*, by means of a micro-injector, completely disrupted subsequent development to adults (Figures 22 and 23).[34] Prepupae and newly formed pupae were less susceptible, and 3-day-old and older pupae were not affected (Table 5). LD_{50} and LD_{90} values of topically applied azadirachtin were 0.1 and 0.4 μg per larva, respectively. Azadirachtin delayed the development of feeding and nonfeeding immature stages. Growth regulating effects were also observed in larvae of the alder leaf beetle, *Agelastica alni.*[60]

A systemic effect of an aqueous and methanolic neem leaf extract was obtained by watering young bean plants. The 4th instar larvae of *Epilachna varivestis* were placed on these treated plants after watering and 48 hr later were transferred to untreated bean leaves. A concentration of 5% of the methanolic extract, for instance, gave 50% mortality in the larval and prepupal instar stages. Some test insects became adults which were partly malformed.[39] A systemic effect, leading to a disturbance of metamorphosis, was also found in the pea aphid, *Acyrthosiphon pisum,*[56] the beet leaf bug, *Piesma quadratum,*[52] various leaf- and planthoppers,[55] the spiny bollworm, *Earias insulana,*[61] the cabbage moth, *Plutella xylostella,*[40] and the leaf miners *Liriomyza sativae* and *L. trifolii.* The first observations on translocation of neem products (methanolic and aqueous SKE, azadirachtin) in bean plants were made in 1971 with *Schistocerca gregaria* as a test insect.[63]

Lepidoptera

IGR effects of neem were studied in detail in species from various families, such as *Ephestia kuehniella, Cnaphalocrocis medinalis, Manduca sexta, Pieris brassicae, Mythimna separata, Plutella xylostella,* and *Spodoptera littoralis.*

In trials with *S. littoralis* larvae weighing 70 to 90 mg at the beginning of the experiment and fed on lucerne treated with a concentration of 0.6% of neem oil extract ("neem extractive"), 75% of the individuals could not molt normally, and at 1%, 85% could not.[64] Aqueous NSKE showed such activity only from 1% and above. The molting process was

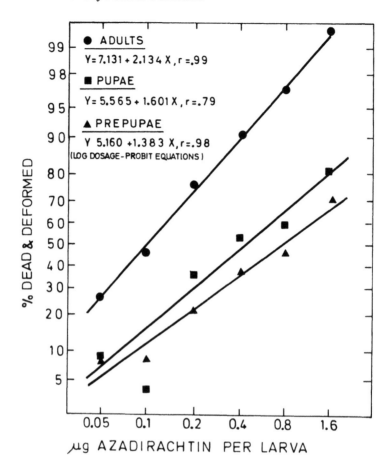

FIGURE 23.   Calculated percentage of dead and deformed Japanese beetles occurring at the indicated stages after topical application of azadirachtin to 3rd instar larvae.

also inhibited in 50% of the 70 to 90 mg larvae and in 15% of the 120 to 150 mg larvae, but no molting inhibition was observed in the groups with higher weights (170 to 190 and 200 to 250 mg). Neonate larvae, fed for 48 hr on lucerne treated with various concentrations of neem extractive failed to molt or pupate. Surprisingly, molting of larvae treated topically, by injection or imbibition of even high concentrations of "neem extractive" was not influenced. After application of a methanolic NSKE at concentrations of 0.3, 0.1, and 0.05% no pupae were obtained. With 0.1 and 0.05%, pupae and adult yields were low, whereas lower concentrations had little or no effect on pupation and adult emergence. The IGR effect seemed to increase with weight of larvae and the closer they were to pupation. Feeding of 30 to 50 mg larvae for 2 days on plants treated with 0.05% of aqueous, methanolic, and ethanolic NSKEs prevented pupation completely or almost completely. An acetonic extract was less effective (Table 6).

Feeding of fully grown larvae of *S. litura* for 6 hr on leaf discs of *Ricinus communis*, treated with various doses of azadirachtin, yielded juvenilizing effects as larval-pupal intermediates and other malformed individuals were obtained.[66]

Topical application of highly purified fractions of NSKE to spinning last instar of *Ephestia kuehniella* (conc. 1 μℓ per larva) resulted in retardation or termination of development, larval-pupal intermediates, malformed pupae, and malformed adults with wing pads or moderately to severely shriveled wings.[26,28]

Table 5
DEAD AND DEFORMED IMMATURE JAPANESE BEETLES FOLLOWING TOPICAL APPLICATION OF 2 μg AZADIRACHTIN/ INSECT (n = 30)[34]

Stage treated	Dead and deformed (%)[b]
Larvae, unfed since storage[a]	100
Larvae, after feeding for 14 days	100
Larvae, feeding completed	92
Prepupae, immobile	92
Pupae, 24 hr after molt	79
Pupae, 72 hr after molt	0
Pupae, "green head"[c]	0

[a] Larvae collected in field (April 1982) and held in soil at 3°C until use.
[b] Corrected by Abbott's formula.
[c] 1—2 days before adult emergence.

Table 6
FEEDING 30—50 MG LARVAE OF *SPODOPTERA LITTORALIS* FOR 2 DAYS WITH LUCERNE TREATED WITH VARIOUS NSKEs — EFFECTS ON ADULT EMERGENCE[64]

Extracting solvents	Dielectric constant (T = 25°C)	Adult emergence % Neem extract in dipping liquid (%)		
		0.05	0.01	0.005
Water	78.5	0	9 + 13 abn.[a]	70
Methanol	32.6	0	8 + 2 abn.	66
Ethanol	42.3	1	28	92
Acetone	20.7	2.5 + 2 abn.	26.5	88

[a] abn. = abnormal adults.

In other experiments with an NSKE, there was the tendency for *E. kuehniella* larvae after uptake of treated food (approximately 4 and 8 ppm/kg) to develop an intermediate additional instar between the 4th and 5th instar.[67] Therefore, the total number of instars was six which is common in some populations but rare or absent in others. The developmental period of treated larvae was prolonged for about 3 days compared to untreated. At concentrations above 2 ppm no adults developed. Depending on the concentration of the NSKE applied, death of the larvae occurred at different molting phases. At the high concentrations of 500 and 250 ppm, all larvae died at the first molting phase which is characterized by splitting of the old larval skin between the head capsule and the first thoracic segment, with subsequent shedding of the head capsule. At 125 and 60 ppm, an increasing number of larvae died in the second phase, expressed by bursting the remaining larval skin on the ventral side and its shedding. With decreasing NSKE concentration death occurred later and later during larval development. At the lower concentrations (4, 2, 1, and 0.5 ppm) the moment of death moved from the molting phases to the feeding phases or into the pupal stage.

In trials with 5th and 6th instar larvae of the rice ear-cutting caterpillar, *Mythimna separata*,

FIGURE 24.   Shrunken larvae of *Mythimna separata* after uptake of NSKE.[68]

partially purified fractions of an ethanolic NSKE as well as methanolic NSKEs were used.[68] Leafcuts of rice served as food after dipping them in solutions of the fractions/extracts (conc. 50, 100, 200, and 500 ppm/ℓ) and of an additional adhesive ("Citowett"). In another test, 6th instar larvae were separately dipped for 3 sec in a 50 ppm/ℓ solution of the partially purified fractions or a 500 ppm/ℓ solution of one of the methanolic NSKEs and afterward offered untreated leaves. *M. separata* 5th or 6th instar larvae which fed shortly before or after molting on leafcuts treated with a fraction which was free of azadirachtin (F_{22}), showed high mortality; a few larvae developed into prepupae which also died about 2 weeks later after continuous shrinking, or developed into larval-pupal intermediates (Figure 24). More or less the same happened after dipping of early 5th instar larvae into a solution of one of the methanolic NSKEs, but none of the larvae developed into the pupal stage. Some treatments delayed prepupal, pupal, and adult stages for about 2 days. The pupal weight was either clearly or only slightly reduced. Moths emerged from pupae in the treatments showed varying degrees of wing deformations. Dipping of 6th instar larvae did not yield visible morphogenetic defects.

Chloroform extracts from chinaberry leaves caused larval mortality when incorporated into a meridic diet fed to caterpillars of the corn earworm, *Heliothis zea,* and the fall armyworm, *Spodoptera frugiperda,* or sprayed on maize seedlings as foodplants.[30] There were also prolonged larval and pupal periods and adults often emerged deformed.

Azadirachtin, sprayed in concentrations of 0.025 and 0.1 mg together with a wetter (Lissapol) on food and insects gave high mortality in the 4th larval instars of the tobacco budworm, *Heliothis virescens* and the cabbage white, *Pieris brassicae.*[45] Only very few deformed pupae developed but no adults. The malformed pupae had swollen, protruding wing sheaths (Figures 25 and 26) and larval heads. As there was no reduction of feeding in the latter species, growth retardation and morphogenetic effects must have been caused solely by azadirachtin.

In the rice leaf-folder, *Cnaphalocrocis medinalis,* treatment of rice leaves serving as food

FIGURE 25.   Dead 5th instar larvae and ''larval-pupal intermediates'' (left) of *Pieris brassicae*, resulting from treatment of
cabbage leaf disks with an enriched NSKE.

FIGURE 26.   Malformed pupae of *Pieris brassicae* as
a result of application of an enriched NSKE in 5th larval
instar.

with different concentrations of neem oil, partially purified fractions of NSKEs and with
methanolic NSKE resulted in the following effects: last instar larvae failed to fold the leaves,
ceased feeding, shrank in size, became darker, and finally died.[42,68] Other larvae developed
into larval-pupal monstrosities, partly covered with the old skin or retaining the hard head
capsule and thoracic legs. Abnormal pupae also occurred. Moths failed sometimes to emerge
from the pupal case or had poorly developed or twisted wings. In the control group, the
larval and pupal mortalities were low and most emerged moths looked normal.

Experimental work with methanolic leaf extracts of *A. indica* and *M. azedarach* revealed
high mortality of 5th instar larvae of the cabbage moth, *Plutella xylostella,* during the first
week after spraying of cabbage leaf disks serving as food.[39] The concentration of the crude
extracts was 2.5 and 5%. Only one individual in the 2.5% treatment with chinaberry leaf
extract reached the pupal stage in which it died. Azadirachtin in a concentration of
10 ppm/ℓ plus an adherent (Citowett), also applied via sprayed cabbage leaf disks, gave
100% mortality in a trial started with 3rd instar larvae. Most larvae died during the next

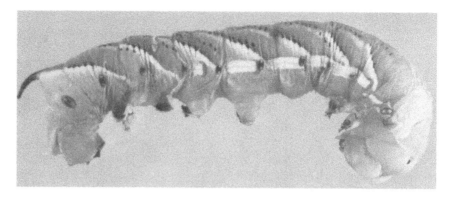

FIGURE 27.   5th instar larvae of *Manduca sexta,* trying to molt into a supernumerary larval instar after topical application of a methanolic NSKE. The old and the new head capsules are visible.[69]

molt, the remaining ones in the 5th instar or, exceptionally, in the pupal stage. Azadirachtin, in a concentration of 0.025, 0.05, and 0.1 mg, sprayed on test insects (5th instar larvae of *Plutella xylostella)* and food at the same time gave high mortality in the larval stage; the remaining insects died in the pupal stage except one which molted into an adult with poorly developed wings.[45]

A comparison of the effect of ethanolic extracts from various parts of the neem tree expressed by mortality of larvae of *P. xylostella,* gave the following order of effectiveness: dried seed > fresh seed > leaf > bark/twig > wood. Third instar larvae were put on sprayed cabbage plants for bioassay. After application of an ethanolic seed extract at concentrations higher than 10^2 ppm/ℓ on potted cabbage plants less than 20% of the larvae placed on them developed into adults compared with 47% in the control.

Incorporation of ground neem seed kernels at concentrations of 0.02, 0.2, and 0.5% into an artificial diet resulted in 100% molting inhibition of 2nd instar larvae of the gypsy moth, *Lymantria dispar.*[31] No molts were observed although some larvae lived up to 27 days. In another trial 1st instar larvae were transferred to oak leaves sprayed with 0.5% of an aqueous NSKE. Four days after the onset of the experiment the test larvae were placed on untreated food. All caterpillars were dead after 15 days and no molts could be observed.

The 5th instar larvae of the tobacco hornworm, *Manduca sexta,* were allowed to feed on an artificial diet containing different concentrations (1, 2, 5, 10, 25, and 50 ppm) of a methanolic NSKE.[69] Caterpillars which fed on a diet containing concentrations between 5 and 50 ppm showed a faded skin, and some areas of their body became brownish. Brown patches appeared on the base of the forelegs and extended gradually into lateral and dorsal areas. Finally, the caterpillars became totally black and died. Larvae which fed on a diet with low concentrations (1 and 2 ppm) of NSKE shrivelled and died in the prepupal stage or molted into malformed, nonviable pupae. Larvae treated topically with 1000 ppm/ℓ of a NSKE in the wandering stage, molted into pupae with the same types of malformation. The malformed individuals could be subdivided in seven groups based on progression of ecdysis. Larval-pupal intermediates with larval head and thorax but pupal abdomen were quite common. Numerous pupae showed more or less severe malformations of the appendages of the head and thorax; the tongue case was the most affected part. Even slightly malformed pupal cases could not be left by moths which developed in them. Many normal-looking pupae molted into adults with crippled wings and a loose proboscis. Furthermore, large areas of the adult cuticle were free of scales.

Surprisingly, some larvae which developed on a diet with 1 ppm NSKE and others from a trial with topical application of NSKE tried to molt or molted into a supernumerary 6th larval instar (Figure 27).[69] Some of these larvae started to shed the cuticle of the 5th instar

Table 7
ACTIVITY OF AZADIRACHTIN ON
ECDYSIS AND GROWTH OF NEWLY
HATCHED LARVAE OF 4
AGRICULTURAL PEST INSECTS
TREATED IN A 10 DAY FEEDING
BIOASSAY[72]

Species	EI$_{95}$ (ppm)[a]	ED$_{50}$ (ppm)[b]
Heliothis zea	2	0.7
H. virescens	2	0.7
Spodoptera frugiperda	1	0.4
Pectinophora gossypiella	10	0.4

[a]   EI$_{95}$ values are the doses for 95% ecdysis activity.
[b]   ED$_{50}$ values are the effective doses for 50% growth inhibition.

but were often not successful; they showed a new larval cuticle underneath the old partially withdrawn exuvia. Those larvae which completed the molt showed abnormal abdominal legs and were not viable. First instar larvae reared on treated diet (conc. 25 or 1 ppm of NSKE) or treated tobacco leaf disks died during the next molt. There was an extension of their developmental period by 1 week.

Growth regulating effects of neem extracts or azadirachtin were also found in some other species of Lepidoptera, for instance, *Boarmia selenaria*,[70] *Earias insulana*,[61] *Mamestra brassicae*,[71] *Heliothis armigera*,[5] *H. zea*,[72] *H. virescens*,[45,72] *Spodoptera frugiperda*,[72,73] *Chilo partellus*,[5] *Ostrinia furnacalis*,[74] *Crocidolomia binotalis*,[75] and *Pectinophora gossypiella*.[72] The EI$_{95}$ values (doses for 95% ecdysis inhibitory activity) of 1st instar larvae of *H. zea* and *H. virescens* were 2, of *S. frugiperda* 1, and of *P. gossypiella* 10 ppm (Table 7).[72]

Diptera

Very little information on effects of neem products in members of this insect order has been available up to now. A methanolic NSKE, incorporated into a semi-liquid rearing medium for the Mediterranean fruit fly, *Ceratitis capitata*, at concentrations of 2.5, 5, 10, 15, and 20 ppm prolonged the larval development by 1 (5 ppm) or 2 (20 ppm) days.[76] The pupal size was negatively affected in a dose-dependent manner. Significant larval mortality was caused by 20 ppm. The eclosion rate of adults from pupae in the 15 ppm treatment was only 16% that of the control group. With rising concentrations, an increasing number of normal-emerged fruit flies were unable to fly; in the 20 ppm group an average percentage of about 78%. The quality of adults was also negatively affected, expressed by startle activity or irritability, response of virgin females to the male sex pheromone, and the mating propensity of sexually mature, virgin flies of both sexes (Table 8). Furthermore, the longevity of adults was drastically reduced, for example, less than 50% of the flies of the 10 ppm group reached sexual maturity.

Application of 0.2 and 0.4% ethanolic NSKEs against the leaf miner *Liriomyza trifolii* on chrysanthemum and tomato resulted in high mortality of 1st and 2nd instar larvae. The pupation rate of treated 3rd instar larvae was not affected, but the emergence of adults was, as few or no flies emerged from treated plants.[77] In trials with the mosquito, *Aedes aegypti*, methanolic extracts from NSKs of individual trees of different origin in concentrations of

Table 8
EFFECTS OF LOW CONCENTRATIONS OF A METHANOLIC NSKE INCORPORATED INTO THE LARVAL DIET OF *CERATITIS CAPITATA*[76]

	Control	2.5 ppm	5 ppm	10 ppm	15 ppm	20 ppm
Pupation rate (%)	76.7	71.4[a]	73.8	76.4	67.2[a]	51.1[a]
Pupal diameter (mm)[b]	2.34	2.28	2.25	2.17	2.12	2.03
Emergence rate (%)	96.4	93.6	89.6[a]	57.7[a]	15.4[a]	4.6[a]
Flight ability index	92.5	90.6	87.9	82.7[a]	67.5[a]	21.7[a]
Startle activity index	49.1	48.9	41.9[a]	38.3[a]	—	—
Olfactometric female response (%)	40.0	26.5[a]	28.4[a]	22.3[a]	—	—
Mating propensity index	57.8	51.3	38.5[a]	31.3[a]	—	—

[a] Significant difference from control ($p = 0.05$).
[b] Pupal diameter shows a linear correlation to conc. ($r = 0.9948$).

about 50 to 100 ppm/ℓ were added to water containing 4th instar larvae. The test larvae were exposed continuously to the treated water.[16] The LC_{95} values (95% mortality mainly due to growth regulating effects) were between 50.7 and 96.2 ppm. The majority of killed specimens died as white, slightly melanized, or normally melanized pupae with adult structures visible through the skin about 6 to 7 days after the onset of the experiment. A small portion of the dead pupae showed protruded mouthparts and deformed wings.

Neem seed kernel extracts (coded AZT-VR-K) caused an extreme prolongation of the larval period when 1st instar larvae were permanently exposed to treated water until adult emergence. The growth-disrupting efficacy of the extracts increased with decreasing polarity of the solvents used for extraction.[78]

Effect on Fecundity, Longevity, and Egg Sterility

A moderate to strong influence on fecundity was observed by various authors in Coleoptera *(Epilachna varivestis, Agelastica alni,* and *Leptinotarsa decemlineata),* Lepidoptera *(Spodoptera littoralis),* Heteroptera *(Dysdercus fasciatus),* and Homoptera *(Nilaparvata lugens, Sogatella furcifera, Acyrthosiphon pisum).* In Diptera such as *Ceratitis capitata,* only high concentrations of neem products fed to adults seem to exert a moderate negative influence on fecundity. Egg fertility was reduced only in *L. decemlineata* and *D. fasciatus* after treatment of females.

Fecundity-reducing effects could be detected for the first time in Mexican bean beetles, *E. varivestis* fed on bean leaves sprayed with azadirachtin (approximately 50 ppm/ℓ) during the first 5 days after molting.[35] Twelve pairs of treated bean beetles started egg laying 8 days later than untreated ones which meant on the 16th day after the onset of the experiment. On the 39th day of the trial 8759 eggs were laid by untreated females but only 704 by treated ones. These figures were somewhat influenced by increasing mortality among treated females. Egg fertility was slightly less in the treatments, namely 65.6% compared to 70.6% in controls but this difference was not significant. The general reproduction rate of the treated beetles was only about 8 or 6% of that of control insects when calculated for the 40th or 55th day, respectively, of the beetle's life. Treatment of males had no negative effect on fecundity.

Further studies employing a methanolic NSKE (0.25%) whose residues were consumed for 3 days by newly molted Mexican bean beetles on bean leaves gave similar results.[37] There was no significant effect on longevity of treated beetles, but fecundity calculated on the basis of female-days and egg fertility were markedly reduced (Figure 28). Eggs laid by treated females were often covered with droplets and exhibited brownish dots. Furthermore, they were sensitive to a fungus which developed and spread rapidly over the egg masses.

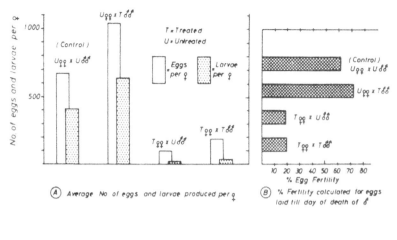

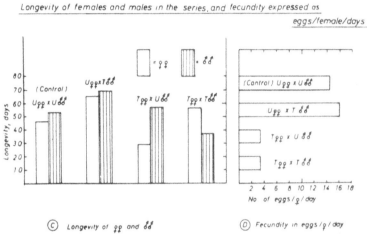

FIGURE 28. Longevity, fecundity, and egg fertility of *Epilachna varivestis* adults exposed before copulation for 3 days to beans sprayed with 0.25% methanolic NSKE.

In a trial with adults of the Colorado potato beetle, *L. decemlineata* aging 10 to 16 days, a purified extract from whole neem fruits (conc. 100 ppm/ℓ) was employed.[36] The beetles, kept in pairs in cages were fed for 5 days on treated potato leaves or leaves treated with solvent only (control). On the 16th day after beginning of the experiment 4,367 eggs in total were laid by untreated females, and somewhat later the number of larvae emerged was 3,645 (emergence rate 83.5%) (Figure 29). On the other hand, the treated females stopped egg laying 1 day after the onset of the trial. Up to this time 172 eggs had been laid, from which 120 larvae hatched (59.3%). From the 9th to the 16th day further egg laying occurred, resulting in 284 eggs with an emergence rate of 41.9%. No eggs were produced after this date by treated females. On the 44th day the total egg production of treated females was 11,551 eggs and 9,301 larvae emerged from them. Mortality of treated beetles had no significant influence in this trial.

Agelastica alni females, fed on alder leaves treated with a methanolic NSKE, reacted also by laying fewer eggs than untreated females.[60] There are some other records on reduced fecundity of test insects in trials with various store pests, mainly beetles, but these effects have not been studied in great detail and were often influenced by high mortality of adults.

In experiments with *Spodoptera littoralis*, larvae of different weights were grouped and

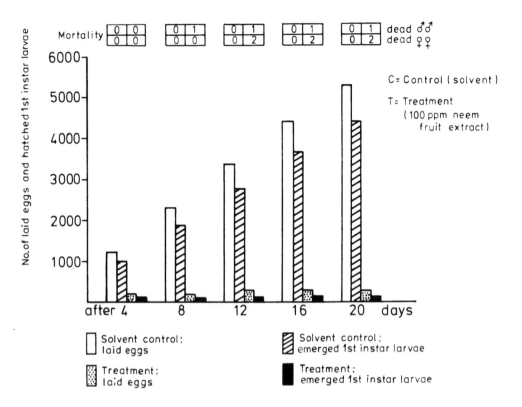

FIGURE 29.   Fecundity of *Leptinotarsa decemlineata* females and emergence rate of 1st instar larvae after consumption of potato leaves treated with an extract from neem fruit.

fed on lucerne leaves treated with low concentrations (0.01, 0.001, and 0.005%) of a methanolic NSKE.[38] Adult moths derived from larvae weighing 2 to 4 mg at the onset of the experiment and fed on lucerne treated with 0.001% NSKE laid only about 12% of the eggs of control adults. At the two higher concentrations, practically no eggs were laid. With 30 to 50 and 100 to 120 mg larvae the negative effect on fecundity was smaller but increased again with 170 to 200 mg larvae (Figure 30). All eggs hatched normally, indicating that there was no effect on sterility. The fecundity-reducing effect exerted by NSKE was on females only.

The fecundity of queens of the ant *Formica polyctena* kept in small colonies (1 ♀ + 100 workers) in the laboratory, was considerably reduced for more than 2 months by application of an enriched NSKE (AZT-VR-K), integrated into a standardized feed at a concentration of 0.5 and 0.25%. Workers were not affected by this treatment.[79]

The fecundity of female leaf miners, *Liriomyza trifolii*, treated at early larval instars (1st and 2nd instar) with ethanolic NSKE at 0.1% was significantly lower than the control but longevity was not influenced. The same applied to females treated in the 3rd larval instar with NSKE at 0.4%. The treatments with 0.4% neem extract significantly altered the longevity of males. The average life expectancy for treated males as 5.8 days as compared to the control with a mean of 10.7 days.[77]

Dysdercus fasciatus females produced after topical treatment of 5th instar nymphs with a methanolic NSKE only 59% (dose 2 μg per insect) of the eggs of control insects.[41] The eggs laid by adults treated with the higher concentration proved to be completely sterile and those laid by the females treated with the lower concentration showed an emergence rate of 33.1%. The emergence rates in the controls were 45.9% (solvent control) and 63.2% (untreated control). Treatment of 3rd instar nymphs had no influence on fecundity.

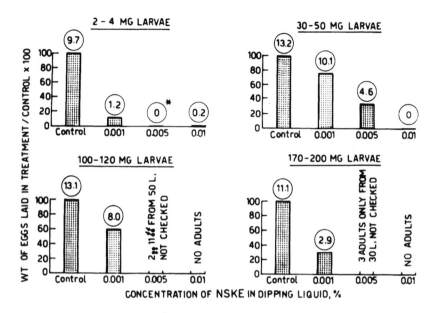

FIGURE 30. Weight of eggs laid by *Spodoptera littoralis* females derived from larvae of different weight classes fed on lucerne dipped into NSKE.

Reproduction and survival of adults of the hoppers, *Nilaparvata lugens, Sogatella furcifera,* and *Nephotettix virescens* were markedly reduced when the insects were caged on neem oil-treated rice plants (Figure 31). More than 6% neem oil was very effective but even 3% reduced fecundity to less than 50% of that of untreated females in all hopper species studied. Higher concentrations of neem oil prolonged the pre-ovipositional period from 2 to 8 days, thereby considerably reducing the oviposition period.[55]

Vicia faba plants were treated with a spray containing a concentration of 20 ppm/ℓ of the extract plus lecithin and sesame oil. First instar nymphs were placed on the host plants. The fecundity and longevity of adults derived from these nymphs were strongly affected as control females produced about 12 times more nymphs and their longevity lasted about 10 days longer than that of treated aphids.[56]

Summarizing neem and chinaberry effects on growth, fecundity, egg sterility, and longevity, it can be concluded that insects may show the following more or less typical reactions after uptake of the active principle(s) *per os*, via cuticle or after injection:

1. A nongustatory antifeedant effect ("secondary antifeedant effect") is observed in most treated insects which has to be separated from a gustatory one ("primary antifeedant effect")
2. The development of immature stages is delayed after treatment of feeding and non-feeding instars/stages
3. Reduced uptake of food due to antifeedant effects may contribute to the delay of postembryonic development but is often not the primary cause of it
4. High mortality may occur between molts, especially after uptake of higher concentrations of active principle(s), otherwise mortality mainly results from disturbances during ecdysis
5. Molts of all instars or stages during postembryonic development are influenced by disturbance of apolysis from the old skin and other processes, leading to various degrees of incomplete ecdysis and also appearance of so-called larval-pupal, nymphal-adult, and pupal-adult intermediates

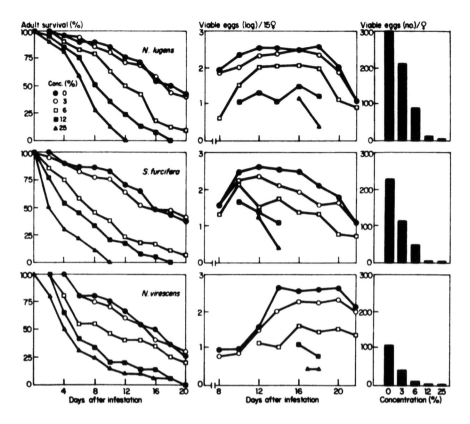

FIGURE 31.   Longevity and fecundity of *Nilaparvata lugens*, *Sogatella furcifera*, and *Nephotettix virescens* confined to neem oil-treated rice plants.[55]

6.   Malformations of wing sheaths in pupae and of wing disks in larvae or nymphs, sometimes are observed so-called "black bodies" in the thoracic region *(Epilachna varivestis)*

7.   Treated larval instars may become "permanent larvae" which are unable to molt due to destruction of their epidermis, probably as a consequence of disturbance of the hormonal system. They may live up to several weeks *(Epilachna varivestis, Locusta migratoria migratorioides, Manduca sexta)*

8.   No definite sensitive phase for IGR effects of neem ingredients exists. However, if the next molt falls during the next 48 hr or less after treatment often no disturbance of ecdysis is observed but the next molt may be negatively influenced

9.   Supernumerary larval instars occur very rarely *(Manduca sexta)* and are not viable

10.   Fecundity is decreased by uptake of active principles either by larvae or nymphs *(Spodoptera littoralis, Dysdercus fasciatus)* or by adults *(Epilachna varivestis, Leptinotarsa decemlineata)* but reduced uptake of food may also lead to decrease of fecundity

11.   Egg sterility is caused by uptake of active ingredients by adults *(Leptinotarsa decemlineata)* or last instar nymphs *(Dysdercus fasciatus)*; eggs laid by treated females may be very susceptible to fungal attack due to a defective chorion *(Epilachna varivestis)*

12.   Laid eggs show a low degree of sensitivity against neem products, if any

13.   Longevity of females and/or males may be shortened after application of higher concentrations of neem extracts or azadirachtin

14. Due to overlapping of antifeedant effects with IGR activities influencing the hormonal system, a complex of factors seems to be responsible for the disturbance of metamorphosis and reduction of fecundity

MODE OF ACTION

Influence on Metamorphosis

Histological Investigations

Histological studies in larvae of *Epilachna varivestis* based on feeding experiments revealed that azadirachtin treatment is followed by a destruction of the epidermis (Figure 32). There are two different processes of degradation of the epidermis which lead to the inability to secrete a new cuticle or to prevention of molting.[80,81] Larvae, treated shortly after molting, show 4 to 6 days later, a retraction of the epidermis from the overlaying cuticle for the most part. The cells remain in contact with each other but a strong flattening is typical. No mitotic activity is observed. At first only few cells degenerate, resulting in holes which are penetrated by hemocytes. The digestion of the epidermis takes place without exception from the apical side. Some days later the epidermis has disappeared.

The second mode of degeneration of the epidermis occurs after treatment of mid-instar larvae. In this case the epidermal cells start mitosis normally but disconnect from each other and divide. Most of them are unable to rebuild a continuous epithelium. Within these cells there is an increasing activity of autophagy, indicated by an increase of acid hydrolase production. Finally, the remnants of the epidermal cells are phagocytosed by hemocytes. Some larvae may develop a new but defective cuticle in several regions of their body. In these cases molting will be severely disturbed. It is striking that such tissues mainly affected are those whose cells show mitotic activities at certain times.

The degeneration of the imaginal wing disks, resulting in so-called "black bodies" is another reason for the appearance of malformed pupae and adults (Figure 20). Either the epithelium degenerates completely, especially after treatment early in the instar, or cells disconnect from each other and break down. Depending on the time of application and amount of active principle taken up by the larva, the wing disk lesions are more or less pronounced. Some disks may be affected, others not. If the lesions are small, they may be repaired by the insect's repair mechanism resulting in a nearly complete but deformed wing.[80]

In addition, various changes in the fat body, the central organ of metabolism, have been detected. In early treated 3rd instar larvae only a separation of the cells within the fat body lobes has been observed. These cells looked very poor and smaller than those of control larvae. There was no sign of glycogen or protein accumulation as in untreated individuals. Consequently, no building material for further tissue synthesis was available. After ripping up the basal membrane, the remnants of the fat cells dispersed through the hemolymph.

Endocrinological Investigations

First attempts to explain the mode of action of growth regulating substances from the neem tree in various species of Heteroptera and Lepidoptera were made in 1972. The effects observed were interpreted as either probably caused by an ecdysone analog[43] or by possible interference of azadirachtin as an ecdysone-like compound with the hormonal balance of insects.[45]

In experiments with 5th instar nymphs of *Oncopeltus fasciatus* topically applied azadirachtin affected both ecdysteroid production and decline of ecdysteroid titers before the attempted adult ecdysis. Ecdysteroid production was delayed and erratic, as determined by radioimmunoassay. The ecdysteroid titers, however, varied strikingly between sexes and also between individuals within the same sex. The similarity between the growth regulating

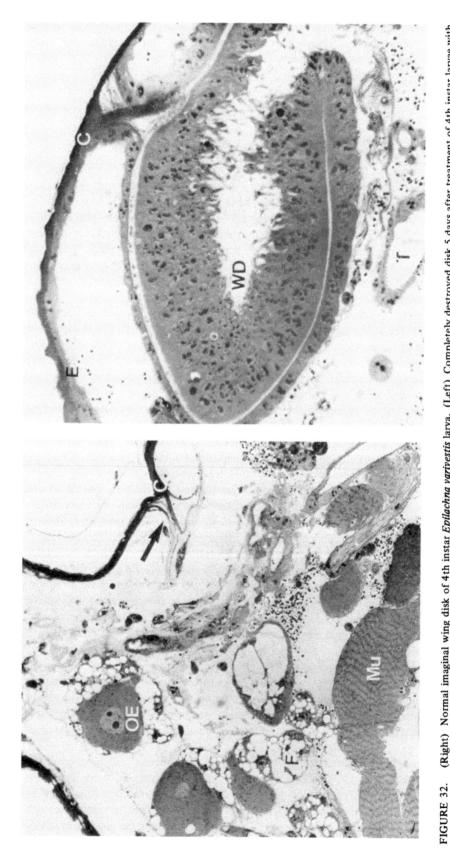

FIGURE 32.   (Right) Normal imaginal wing disk of 4th instar *Epilachna varivestis* larva. (Left) Completely destroyed disk 5 days after treatment of 4th instar larvae with azadirachtin. The epidermis is also totally destroyed. WD = wing disk, C = cuticle, E = epidermis, F = fat body, OE = oenocyte, Mu = muscle, T = trachea, ↘ = peduncle of wing disk.

effects of diflubenzuron (Dimilin) and azadirachtin suggested a blockage of chitin synthesis by azadirachtin. Diflubenzuron acts by blocking the terminal polymerization step in chitin formation. In addition, it was concluded that suppression of the feedback inactivation of ecdysteroids following their peak production and blocking of the suppression of eclosion hormone release could be some of the azadirachtin effects.

Detailed endocrinological investigations to enlighten the physiological mode of action of azadirachtin were carried out mainly by employing the African migratory locust, *Locusta migratoria migratorioides*, as test insects.[47,48]

The ecdysteroid titer of untreated 5th instar nymphs showed two distinct peaks (Figure 33). A small ecdyson peak built up on the 5th day after the preceding molt. Afterwards, the ecdysteroid titer increased steadily within hours, reaching a peak value of about 12 ng/ $\mu\ell$ hemolymph on the 7th day, 2 days before ecdysis, and then it decreased rapidly. Before ecdysis only small amounts of ecdysteroids were left. Nymphs injected with azadirachtin (2.5 μg/g), which completed synthesis of cuticle but were unable to molt on the 18th day after the last ecdysis, showed a maximum hemolymph ecdysteroid concentration similar to that of control nymphs on the 13th day. However, the gradual increase and decline of ecdysteroid concentration was much slower in the treated nymphs and a distinct peak was not detectable. Only small bursts of ecdysteroid concentration could be found in an azadirachtin-treated larva with complete inhibition of ecdysis. In other words, the ecdysteroid titers in *L. m. migratorioides* were either altered or abolished, showing a close correlation with morphogenetic effects.

The effect of azadirachtin on the molting hormone titer could be explained as an interference with the neuroendocrine system controlling ecdyson (prothoracotropic hormone) synthesis. Histological staining of the neurosecretory material from the brain of *L. m. migratorioides* revealed a high concentration of stainable material in the corpus cardiacum. It is assumed that the release of neuropeptides (prothoracotropic and allatotropic hormones) disrupts the control of insect metamorphosis and behavior on the level of the molting and the juvenile hormones (JH), finally leading to inhibition of ecdysis.

The appearance of a supernumerary 6th larval instar in *Manduca sexta* after uptake of a methanolic NSKE or azadirachtin is difficult to explain because such additional instars are normally not formed after application of neem products. The reason could be that the 5th instar larvae do not reach the critical weight of about 5 g, which is necessary to inactivate the corpora allata and consequently JH production in this species.[69] Treated larvae weighing more than 5 g molt into pupae, those between 3 and 5 g become prepupae, and others with a weight of less than 3 g molt into 6th instar larvae with pupal imaginal disks.[82]

Another hypothesis on the mode of action of azadirachtin was propounded recently.[83] Tissue of the blowfly, *Calliphora vicina*, was employed for studies of the role of ecdysteroid receptors in insect development. Ecdysteroid binding was observed in larval epidermis, fat body, brain (including several imaginal disks), and hemolymph. Azadirachtin reduced ponasterone A-binding to ecdysteroid receptors in vitro, a result leading to the suggestion that the steroid-like inhibitor apparently blocks the binding sites for the hormone and thus shows the properties of an anti-ecdysteroid.

Influence on Fecundity and Egg Fertility

Histological Investigations

The ovaries of females of *E. varivestis*, treated with a methanolic NSKE or azadirachtin, are much smaller than those of untreated beetles and also look transparent.[84,85] Ovarioles, in which oocytes are already developed, degenerate slowly and show a tendency of normal oocyte growth of the following younger oocytes, where no oocyte development has started. These differences are not azadirachtin-dose dependent.

The first cytopathological alterations take place especially in the germarium and are limited

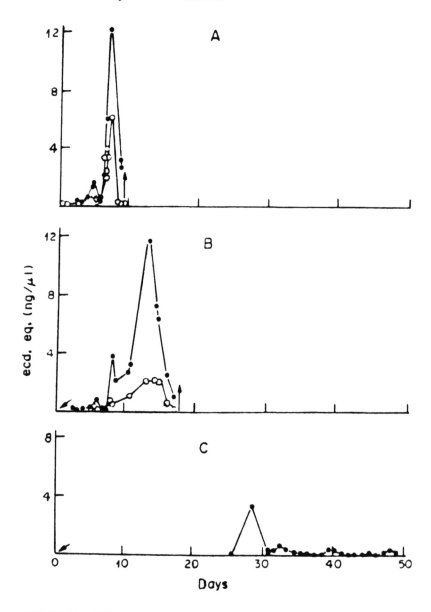

FIGURE 33.   Effect of azadirachtin injection (2.5 μg/g) on the total ecdysteroid (●) and
20-hydroxyecdysone (○) titer of 5th instar *Locusta migratoria migratorioides*. The results
are based on hemolymph measurements of individual nymphs staged with an accuracy of ±
hr. (A) Ecdysteroid titer of control nymphs, ↑: time of ecdysis. (B) Ecdysteroid titer of
azadirachtin-treated larvae with attempted ecdysis, ↑ . (C) Ecdysteroid titer of an azadirachtin-
treated nymph with complete molt inhibition, ↙: time of injection.

to localized areas of the follicular epithelium. In oocytes and surrounding prefollicle cells
mitochondria are first dissolved and vacuolized. Further pathological alterations are loss of
homogenity of the ooplasm of previtellogenic oocytes, encapsulation of the mitochondria
inside the ooplasm, formation of autophagic vacuoles, change of the shape of the nuclei of
oocytes, and partial destruction of the prefollicular epithelium (Figure 34).

In oocytes of beetles reaching the stage of vitellogenesis minimal yolk deposition is
observed. The follicular epithelium shows an irregular development and various stages of

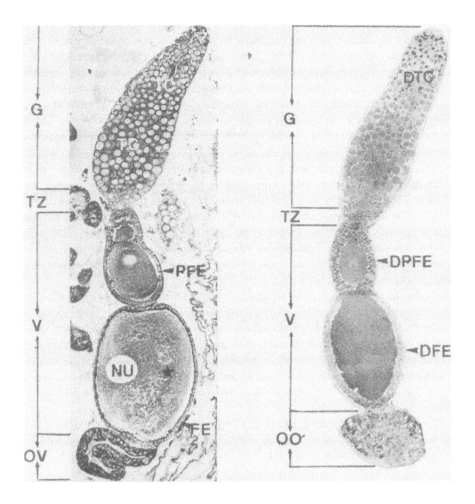

FIGURE 34. (Left) The ovariole of a 6-day-old untreated female of *Epilachna varivestis*. The trophocytes (TC) are well developed. Oogenesis takes place in the vitellarium (V). The terminal oocyte (*) is fully engaged in yolk deposition. FE = follicular epithelium; G = germarium; NU = nucleus; OV = oviduct; PFE = prefollicular epithelium; TZ = transition zone. (Magnification × 50) (Right) Ovariole of a 14-day-old treated beetle *(Epilachna varivestis)*. The apical trophocytes (DTC) in the germarium (G) are destroyed. The ovariole sheath is infolded and the prefollicular tissue (DPFE) and the follicular tissue (DFE) are degenerated. Proximal in the vitellarium (V) oosorption (OO) takes place. TZ = transition zone. (Magnification × 80)

destruction. Beside other pathological changes, the nucleus of oocytes is indistinct or totally lacking. Oocytes which have been partially destroyed and deformed as described above are dissolved in the egg chamber and resorbed. These effects are not reversible. Different reactions of females treated in different physiological states, e.g., before and after beginning of the oviposition period, indicate the existence of a sensitive period for azadirachtin effects.

Eggs which have been laid by treated females show different degrees of destruction of the characteristic chorionic surface.[85] Furthermore, the exochorion is perforated and the micropyles are often cemented. These defects may be caused by the degenerated follicular epithelium in the vitellarium and/or by an insufficient production of the chorion precursor material in the follicle cells.

After azadirachtin treatment of males total lysis of the apical testis cells is observed about 14 days after treatment.[85] Spermatids and spermatozoa are not or only slightly affected. However, these alterations could also be secondary effects.

Endocrinological Investigations

The influence of azadirachtin on the hormonal regulation of oogenesis was studied in *Locusta migratoria migratorioides*.[23,86,87] After azadirachtin injection between 2 and 10 days after the last molt, follicle growth was inhibited. This is explainable as a possible consequence of interference of the active principle with vitellogenin synthesis and/or with incorporation into the oocytes either directly or indirectly through endocrine control. Whereas in untreated adult females the JH titer in the hemolymph increases about 8 days after the last molt, inducing vitellogenin biosynthesis in the fat body and consequently oogenesis, azadirachtin injection prevents JH production, and therefore, no vitellogenin and no eggs are produced.

The concentration of ecdysteroids in ovaries of untreated females reached its maximum toward the end of vitellogenesis. After injection of azadirachtin at the end of oogenesis (10th to the 13th day after preceding molt) the total amount of ovarian ecdysteroids was drastically reduced (27.9 + 4.35 ng/mg compared to 6 + 15.6 ng/mg; n = 10). Ovarian ecdysteroid synthesis is probably under direct control of the brain hormone. Azadirachtin most probably reduced the total amount and concentration of ecdysteroids relative to untreated individuals by acting on the neuroendocrine system which may also explain the decrease in body weight and inhibition of oviposition.

On the other hand, histological investigations revealed histopathological alterations in the corpora allata and corpora cardiaca of *Epilachna varivestis*.[85] Therefore, the production of JH may be disturbed, leading to consequences regarding vitellogenin production and consequently fecundity. Young treated females show striking alterations in their protein spectrum in the hemolymph and generally do not produce vitellogenin. Females which are 8 days and older during treatment demonstrate a nearly total lack of vitellogenin in the ovaries and a great reduction of fractions in the hemolymph. In some cases vitellogenin is detected in the hemolymph but not in the ovaries.

Surprisingly, so-called "permanent nymphs" were able to synthesize vitellogenin 40 days after an azadirachtin injection.[23] Normally this compound is produced in adults only. Vitellogenin is under the control of the JH.

METABOLISM

The metabolism of azadirachtin A, hydrogenated to [22,23-^{3}H$_2$] dihydroazadirachtin and to the corresponding tritiated dihydroazadirachtin A was studied in *L. m. migratorioides*.[23] After injection, the bulk of radioactivity was excreted during the first 24 hr. Most of the excreted material proved to be unchanged dihydroazadirachtin A. Afterwards, this compound disappeared and more polar compounds were extracted. An analysis of Malpighian tubules 6 days after injection of the labeled compound yielded three main products which were not identical with the substance applied.

PRACTICAL APPLICATION

The pure compound azadirachtin cannot be applied against field or store pests on a commercial scale because its purification or synthesis would be complicated and therefore too expensive. Consequently, only extracts from the seeds, and perhaps to a certain extent also the leaves, have a real chance of being used in field and store pest control, preferably in those countries where neem and chinaberry trees are abundant (parts of southern and Southeast Asia, Africa, and the Americas).

For a long time, parts of neem and chinaberry have been used in a very simple way for pest control in stores in some areas of Asia (India, Pakistan) and Africa (Togo, Cameroon, etc.). A so-called "sandwich method" was often applied which means layers of stored products and neem or chinaberry leaves were put one after another in the store huts. However,

Table 9
DAMAGE (%) DUE TO *CALLOSOBRUCHUS MACULATUS* IN
LEGUMINOUS SEEDS MIXED WITH DIFFERENT QUANTITIES OF
NEEM KERNEL POWDER (NKP)[5]

Treatment g NKP/100 g seeds	Average pecentage[a] damage to			
	Mung beans (after 8 months)	Gram (after 11 mo.)	Cowpea (after 9 mo.)	Pea (after 9 mo.)
Control	58.97	90.00	90.00	63.30
0.5	18.90	31.13	30.13	31.23
1.0	11.93	6.07	16.40	15.20
2.0	9.77	2.33	3.13	19.00
S.E.M. ±	2.088	2.784	11.475	6.079
C.D. at 1%	9.90	13.21	54.44	28.84

[a] Values obtained by angular transformation.

more recent findings revealed that the effect of this measure against some major store pests was only moderate if not negligible. Some people in India try to protect leather goods and woolen clothes by using neem and chinaberry leaves.[5]

Numerous experiments in Asia (India, Pakistan) and Africa (Togo, Nigeria, Sudan) with ground NSKs, ground neem leaves, and neem oil mixed with stored products, resulted in satisfactory, sometimes excellent control of some important store pests.[5,9] Various species of moths, for instance *Ephestia kuehniella,* and bruchids, such as *Callosobruchus maculatus,* seem to be very sensitive to neem products. Mixing of cowpeas or green gram with ground neem seed kernels (1 to 4% w/w) or with 2 to 4 mℓ neem oil per kilogram legume seeds gave very good, long-lasting protection against *C. maculatus* in India and West Africa (Table 9).

Some store pests are less sensitive to neem products than other closely related species, for instance *Sitophilus zeamais* compared with *S. oryzae* and *C. chinensis* compared with *C. maculatus.*[88,89] In many cases it may be rather difficult to separate antifeedant and/or repellent effects from growth regulating and fecundity reducing effects of neem products because there is some overlapping. Especially under the conditions in stores all these factors may play a role of varying importance.

Field application of neem is practiced in India in tobacco nurseries. It seems to be increasingly popular among tobacco growers due to its cheapness and effectiveness.[90] In this case aqueous NSKEs are employed which are obtained by a very simple method of extraction: NSKs have to be crushed or ground and then put in a cloth which is placed in a kettle or pot filled with water. The water will extract the active principles overnight. On the next day the aqueous extract has to be filtered through a cloth or close-meshed sieve to remove bigger particles. After that the spray is ready for use in ordinary knapsack or other sprayers. Watering cans may also be used. The spray should be applied as soon as possible, otherwise bacteria may destroy the active ingredient(s) within a few hours or days. Normally about 100 or 50 g of ground or crushed NSKs are needed per liter of water,[91] but sometimes even 25 g/ℓ are sufficient if fresh seed kernels are available.

Numerous successful field trials with neem oil, chinaberry oil, neem and chinaberry seed cake, methanolic neem and chinaberry SKEs, and enriched NSKEs have been carried out mainly in Third World countries against a wide range of field pests (Orthoptera, Heteroptera, Homoptera, Hymenoptera, Lepidoptera, Coleoptera, and Diptera),[5] and it is hoped that the practical use of neem products will become more popular among farmers in the future.

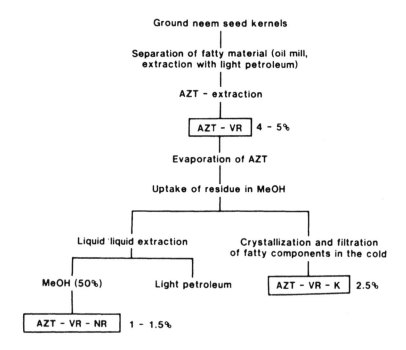

FIGURE 35. Flow diagram of a solid/solid extraction using azetropic mixture of methanol and tertiary butyl methyl ether (AZT) as extract solvent.[92]

However, this also depends on the establishment of small-scale industries in areas with abundant neem and chinaberry trees. Such neem and chinaberry factories are planned in several Asian countries (Burma, China).

Effective enriched formulated NSKEs have been developed in recent years by research workers and also by a private firm in the U.S. (Figure 35).[71,92-94] Neem oil and neem oil blended with enriched NSKEs are suitable for application with ULV sprayers (about 10 or 20/ha). These oils might be phytotoxic depending on concentration, weather conditions, and plant species treated. Due to sensitivity of azadirachtin to sunlight, the enriched and formulated neem seed kernel extracts must be sprayed at concentrations ranging from about 300 to 500 ppm/ℓ.[71] Under tropical conditions 1000 to 2000 ppm/ℓ may be necessary. Weekly applications are needed in case of continuous immigration of pests into treated fields.

The azadirachtin content of crude neem and chinaberry extracts can be estimated by application of a liquid chromatographic reversed-phase procedure. An estimation of the azadirachtin content is feasible through the use of an external azadirachtin standard and valley-to-valley integration.[95] A rough estimation for practical purposes is also possible by applying TLC (Figure 36).[92] Extraction of neem seed kernels with 95% ethanol gave the best yield of azadirachtin, followed by a methanol-H$_2$O mixture (85:15) (Table 10).

Neem seed cake, which proved to be very effective as a natural fertilizer in paddy fields, may also be used on a larger scale in coming years. It may show insecticidal effects by systemic action and reduction of ostracods, the worst enemies of nitrogen-fixing blue-green algae *(Anabaena* sp., *Nostoc* sp.), leading to a considerable increase of rice yields.[97]

Regarding the practical use of neem products against insect pests, farmers may be skeptical in the beginning due to a delayed lethal effect, especially if they are accustomed to synthetic pesticides with quick contact and knockdown effects. Therefore, farmers should be well informed about the slow mode of action by extension staff members when neem products are used in practice.

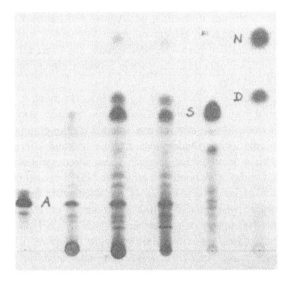

FIGURE 36.   Thin-layer chromatogram of NSKE in the following order (from left to right): azadirachtin, water, methanol,
AZT, salannin, and desacetyl-nimbin. A = azadirachtin, S =
salannin, D = desacetyl-nimbin, N = nimbin.

Table 10
AZADIRACHTIN CONTENT
DETERMINED BY HPLC IN NEEM SEED
KERNEL EXTRACTS PREPARED WITH
VARIOUS SOLVENTS[96]

Solvent	Azadirachtin (μg/10 mℓ)
Ethanol (95%)	2.80
Methanol-H$_2$O (85:15)	2.60
Methanol	2.19
Methylene chloride	1.73
Ethyl ether	1.28
Acetone	0.74

SIDE EFFECTS ON BENEFICIAL ARTHROPODS

Azadirachtin and other related triterpenoids from the neem and chinaberry trees exhibit
a broad insecticidal spectrum in larvae/nymphs and pupae due to growth regulating and other
effects which are observed in various orders of insects. Due to this broad spectrum of
efficacy, side effects on beneficial insects, especially on exposed larvae/nymphs of predators,
have to be taken into consideration.

Growth-regulating compounds, acting by disturbance of the hormonal system of arthropods, do not normally harm adult insects unless they are phytophagous. In the latter case,
they may reduce fecundity and longevity and increase egg sterility as already mentioned in
a preceding chapter.

Sparing of adult beneficial insects (parasites, predators, pollinators) and spiders is very
important from an ecological viewpoint in integrated pest control. Thus far, such selective
properties have been observed mainly or exclusively in growth-regulating natural and syn-

thetic compounds, for instance, juvenoids, ecdysoids, and benzoyl phenyl ureas, such as diflubenzuron. However, negative effects on fecundity and/or egg fertility may not be completely ruled out in adults of parasitic, predaceous, and other nonphytophagous arthropods if the active principles penetrate the cuticle (contact effect) or are taken up by consumption of contaminated prey.

Neem and chinaberry oil (conc. about 50 μg per insect) proved to be only slightly toxic to the predaceous mirid bug, *Cyrtorhinus lividipennis*, but nontoxic to the spider, *Lycosa pseudoannulata*.[98] In rice plots treated several times with the enriched NSKE-AZT-VR-K at 500 ppm/ℓ more predators *(C. lividipennis, Lycosa* sp., and *Microvelia douglasi atrolineata)* were counted than in untreated plots.[99] These three species are important natural enemies of rice pests. On the other hand, a methanolic NSKE caused by contact an increased mortality during molts of larvae and pupae of the coccinellid beetle *Coccinella septempunctata* in laboratory experiments.[100] Third instar larvae of the hover fly, *Episyrphus balteatus*, treated with 100 ppm/ℓ of an enriched NSKE (MTB/H$_2$O-VR-K) + sesame oil (ratio 1:4) died in most cases sometime after spraying, however, a few individuals were able to molt into adults. Adults of *Diaraetiella rapae* and *Ephedrus cerasicola*, which are important parasites of the green peach aphid, *Myzus persicae* in greenhouses, emerged normally after spraying of aphid mummies, containing larvae or pupae of the parasites, with an enriched methanolic NSKE (500 and 100 ppm/ℓ) + sesame oil (ratio 1:2).[101]

An interesting effect of a 20% aqueous neem seed kernel suspension was observed in the egg parasite *Telenomus remus* (Hymenoptera, Scelionidae).[102] Eggs of the noctuid moth, *Spodoptera litura*, were normally parasitized by *T. remus* after application of the neem seed kernel suspension. The treatment accelerated the development of the parasite whose adults hatched normally from the eggs of the host, but the longevity of these adults was significantly reduced. On the other hand, neem treatment of *S. litura* eggs after egg deposition by the parasite resulted in a prolonged development period, a normal emergence of the wasps, and an increased adult longevity (Table 11).

Neem oil and partially purified fractions of a methanolic NSKE did not inhibit development and emergence of internal solitary and gregarious parasites of the rice leaf folder, *Cnaphalocrocis medinalis*.[42,68] After using 50% neem oil sprays under field conditions even an increased rate of parasitization was observed. In sprayed plots 35% of the leaf folder larvae were parasitized which was twice that in the controls. This result was explained by the fact that spinning activity of the leaf folder larvae was considerably reduced by uptake of active neem oil principles. For this reason, the leafrolls could not be closed completely. Consequently, the parasites, exclusively Hymenoptera of various families, were able to trace their hosts easily and to deposit numerous eggs in them.

Larvae of honey bees proved to be sensitive to azadirachtin in laboratory experiments.[58] However, in field cages in which plots of flowering rape, mustard, and *Phacelia tanacetifolia* plants were repeatedly (two to three times) treated with a relatively high concentration (500 ppm/ℓ) of an enriched formulated NSKE (AZT-VR-K), typical growth-regulating effects were observed only in very small bee colonies consisting of a queen and about 200 to 250 workers.[59] In larger bee colonies consisting of a queen and about 2000 to 2500 workers no negative influence was detectable. In the affected small bee populations a few pupae died and numerous workers could not leave the cells in which they developed after they bit off the lid; those which succeeded showed malformed wings or parts of the pupal skin adhering to the abdomen. They disappeared after about 1 week from the colonies. The lack of any growth-regulating effect in the bigger bee populations may justify the suggestion that the active principles of neem were either strongly diluted and/or inactivated by enzymes of the bee workers.

Table 11
EFFECT OF 2% NEEM SEED KERNEL SUSPENSION ON PERCENT PARASITISM, DEVELOPMENT, AND LONGEVITY OF *TELENOMUS REMUS* IN *SPODOPTERA LITURA* EGGS[102]

	No. of *S. litura* eggs exposed		*T. remus* emergence (%)		Development period (days)		Adult period (days)	
	T	C	T	C	T	C	T	C
S. litura Eggs Treated Before Parasitization								
Mean	322.5	351.8	87.6	88.0	10.0	11.0	18.5	19.0
1. Calculated "t" value			0.3		3.0[a]		3.0	
Tabulated "t" value					2.3		2.3	
S. litura Eggs Treated After Parasitization								
Mean	236.7	89.7	90.9	11.2	10.0	21.0	20.0	
1. Calculated "t" value	1.4		0.6		9.0[b]			
Tabulated "t" value					3.2			
2. Calculated "t" value	12.2[b]		0.8		3.3[b]			
Tabulated "t" value	3.2				3.2			

Note: T = treatment, C = control. (1) Comparison between treatment and control. (2) Comparison between treatments.

[a] Significant at the 5% level.
[b] Significant at the 1% level.

RESISTANCE

At present no information is available on possible development of resistance of insect pests against azadirachtin or other neem products because they have not been applied in fields or stores for a long period and/or on a large scale. Theoretically it should be more difficult for insects to develop resistance against a mixture of compounds with perhaps different modes of action in comparison to a single more or less pure compound. However, the presence of a number of insects on neem trees indicates that specialized species may adapt after a certain period of time.

PESTS AND DISEASES OF NEEM AND CHINABERRY

Interestingly, the neem tree is attacked by a number of pests under natural conditions. Especially in its centers of origin (Burma, India, Indonesia) several insects and spider mites feed on different parts of the plant. Some species of scale insects, for instance *Aonidiella orientalis, Coccus hesperidum, Ceroplastes* sp., and *Pulvinaria* sp. have been recorded in Asia and Africa, but also various Lepidoptera and Coleoptera. In northern India, neem trees may be defoliated by large numbers of chafers (*Holotrichia consanguinea, Lachnosterna serrata,* and others),[101] and in Nicaragua and India young trees by the leaf-cutting ant *Acromyrmex octospinosus* and *Solenopsis* sp., respectively. Nematodes, such as *Xiphinema basiri,* attack the neem roots in the Sudan.[102] Fruit bats *(Eidolon helvum helvum)* and children suck the pulp of ripe fruit. Surprisingly, even the seed kernels are attacked by beetles, such as *Oryzaephilus surinamensis, O. acuminatus,* and *Rhyzopertha dominica,* which may even breed in them successfully.[103]

A disease causing a dieback of young shoots especially of older neem trees is common in Upper Burma.

Chinaberry trees are also attacked by a number of pests belonging to various orders of insects and spider mites.

PERSISTENCE

The chemical structure of azadirachtin confirms the presence of a dihydrofuran ring, a ketal group, and two tertiary hydroxyl groups which are all sensitive to strong acid conditions.[13] There are, in addition, four ester groups which are sensitive to alkali. Under neutral conditions azadirachtin proved to be quite stable. The persistence of the compound and perhaps of other insect growth-regulating triterpenoids from neem and chinaberry may, therefore, be affected by acid or alkaline conditions of the treated plant surfaces, in plants (systemic effect), and/or by UV light, air, and rainfall.

In trials for studies of the persistence of neem products, including azadirachtin, the larval feeding activity (expressed by weight gain) on neem-treated leaves in comparison to untreated ones served as criterion. A 0.3% aqueous NSKE sprayed on avocado leaves was highly active against *Boarmia selenaria* larvae when treated plants were kept in the laboratory or outside in the shade. On the oher hand, the activity was much less on treated plants kept in the sun.[70] An aqueous NSKE, sprayed on oak leaves in the field in a concentration of 0.5%, showed a fairly long residual effect.[31] Twelve days after spraying, the residues were still effective and prevented successful molting of 2nd instar larvae of *Lymantria dispar* despite heavy rainfall during the test period. In field experiments sugar beet, lucerne, and cotton leaves sprayed with various neem products (aqueous seed suspension, neem oil, neem extractive) were collected 1, 4, 6, 8, and 10 days after treatment and offered to larvae of the Egyptian cotton leafworm, *Spodoptera littoralis,* in the laboratory. Aqueous neem seed suspensions (0.2 and 0.6%) sprayed on sugar beet leaves showed the highest residual activity. Concentrations of 1% of the various products on sugar beet leaves 1 day after treatment gave the highest antifeedant activity and also gave the aqueous seed kernel suspension some activity for 8 days after treatment.[104] After uptake of a NSKE by roots of *Brassica rapa* plants following soil treatment, the active principles were effective in a purely systemic manner for about 10 days against the cabbage moth, *Plutella xylostella.*[40]

Exposure of neem oil-treated rice plants to sunlight for 2 to 4 days led to a degradation of the oil's activity.[42] In other experiments, exposure of azadirachtin for 16 days to sunlight resulted in destruction of the compound, which was confirmed by HPLC analyses.[95,96] More than 50% of the antifeedant potency against newly emerged 1st instar larvae of the fall armyworm, *Spodoptera frugiperda,* were lost after 7 days of exposure. The UV-absorbing additive *p*-aminobenzoic acid gave slight protection to azadirachtin from photodegradation. Various plant oils (neem, angelica, castor, and calamus) provided moderate protection (<25%) from degradation by sunlight. Continuous irradiation of an enriched NSKE in the laboratory with UV light led to loss of about 46% of the growth-regulating activity after 24 hr, expressed by the mortality rate of 4th instar larvae of the test insect *Plutella xylostella.* There was no further degradation during the next 2 days but after 4 days 56% of the original activity had vanished (Figure 37). Addition of the synergistic compound piperonyl butoxide to an enriched NSKE resulted in a slower reduction of the growth-regulating effect in larvae and pupae of *P. xylostella* under field conditions. This was found in trials with and without rainfall during the testing period.[105]

A purified fraction from an extract of whole neem fruit was sprayed in concentrations of 100 and 1000 ppm/ℓ on bean plants in two experiments. Then the plants were kept outdoors for 6 days and afterwards brought back to the laboratory to place 3rd instar larvae of *Epilachna varivestis* on them for 48 hr. Next the larvae were transferred to untreated bean leaves.

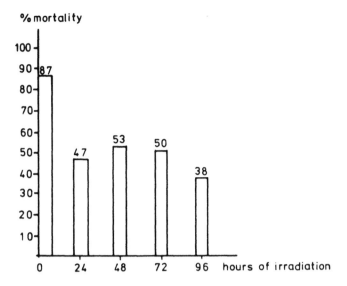

FIGURE 37.   Decrease of IGR activity of an enriched NSKE (AZT-VR-K) caused by permanent UV radiation.

During the first trial the temperature rose to 28 to 30°C and the period of sunshine per day was about 8 hr. Rainfall was relatively poor: 0.3 ℓ/m^2. In the second experiment the temperature reached 25°C, the period of sunshine per day was 5.2 hr, and the rainfall was 28.4 ℓ/m^2. According to the growth-regulating effects exhibited by the test insects the active principles in sprays containing 100 ppm were practically completely destroyed within 6 days.[39] On the other hand, larvae kept on beans treated with 1000 ppm molted into the 4th instar, during which time they suffered heavy mortality. Some surviving individuals molted into malformed or normal-looking adults.

The systemic effect of an enriched NSKE (AZT-VR-K) on plant and leaf hoppers of rice (*Nephotettix virescens, Nilaparvata lugens, Sogatella furcifera*) lasted at least 12 days after uptake of the active material by the roots of the treated rice plants in laboratory experiments.[55]

Solar radiation degraded the active principle(s) in an AZT-VR-K extract (200 ppm/ℓ) with the age of the residues to about 25% of the original effect. After application of 2000 ppm/ℓ, however, no reduction of the efficacy, expressed by mortality of *Spodoptera frugiperda* larvae, was observed after 1 week.[73]

SYNERGISM

Very few papers deal with the effect of some insecticide synergists when combined with neem extracts. The best synergistic effects were obtained with piprotal (Tropital) under laboratory conditions.[106] This compound increased the growth-regulating activity of a methanolic NSKE (ratio 2:1, 60 + 30 ppm/ℓ) in 4th instar larvae of the Mexican bean beetle, *Epilachna varivestis,* about sixfold. In 4th instar larvae and pupae of the cabbage moth, *Plutella xylostella,* piprotal and sesoxane improved the efficacy of the above-mentioned extract (ratio 2:1, 60 + 30 ppm/ℓ) approximately threefold. Mortality was observed primarily in the pupal stage.

Piperonyl butoxide, a synergist which is mainly used to increase the effect of pyrethrum, distinctly increased the activity of various crude and enriched NSKE under laboratory conditions. The ratio of synergist:extract was mostly 5:1 (100 + 20 ppm/ℓ) or 10:1 (40 + 4 ppm/ℓ). Test insects were Lepidoptera *(Plutella xylostella, Pieris brassicae)* and Coleoptera *(Leptinotarsa decemlineata).*[105]

Table 12
EFFECTS OF NEEM AND *BACILLUS THURINGIENSIS* (B. t.) MIXTURES ON 10-DAY-OLD *SPODOPTERA FRUGIPERDA* LARVAE[73]

Treatment	Mortality (%)	LT_{50}, days	LT_{100}, days	Av. pupal wt (mg)	Days to 50% pupation
Control	20.0	—	—	163	13
B. t. 0.1%	70.0	12	—	119[a]	14
Neem 10 ppm	83.8	12	—	93.6[a]	18[a]
Neem 10 ppm + B. t. 0.1%	100	5	15	—	—
Neem 20 ppm	100	11	15	—	—
Neem 20 ppm + B. t. 0.1%	100	4	8	—	—
Neem 50 ppm	100	7	12	—	—
Neem 50 ppm + B. t. 0.1%	100	4	8	—	—

[a]   Significant difference from control at $p = 0.01$.

Field experiments with combinations of piperonyl butoxide and an enriched, formulated NSKE (MTB-H$_2$O-VR) resulted in a limited improvement of the effect.[105] Larvae of *P. xylostella* proved to be more sensitive than those of *L. decemlineata*. The improvement was mainly expressed by an acceleration of the growth-regulating activity and to a lesser extent by increased mortality.

NSKE showed an improved effect against the store pests *Callosobruchus chinensis* and *Rhyzopertha dominica* when combined with oleoresin extracted from the flowers of *Chrysanthemum cinerariaefolium*.[107] A combination of an enriched NSKE (AZT-VR-K), incorporated into an artificial diet at a concentration of 10 ppm, with *Bacillus thuringiensis* (Dipel; 0.1%) gave 100% mortality of 10-day-old larvae of *Spodoptera frugiperda* whereas only 70% mortality occurred in control larvae treated with Dipel alone (Table 12). Neem alone (10 ppm) gave 83% mortality. The neem + *B. thuringiensis* combination also accelerated the lethal effect compared to that of both components alone. Combinations of enriched NSKE with lecithin and sesame oil improved the efficacy against the pea aphid *Acrythosiphon pisum* compared to NSKE alone under laboratory conditions.[56] The addition of wetting agents, such as Citowett, improves the effect of various neem products up to a certain extent.[32]

Combination of custard apple seed extract with neem seed extract gave a clear synergistic effect against the three household pests *Callosobruchus chinensis*, *Rhyzopertha dominica*, and *Musca domestica nebulo*.[107]

TOXICOLOGY

Reliable information on toxicity of neem and chinaberry extracts or azadirachtin is still rather scarce.

No oral toxicity of a methanolic NSKE was found in rats at 8.5 mg/kg, and no dermal toxicity or reaction of treated eyes could be detected in rabbits.[108] An enriched NSKE (AZT-VR-K) showed no oral toxicity at 5 mg/kg (rat).[108] Margosan A, an enriched and formulated NSKE, produced by Vickwood Ltd. (U.S.), proved to be a safe pesticide in various toxicological tests. Tests with the product Neemrich I, obtained from a fraction of neem oil, yielded the following LD_{50} (mg/kg) values: rat (dermal) 11.220, rat (oral) 8.705, mouse (oral) 6.760, pigeon (oral) 6.309, chicken (oral) 39,856, fish (exposure method) 37.5 ppm, and honey bee (topical) 73.5 mg/g. Neemrich I was also a nonirritant to skin and eyes (mucous membrane) of rabbits.[94] The LD_{50} of expressed neem oil to the red-winged blackbird *Agelaius phoeniceus* was >1000 mg/kg and that of extracted oil 1000 mg/kg.[96] Guppies tolerated 100 ppm/ℓ of an enriched NSKE (AZT-VR-K), whereas larvae and pupae of mosquitoes died at concentrations between 10 and 20 ppm/ℓ. On the other hand, neem oil extractive proved to be toxic to fish *(Gambusia* sp.) and tadpoles; in a concentration of 0.04% it caused 100% mortality within 1 to 2 days. Neem seed cake fed to rats for 9 months showed no carcinogenic effect.[109] The standard Ames test carried out with 100 to 3000 μg of azadirachtin showed no mutagenic activity on four strains of *Salmonella typhimurium.*[110]

No negative consequences are observed in drinking extracts from neem leaves and roots against feverish diseases, such as malaria. This is a widespread practice in West Africa. The same is true after drinking "tea" prepared from neem leaves in India. Young leaves and influorescenses from edible neem, *A. indica* var. *siamensis,* are widely utilized as some sort of vegetable in Burma and Thailand, apparently without negative consequences. The fruits of the edible neem tree seem to contain the same active principle(s) as those of the typical (Indian) neem tree.[111] Neem twigs are widely used as toothbrushes in Asian and African countries, probably because of an antibacterial effect. Extracts from neem bark are used in toothpastes sold on the European and Indian markets. Crushed neem seed kernels are employed to get children free of intestinal worms. Many parts of neem trees are used for medicinal purposes by Ayurveda medicine in India.

However, toxicological problems may arise if neem oil is pressed or extracted from neem seeds infested by fungi. In this case the oil may contain the toxic and highly carcinogenic aflatoxin.[112,113] The growth of fungi on neem seed kernels may be prevented by careful drying soon after harvest and cleaning, especially under the conditions of the humid tropics. Fungal contamination seems to play no role under arid conditions. Neem seeds should never be stored or transported in plastic bags or sacs which favor the growth of fungi. If fungal growth cannot be avoided, infested seeds should be selected and discarded before oil-pressing or extraction of oil and pesticides.

In contrast to neem, chinaberry is considered to be a toxic plant.[114] Recently, four toxic tetranortriterpenoids were identified in its seed kernels in Australia.[19] According to various records, the fruits of *M. azedarach* are toxic to livestock. Pigs are often mentioned as rather sensitive, but cases of poisoning of cattle, goats, sheep, and poultry have also been reported. Experimental feeding of the fruit to pigs and sheep established its toxicity as close to 0.5% of the animal's body weight.[19] In other experiments the approximate LD_{50} of the toxic principles (meliatoxins) for orally dosed pigs was found to be 6.4 mg/kg. Fruit from trees in some parts of New South Wales were found to be nontoxic whereas fruit from trees from other parts of New South Wales were toxic. Aborigines in Australia use leaf extracts of *M. azedarach* as a fish poison.

CONCLUSIONS

Undoubtedly, the neem and the chinaberry tree are highly promising sources of biologically active natural pesticides, especially of the triterpenoid azadirachtin, which is a potent antifeedant, egg laying deterrent, IGR, and sterilant at the same time. Due to the effects of the active principle(s) on all instars/stages of many insect pests with the exception of the egg stage, neem and chinaberry products have some important advantages compared with other IGR, such as JHs and their synthetic analogs or mimics. Despite their effects in many insect species from various orders, neem products are selective by sparing adults, which however may show a reduced fecundity under certain conditions.

For this and other reasons they may not be very harmful to many parasites and some predators and also rather harmless to honey bees. Most neem products seem to be not directly toxic but indirectly. Toxicity for warm-blooded animals including men seems to be very low, if not negligible.

Due to economic reasons, the pure compound azadirachtin cannot be used for practical pest control measures. The only way is the application of crude extracts or of differently formulated enriched products. Such enriched ''cocktails'' have already been developed and formulated in recent years; they gave good control of a wide range of insect pests in many field trials in tropical and temperate climates.[71] The application of neem oil, employing ULV sprayers, may also give satisfactory results, probably by a joint effect of antifeedant and IGR activities so that oil application is also feasible. Aqueous NSKEs are very cheap and easy to prepare, consequently they are very suitable especially for small subsistence farmers in Third World countries.

One of the main problems of neem application in the field is the sensitivity of the active principle(s) to UV light. The possibilities of increasing the half-life of azadirachtin are still limited. However, this problem can be partly overcome by application of higher doses, for instance 300 to 500 ppm/ℓ of enriched products, but this makes control measures more expensive. The same applies when synergists, such as piperonyl butoxide, are used for stabilization. Another problem of neem application is a psychological one, due to the delayed action of neem. Farmers used to conventional pesticides with a quick knockdown and toxic action may be skeptical if they cannot recognize a convincing result of their pest control measures soon after they have been carried out. However, this is a common problem of IGRs.[115]

The best way to utilize neem or chinaberry products for pest control and other purposes (fertilizer, feed for domestic animals, soap production, etc.) in the near future is perhaps the establishment of small cottage industries in areas where abundant neem and chinaberry trees occur, for instance in India, Burma, and southern China (chinaberry);[116] in other countries these trees should be planted in great numbers along roadsides or in agroforestry systems and exploited as soon as fruit production takes place.

It is hoped that neem products will soon become applicable, safe natural pesticides, first of all in developing countries where they should be used in a careful way primarily in integrated pest control programs to avoid or postpone the buildup of insect resistance. However, environmentally sound pesticides based on neem should also be suitable for industrialized countries, especially for home gardening.

REFERENCES

1. **Sláma, K. and Williams, C. M.,** ''Paperfactor'' as an inhibitor of embryonic development of the European bug, *Pyrrhocoris apterus, Nature (London),* 210, 329, 1966.
2. **Nakanishi, K., Koreeda, M., and Schooley, D. A.,** Recent studies on ecdysones, in *Invertebrate Endocrinology and Hormonal Heterophylly,* Burdette, W. J., Ed., Springer-Verlag, Basel, 1974, 204.
3. **Rees, H. H.,** Ecdysones, in *Aspects of Terpenoid Chemistry and Biochemistry,* Goodwin, T. W., Ed., Academic Press, London, 1971, 181.
4. **Bowers, W. S., Ohta, W., Cleere, J. S., and Marsella, P. A.,** Discovery of insect anti-juvenile hormones in plants. Plants yield a potential fourth-generation insecticide, *Science,* 193, 542, 1976.
5. **Anon.,** Neem in agriculture. II. Neem in pest control, *IARI Res. Bull.,* 40, 18, 1983.
6. **Pradhan, S., Jotwani, M. G., and Rai, B. K.,** The neem seed deterrent to locusts, *Indian Farming,* 12, 7, 1962.
7. **Volkonsky, M.,** Sur l'action acridifuge des extraits de feuilles de *Melia azedarach, Arch. Inst. Pasteur Alg.,* 15, 427, 1937.
8. **Butterworth, J. H. and Morgan, E. D.,** Isolation of a substance that suppresses feeding in locusts, *Chem. Commun.,* p. 23, 1968.
9. **Ketkar, C. M.,** Utilization of Neem *(Azadirachta indica* A. Juss) and its By-Products, Final Technical Report, Khadi & Village Industrial Commission, Bombay, 1976.

10. **Tirimanna, A. S. L.,** Surveying the chemical constituents of the neem leaf by a simple method: 2-dimensional thin layer chromatography, *Proc. 2nd. Int. Neem Conf.,* (Rauischholzhausen, 1983), 67, 1984.

11. **Pliske, T. E.,** The establishment of neem plantations in the American tropics, *Proc. 2nd Int. Neem Conf.,* (Rauischholzhausen, 1983), 521, 1984.

12. **Radwanski, S. A.,** Multiple land utilization in the tropics: an integrated approach with proposals for an international neem tree research and development program, *Proc. 1st Int. Neem Conf.,* (Rottach-Egern, 1980), 43, 1981.

13. **Morgan, E. D.,** Strategy in the isolation of insect control substances from plants, *Proc. 1st Int. Neem Conf.,* (Rottach-Egern, 1980), 43, 1981.

14. **Cramer, R.,** Neue Tetra- und Pentanortriterpenoide aus *Azadirachta indica* A. Juss (Meliaceae) mit insektenfraβhemmender Wirkung, Ph.D. thesis, University of Tübingen, West Germany, 1979.

15. **Kraus, W., Cramer, R., and Sawitzki, G.,** Tetranortriterpenoids from the seeds of *Azadirachta indica, Phytochemistry,* 20, 117, 1981.

16. **Schmutterer, H. and Zebitz, C. P. W.,** Standardized methanolic extracts from seeds of single neem trees of African and Asian origin: the effect on Coleoptera and Diptera, *Proc. 2nd Int. Neem Conf.,* (Rauischholzhausen, 1983), 83, 1984.

17. **Ermel, K., Pahlich, E., and Schmutterer, H.,** Comparison of the azadirachtin content of neem seeds from ecotypes of Asian and African origin, *Proc. 2nd Int. Neem Conf.,* (Rauischhozzhausen, 1983), 91, 1984.

18. **Morgan, E. D. and Thornton, M. D.,** Azadirachtin in the fruit of *Melia azedarach, Phytochemistry,* 12, 391, 1973.

19. **Oelrichs, P., Hill, M. W., Vallely, P. J., Macleod, J. K., and Molonski, T. F.,** Toxic tetranortriterpenes of the fruit of *Melia azedarach, Phytochemistry,* 22, 531, 1983.

20. **Mwangi, R. W.,** Locust antifeedant activity in fruits of *Melia volkensii. Ent. Exp. Appl.,* 32, 277, 1982.

21. **Butterworth, J. H., Morgan, E. D., and Percy, G. R.,** The structure of azadirachtin; the functional groups, *J. Chem. Soc. Perkin Trans.,* 1, 2445, 1972.

22. **Zanno, P. R., Muira, I., Nakanishi, K., and Elder, D. L.,** Structure of the insect phagorepellent azadirachtin. Application of PRFT/CWD carbon-13 nuclear magnetic resonance, *J. Am. Chem. Soc.,* 97, 1975, 1975.

23. **Butterworth, J. H. and Morgan, E. D.,** Investigation of the locust feeding inhibition of the seeds of the Neem tree *(Azadirachta indica), J. Insect Physiol.,* 17, 969, 1971.

24. **Uebel, E. C., Warthen, J. D., Jr., and Jacobson, M.,** Preparative reversed-phase liquid chromatographic isolation of azadirachtin from neem kernels, *J. Liquid Chromatogr.,* 2, 875, 1979.

25. **Rembold, H., Sharma, G. K., and Czoppelt, Ch.,** Growth-regulating activity of azadirachtin in two holometabolous insects, *Proc. 1st Int. Neem Conf.,* (Rottach-Egern, 1980), 121, 1981.

26. **Rembold, H., Forster, H., Czoppelt, Ch., Rao, P. J., and Sieber, K.-P.,** Azadirachtin, a group of insect growth regulators from the neem tree, *Azadirachta indica* A. Juss, *Proc. 2nd Int. Neem Conf.,* (Rauischholzhausen, 1983), 153, 1984.

27. **Forster, H.,** Isolierung von Azadirachtin aus Neem *(Azadirachta indica)* und radioaktive Markierung von Azadirachtin A, Diploma thesis, University of Munich, 1982.

28. **Rembold, H., Sharma, G. K., Czoppelt, Ch., and Schmutterer, H.,** Evidence of growth disruption in insects without feeding inhibition by neem seed fractions, *Z. Pflkrankh. Pflschutz,* 87, 290, 1980.

29. **Schmutterer, H. and Rembold, H.,** Zur Wirkung einiger Reinfraktionen aus Samen von *Azadirachta indica* auf Fraβaktivität und Metamorphose von *Epilachna varivestis* (Col., Coccinellidae), *Z. Angew. Entomol.,* 89, 179, 1981.

30. **McMillian, W. W., Bowman, M. C., Burton, R. L., Starks, K. I., and Wiseman, B. R.,** Extract of chinaberry leaf as feeding deterrent and growth retardant for larvae of the corn earworm and fall armyworm, *J. Econ. Entomol.,* 62, 70, 1969.

31. **Skatulla, U. and Meisner, J.,** Labor-Versuche mit Neemsamenextrakt zur Bekämpfung des Schwammspinners, *Lymantria dispar* L., *Anz. Schaedlingskd. Pflanz. Umweltschutz,* 48, 38, 1975.

32. **Steets, R.,** Die Wirkung von Rohextrakten aus den Meliaceen *Azadirachta indica* und *Melia azedarach* auf verschiedene Insektenarten, *Z. Angew. Entomol.,* 77, 306, 1976.

33. **Redfern, R. E., Kelly, T. J., Borkovec, A. B., and Hayes, D. K.,** Ecdysteroid titers and molting alterations in last stage *Oncopeltus* nymphs treated with insect growth regulators, *Pest. Biochem. Physiol.,* 18, 351, 1982.

34. **Ladd, T. L.,** The influence of azadirachtin on the growth and development of immature forms of the Japanese beetle, *Popillia japonica, Proc. 2nd Int. Neem Conf.,* (Rauischholzhausen, 1983), 425, 1984.

35. **Steets, R. and Schmutterer, H.,** Einfluβ von Azadirachtin auf die Lebensdauer und das Reproduktionsveermögen von *Epilachna varivestis* Muls. (Coleoptera, Coccinellidae), *Z. Pflankzenkr. Pflanzenschutz,* 82, 176, 1975.

36. **Steets, R.,** Zur Wirkung eines gereinigten Extraktes aus *Azadirachta indica* A. Juss auf *Leptinotarsa decemlineata* Say (Coleoptera, Chrysomelidae), *Z. Angew. Entomol.,* 82, 169, 1976/77.

37. **Ascher, K. R. S.**, Some physical (solubility) properties and biological (sterilant for *Epilachna varivestis* females) effects of dried methanolic neem *(Azadirachta indica)* seed kernel extract, *Proc. 1st Int. Neem Conf.*, (Rottach-Egern, 1980), 63, 1981.

38. **Ascher, K. R. S., Eliyahu, M., Nemny, N. E., and Meisner, J.**, Neem seed kernel extract as an inhibitor of growth and fecundity in *Spodoptera littoralis*, *Proc. 2nd Int. Neem Conf.*, (Rauischholzhausen, 1983), 331, 1984.

39. **Steets, R.**, Zur Wirkung von Inhaltsstoffen aus Meliaceen und Anacardiaceen auf Coleopteren und Lepidopteren, Ph.D. thesis, University of Giessen, W. Germany, 1976.

40. **Mong Ting Tan and Sudderuddin, K. I.**, Effects of a neem tree *(Azadirachta indica)* extract on diamondback moth *(Plutella xylostella* L.), *Mal. Appl. Biol.*, 7, 1, 1978.

41. **Ochse, L.**, Zur Wirkung von Rohextrakten aus Samen des Neembaumes *(Azadirachta indica* A. Juss) auf die Baumwollrotwanze *(Dysdercus fasciatus* Sign.), Dipl. thesis, University of Giessen, W. Germany, 1981.

42. **Saxena, R. C., Waldbauer, G. P., Liquido, N. J., and Puma, B. C.**, Effects of neem seed oil on the rice leaf folder, *Cnaphalocrocis medinalis*, *Proc. 1st Int. Neem Conf.*, (Rottach-Egern, 1980), 189, 1981.

43. **Leuschner, K.**, Effect of an unknown substance on a shield bug, *Naturwissenschaften*, 59, 217, 1972.

44. **Leuschner, K.**, Wirkung von Juvenilhormon-Analogen und Phytoecdysoiden auf Entwicklung, Fortpflanzung, Paarungsverhalten und Eiparasiten der ostafrikanischen Kaffeewanzen *Antestiopsis orbitalis bechuana* Kirk. und *Antestiopsis orbitalis ghesquierei* Car. (Heteroptera:Pentatomidae), Ph.D. thesis, University of Giessen, W. Germany, 1974.

45. **Ruscoe, C. N. E.**, Growth disruption effects of an insect antifeedant, *Nature (London) New Biol.*, 236, 159, 1972.

46. **Quadri, S. H. and Narsaiah, J.**, Effect of azadirachtin on the moulting process of last instar nymphs of *Periplaneta americana*, *Indian J. Exp. Biol.*, 16, 141, 1978.

47. **Sieber, K.-P. and Rembold, H.**, The effects of azadirachtin on the endocrine control of moulting in *Locusta migratoria*, *J. Insect Physiol.*, 29, 523, 1983.

48. **Rembold, H., Forster, H., Czoppelt, Ch., and Sieber, K.-P.**, Studies on the mode of action of azadirachtin, *Proc. 2nd Int. Neem Conf.*, (Rauischholzhausen, 1983), 153, 1984.

49. **Warthen, J. D. and Uebel, E. C.**, Effect of azadirachtin on house crickets, *Acheta domesticus*, *Proc. 1st Int. Neem Conf.*, (Rottach-Egern, 1980), 137, 1981.

50. **Abraham, C. C. and Ambika, B.**, Effect of leaf and kernel extracts of neem on moulting and vitellogenesis in *Dysdercus cingulatus* Fabr. (Heteroptera, Pyrrhocoridae), *Curr. Sci.*, 48, 554, 1979.

51. **Koul, O.**, Azadirachtin: I-interaction with the development of red cotton bugs, *Ent. Exp. Appl.*, 36, 85, 1984.

52. **Schmutterer, H.**, Some properties of components of the neem tree *(Azadirachta indica)* and their use in pest control in developing countries, *Meded. Landbouwhogesch. Rijksuniv. Gent*, 46, 39, 1981.

53. **Redfern, R. E., Warthen, J. D., Jr., Uebel, E. C., and Mills, G. D., Jr.**, The antifeedant and growth-disrupting effects of azadirachtin on *Spodoptera frugiperda* and *Oncopeltus fasciatus*, *Proc. 1st Int. Neem Conf.*, (Rottach-Egern, 1980), 121, 1981.

54. **Redfern, R. E., Warthen, J. D., Jr., Mills, G. D., and Uebel, E. C.**, Molting inhibitory effects of azadirachtin on large milkweed bug, Agric. Res. Results No. ARR-NE-5, U.S. Department of Agriculture, Washington, D.C., 1979.

55. **von der Heyde, J., Saxena, R. C., and Schmutterer, H.**, Neem oil and neem extracts as potential insecticides for control of homopterous rice pests, *Proc. 2nd Int. Neem Conf.*, (Rauischholzhausen, 1983), 377, 1984.

56. **Schauer, M.**, Effects of variously formulated neem seed extracts on *Acyrthosiphon pisum* and *Aphis fabae* (Homoptera:Aphididae), *Proc. 2nd Int. Neem Conf.*, (Rauischholzhausen, 1983), 141, 1984.

57. **Saxena, R. C., Liquido, N. J., and Justo, H. D.**, Neem seed oil, a potential antifeedant for the control of the rice brown planthopper, *Nilaparvata lugens*, *Proc. 1st Int. Neem Conf.*, (Rottach-Egern, 1980), 171, 1981.

58. **Rembold, H., Sharma, G. K., and Czoppelt, Ch.**, Growth-regulating activity of azadirachtin in two holometabolous insects, *Proc. 1st Int. Neem Conf.*, (Rottach-Egern, 1980), 121, 1981.

59. **Schmutterer, H. and Holst, H.**, unpublished data.

60. **Speckbacher, U.**, Zur Wirkung von Inhaltsstoffen aus *Azadirachta indica* A. Juss (Meliaceae) auf einige Insektenarten, M.Sc. thesis, University of Munich, 1977.

61. **Meisner, J., Kehat, M., Zur, M., and Eizick, N. F.**, Response of *Earias insulana* larvae to neem *(Azadirachta indica* A. Juss) kernel extract, *Phytoparasitica*, 6, 85, 1978.

62. **Webb, R. E., Hinebaugh, M. A., Lindquist, R. K., and Jacobson, M.**, Evaluation of aqueous solution of neem seed extract against *Liriomyza sativae* and *L. trifolii* (Diptera:Agromyzidae), *J. Econ. Entomol.*, 76, 357, 1983.

63. **Gill, J. S. and Lewis, G. T.**, Systemic action of an insect feeding deterrent, *Nature (London)*, 232, 402, 1971.

64. **Ascher, K. R. S., Eliyahu, M., Nemny, N., and Meisner, J.,** Neem extracts as inhibitor of growth and fecundity in *Spodoptera littoralis, Proc. 2nd Int. Neem Conf.,* (Rauischholzhausen, 1983), 331, 1984.

65. **Meisner, J. and Ascher, K. R. S.,** Insect growth regulating (IGR) effects of neem products on *Spodoptera littoralis, Proc. 2nd Int. Neem Conf.,* (Rauischholzhausen, 1983), 345, 1984.

66. **Guyar, G. T. and Methrotra, K. N.,** Juvenilizing effect of azadirachtin on a noctuid moth, *Spodoptera litura* Fabr., *Indian J. Exp. Biol.,* 21, 292, 1983.

67. **Maurer, G.,** Effect of a methanolic extract of neem seed kernels on the metamorphosis of *Ephestia kuehniella, Proc. 2nd Int. Neem Conf.,* (Rauischholzhausen, 1983), 365, 1984.

68. **Schmutterer, H., Saxena, R. C., and von der Heyde, J.,** Morphogenetic effects of some partially-purified fractions and methanolic extracts of neem seeds on *Mythimna separata* (Walker) and *Cnaphalocrocis medinalis* (Guenée), *Z. Angew. Entomol.,* 95, 230, 1983.

69. **Haasler, C.,** Effects of neem seed extract on the post-embryonic development of the tobacco hornworm, *Manduca sexta* (Joh.) (Lepidoptera:Sphingidae), *Proc. 2nd Int. Neem Conf.,* (Rauischhoizhausen, 1983), 321, 1984.

70. **Meisner, J., Wysoki, M., and Ascher, K. R. S.,** The residual effect of some products from neem *(Azadirachta indica* A. Juss) seeds upon larvae of *Boarmia (Ascotis) selenaria* (Schiff.) in laboratory trials, *Phytoparasitica,* 4, 185, 1976.

71. **Feuerhake, K. and Schmutterer, H.,** Einfache Verfahreen zur Gewinnung und Formulierung von Neem-samenextrakten und deren Wirkung auf verschiedene Schadinsekten, *Z. Pflanzenkr. Pflanzenschutz,* 89, 737, 1982.

72. **Kubo, I. and Klocke, J. A.,** Azadirachtin, insect ecdysis inhibitor, *Agric. Biol. Chem.,* 46, 1951, 1982.

73. **Hellpap, C.,** Effects of neem kernel extracts on the fall armyworm, *Spodoptera frugiperda, Proc. 2nd Int. Neem Conf.,* (Rauischholzhausen, 1983), 353, 1984.

74. **Chiu Shin-Foon,** Recent research findings on Meliaceae and other promising botanical insecticides in China, *Z. Pflkranzenkr. Pflanzenschutz,* 92, 310, 1985.

75. **Fagoonee, I. and Laugé, G.,** Noxious effects of neem extracts on *Crocidolomia binotalis, Phytoparasitica,* 9, 111, 1981.

76. **Steffens, R. and Schmutterer, H.,** The effect of a crude methanolic neem *(Azadirachta indica)* seed kernel extract on metamorphosis and quality of adults of the mediterranean fruit fly, *Certitis capitata* (Diptera, Tephritidae), *Z. Angew. Entomol.,* 14, 98, 1982.

77. **Stein, U.,** The Potential of Neem for Inclusion in a Pest Management Program for the Control of *Liriomyza trifolii* on Chrysanthemum, M.Sc. thesis, University of California, Berkeley, 1984.

78. **Zebitz, C. P. W.,** Effect of some crude and azadirachtin-enriched neem *(Azadirachta indica)* seed kernel extracts on larvae of *Aedes aegypti, Entomol. Exp. Appl.,* 35, 11, 1984.

79. **Schmidt, G. H.,** unpublished data.

80. **Schlüter, U.,** Histological observations on the phenomenon of black legs and thoracic spots: effects of pure fractions of neem kernel extracts on *Epilachna varivestis, Proc. 1st Int. Neem Conf.,* (Rottach-Egern, 1980), 97, 1981.

81. **Schlüter, U. and Schulz, W.-D.,** Structural damages caused by neem in *Epilachna varivestis.* A summary of histological and ultrastructural data. I. Tissues affected in larvae, *Proc. 2nd Int. Neem Conf.,* (Rauischholzhausen, 1983), 227, 1984.

82. **Bidmon, H.-J.,** Entwicklung und ultrastrukturelle Veränderungen der Protorakaldrüsen unbehandelter und mit Azadirachtin behandelter *Manduca sexta* L. (Lepidoptera, Sphingidae), Dipl. thesis, University of Giessen, W. Germany, 1984.

83. **Käuser, G. and Koolman, J.,** Ecdysteroid receptors in tissues of the blowfly *Calliphora vicina,* Abstr. Int. Cong. Invertebrate Reproduction, Tübingen, W. Germany, 1983.

84. **Schulz, W.-D.,** Pathological alterations in the ovaries of *Epilachna varivestis* induced by an extract from neem kernels, *Proc. 1st Int. Neem Conf.,* (Rottach-Egern, 1980), 81, 1981.

85. **Schulz, W.-D. and Schlüter, U.,** Structural damages caused by neem in *Epilachna varivestis.* A summary of histological and ultrastructural data. II. Tissues affected in adults, *Proc. 2nd Int. Neem Conf.,* (Rauischholzhausen, 1983), 237, 1984.

86. **Rembold, H. and Sieber, K.-P.,** Effect of azadirachtin on oocyste development in *Locusta migratoria migratorioides, Proc. 1st Int. Neem Conf.,* (Rottach-Egern, 1980), 75, 1981.

87. **Rembold, H. and Sieber, K.-P.,** Effect of azadirachtin on oocyste development in *Locusta migratoria migratorioides* (R. + F.), *Z. Naturforsch.,* 36C, 466, 1981.

88. **Pereira, J. and Wohlgemuth, R.,** Neem *(Azadirachta indica* A. Juss) of West African origin as a protectant of stored maize, *Z. Angew. Entomol.,* 94, 208, 1982.

89. **Yadav, T. D.,** Studies on the Insecticidal Treatment Against Bruchids, *Callosobruchus maculatus* (Fab.) and *C. chinensis* (Linn.) Damaging Stored Leguminous Seeds, Ph.D. thesis, Agra University, Agra, India, 1973.

90. **Joshi, B. G., Ramaprasad, G., and Sitaramaiah, S.,** Neem seed kernel suspension protects tobacco nurseries, *Indian Farm.,* 28, 1, 1978.

91. **Dreyer, M.,** Effects of aqueous neem seed extracts and neem oil on the main pests of courgette, *Cucurbita pepo* L., in Togo, *Proc. 2nd Int. Neem Conf.,* (Rauischholzhausen, 1983), 435, 1984.

92. **Feuerhake, K.,** Effectiveness and selectivity of various solvents for the extraction of neem seed components with insecticidal activity put to practical use, *Proc. 2nd Int. Neem Conf.,* (Rauischholzhausen, 1983), 103, 1984.

93. **Sharma, H. C., Leuschner, K., Sankaram, A. V. B., Gunasekmar, D., Marthandamurthi, M., Bhaskariam, K., and Subramanyam, M.,** Insect antifeedant and growth regulators for *Mythimna separata* from fractions isolated from *Azadirachta indica* extracts, *Proc. 2nd Int. Neem Conf.,* (Rauischholzhausen, 1983), 291, 1984.

94. **Sharma, R. N., Nagasampagi, B. A., Bhosale, A. S., Kulkarni, M. M., and Tungikar, K. B.,** "Neemrich": the concept of enriched particular fractions from neem for behavioral and physiological control of insects, *Proc. 2nd Int. Neem Conf.,* (Rauischholzhausen, 1983), 115, 1984.

95. **Stokes, J. B. and Redfern, R. E.,** Effect of sunlight on azadirachtin: antifeedant potency, *J. Environ. Sci. Health,* A17, 57, 1982.

96. **Jacobson, M., Stokes, J. B., Warthen, J. D., Jr., Redfern, R. E., Reed, D. K., Webb, R. E., and Telek, L.,** Neem research in the U.S. Department of Agriculture: an update, *Proc. 2nd Int. Neem Conf.,* (Rauischholzhausen, 1983), 31, 1984.

97. **Grant, I. F.,** Increasing biological nitrogen fixation in flooded rice using neem *(Azadirachta indica), Proc. 2nd Int. Neem Conf.,* (Rauischholzhausen, 1983), 493, 1984.

98. **Saxena, R. C., Epino, P. B., Tu Cheng-Wen, and Puma, B. C.,** Neem, chinaberry, custard apple: antifeedant and insecticidal effects of seed oils on leafhopper pests of rice, *Proc. 2nd Int. Neem Conf.,* (Rauischholzhausen, 1983), 403, 1984.

99. **von der Heyde, J.,** Zur Wirkung von Niemprodukten auf Reiszikaden unter Labor-, Gewächshaus- und Feldbedingungen in den Philippinen, Ph.D. thesis, University of Giessen, W. Germany, 1985.

100. **Schmutterer, H.,** Ten years of neem research in the Federal Republic of Germany, *Proc. 1st Int. Neem Conf.,* (Rottach-Egern, 1980), 21, 1981.

101. **Schauer, M.,** unpublished data.

102. **Joshi, B. G., Ramaprasad, G., and Sitaramaiah, S.,** Effect of neem kernel suspension on *Telenomus remus,* an egg parasite of *Spodoptera litura, Phytoparasitica,* 10, 61, 1982.

103. **Sarup, P. and Srivastava, V. S.,** Observations on the damage of neem *(Azadirachta indica* A. Juss) seed kernel in storage by various pests and efficacy of the damaged kernel as an antifeedant against the desert locust *Schistocerca gregaria* Forsk., *Indian J. Entomol.,* 33, 228, 1971.

104. **Meisner, J., Ascher, K. R. S., and Aly, R.,** The residual effects of some products of neem seeds on larvae of *Spodoptera littoralis* in laboratory and field trials, *Proc. 1st Int. Neem Conf.,* (Rottach-Egern, 1980), 157, 1981.

105. **Lange, W.,** Piperonyl butoxide: synergistic effects on different neem seed extracts and influence on the degradation by UV light of an enriched extract, *Proc. 2nd Int. Neem Conf.,* (Rauischholzhausen, 1983), 129, 1984.

106. **Lange, W. and Schmutterer, H.,** Versuche mit Synergisten zur Steigerung der metamorphosestörenden Wirkung eines methanolischen Rohextraktes aus Samen des Niembaumes *(Azadirachta indica), Z. Pflanzenkr. Pflanzenschutz,* 89, 258, 1982.

107. **Quadri, S. S. and Rao, B. B.,** Effect of combining some indigenous plant seed extracts against household insects, *Pesticides,* 11, 21, 1977.

108. **Schmutterer, H.,** Neem research in the Federal Republic of Germany since the First International Neem Conference, *Proc. 2nd Int. Neem Conf.,* (Rauischholzhausen, 1983), 21, 1984.

109. **Sardeshpande, P. D.,** Carcinogenic potency of neem seed cake in rat, vide letter 423/3/Pathological Research/76, Department of Pathology, Bombay Veterinary College, Parel, Bombay, 1976.

110. **Jacobson, M.,** Neem research in the U.S. Department of Agriculture: chemical, biological and cultural aspects, *Proc. 1st Int. Neem Conf.,* (Rottach-Egern, 1980), 33, 1981.

111. **Sombatsiri, K. and Tigvattanont, S.,** Effect of neem extracts on some insect pests of Thailand, *Proc. 2nd Int. Neem Conf.,* (Rauischholzhausen, 1983), 95, 1984.

112. **Sundaravalli, A. B., Bhaskar Raju, B., and Krishnamoorthy, K. A.,** Neem oil poisoning, *Indian J. Pediat.,* 49, 357, 1952.

113. **Sinnjah, D., Varghese, G., Baskaran, G., and Koo, S. H.,** Fungal flora of neem *(Azadirachta indica)* seeds and neem oil toxicity, *Malays. Appl. Biol.,* 12, 1, 1983.

114. **Kingsbury, J. M.,** *Poisonous Plants of the United States and Canada,* Pretice-Hall, Englewood Cliffs, N.J., 1964.

115. **Staal, G. B.,** Insect growth regulators with juvenile hormone activity, *Ann. Rev. Entomol.,* 20, 417, 1975.

116. **Chiu Shin-Foon,** unpublished data.

117. **Kraus, H., Bekel, M., Klenk, A., and Pöhnl, H.,** The structure of azadirachtin and 22, 23-dihydro-23β-methoxyazadirachtin, *Tetrahedron Lett.,* 26, 6435, 1985.

INDEX

A

Abies balsamea, 4
Abscisic acid, 2, 4
Acalymma vittatum, 86, 106
Acarina, 1
1-Acetoxy-2-hydroxy-4-oxoheneicosa-12,15-diene,
    55—57
1-Acetoxy-2-hydroxy-4-oxoheneicosane, 55, 57
2'-Acetylneriifolin, 86, 87
Acheta domesticus, 36, 127—128
Achillea millefolium, 82
Acorus calamus, 4
Acromyrmex octospinosus, 161
Acrythosiphon pisum, 11, 104, 131, 139, 146, 164
Actidione, see Cycloheximide
Actinomycin D, see Dactinomycin
Aculus cornutus, 8
Adoxophyes orana, 11
Aedes
    aegypti
        azarachtidin in, 145—146
        compounds affecting growth and development
            of, 28, 32, 52
        compounds affecting mortality of, 84, 90, 101
        compounds affecting sterility of, 4, 8, 11
    triseriatus, 82
Affinin, see Spilanthol
Aflatoxin, 6, 8, 165
Agelaius phoeniceus, 165
Agelastica alni, 139, 146—147
Ageratum houstonianum, 4, 119
Agrotis ipsilon, 24, 52
Alanosine, 6, 8
Albizia julibrissin, 24
Alkaloids
    chemical data for, 42, 76—78
    effect on insect growth and development, 41
    effect on insect mortality, 68, 73—75
    structure of, 40, 72—73
Alkylating agents, 1
Allelochemics, see also specific types, 21—57
    alkaloids, 41—42
    amino acids and amines, 24—25
    benzoic acid, 52—54
    benzoxazolinones, 48—49
    benzyl alcohol, 52—54
    enzyme inhibitors, 49
    flavonoids, 46—47
    gossypol, 38—39
    phenolics, 52—54
    phenylpropane, 28—29
    tannins and lignins, 43
    terpenoids, 32—33, 36—37
Allium sativum, 104
Allylglucosinolate, 55—57
Amino acid analogs
    chemical data for, 25, 80

effect on insect growth and development, 24
effect on insect mortality, 68, 79
structure of, 23, 79
p-Aminobenzoic acid, 162
γ-Aminobutyric acid, 12
Amphotericin A, 6, 8
Anabaena, 158
Anabasine, 72, 75, 77
Anacridium melanorhodon, 21
Anasa tristis, 74
Anastrepha
    ludens, 5, 8—9, 11
    suspensa, 104
trans-Anethole, 101, 102
Anethum, 32
    graveolus, 90, 101
Angalin, see *N*-Methyltyrosine
Angeliastica alni, 99
Angelica oil, 162
Angelicin, 26, 28, 29
Animals, chemosterilants from, 11
Anopheles quadrimaculatus, 82
Antestiopsis orbitalis bechuana, 127—129
Anthonomus
    grandis, 4—5, 8—9, 24, 43
    grandis grandis, 12
    grandis thuberiae, 90
Anthramycin, 6, 8, 9, 12
Anthrenus flavipes, 4
Antibiotics, 71
Antifeedant, 21, 67, 119, 149
Anuraphis rosae, 99
Aonidiella orientalis, 161
Aphidus phorodontis, 75
Aphis
    craccivora, 9
    fabae, 8—9
    forbesii, 74
    gossypii, 74, 90, 106
    persicae-niger, 99
    pomi, 8, 99
    rumicis, 74—75, 84, 98—99
Apiol, 101, 102
Apis mellifera, 11, 74—75, 98, 132, 134—135
Areca, 74
    semen, 74
Arecoline, 72, 74, 77
Aristolochia bracteata, 4
Aristolochic acid, 2, 4, 9
β-Asarone, 2, 4, 101, 102
Ascorbic acid, 2, 4, 9
L-Asparaginase, 11
Aspergillus
    flavus, 8
    parasiticus, 8
Astilbin, 44, 46, 47
Astragalus, 24
    radix, 79

Atranorin, 51, 52, 54
Atropine, 72, 74, 76
Atteva fabriciella, 5
Aulocara elliotti, 4, 8
Aureomycin, see Chlortetracycline
Avermectin B₁, 6, 8, 12
Azadirachta indica, see Neem tree
Azadirachtin, 119
    antifeedant activity of, 119
    bioassay of, 125—126
    chemical data for, 37
    content determination, 158—159
    effect on Coleoptera, 132—139, 146
    effect on Diptera, 145—146
    effect on egg fertility, 146—151, 153—156
    effect on embryonic development, 126—127
    effect on fecundity, 146—151, 153—156
    effect on Heteroptera, 128—131, 146
    effect on Homoptera, 131—132, 146
    effect on Hymenoptera, 132
    effect on insect growth and development, 36
    effect on Lepidoptera, 139—146
    effect on longevity, 146—151
    effect on metabolism, 156
    effect on metamorphosis, 151—153
    effect on Orthoptera, 127—128
    effect on postembryonic development, 127—146
    isolation of, 125—126
    mode of action of, 151—156
    persistence of, 162—163, 166
    practical application of, 156—159
    resistance to, 161
    side effects on beneficial arthropods, 159—161
    structure of, 35, 123—125
    synergisms with other insecticides, 163—164
    toxicology of, 164—165
    yield from neem seeds, 121
Azetidine-2-carboxylic acid, 79, 80
Aziridines, 1

B

Bacillus
    brevis, 9
    thuringiensis, 164
Bacteria, 1
Bambermycin, 8
Bead tree, see Chinaberry tree
Behavioral effects, of allelochemics, 21—22
Beneficial arthropods, 159—161
1,2,4-Benzenetriol, 50, 52, 54
Benzoic acid, 50, 52, 53
p-Benzoquinone, 50, 52, 53
Benzoxazolinones, 48, 49
Benzyl alcohol derivatives, 50—53
Berberine
    chemical data for, 42, 77
    effect on insect growth and development, 41
    effect on insect mortality, 74
    structure of, 40, 72

Betula pubescens, 49
Bioassay
    for allelochemics, 22—23
    for azadirachtin, 125—126
Biotin, 9—11
Black body, 150—151
Black spots, 134
Blatella germanica, 4, 75, 84
Blue-green algae, 158
Boarmia selenaria, 145, 162
Bombyx mori, 8
    compounds affecting growth and development of,
        24, 28, 32, 36, 41, 56
    compounds affecting mortality of, 74—75, 79,
        82, 86, 90, 97—98
Bracon hebetor, 8
Brassica, 56
    rapa, 104
Bregmatothrips iridis, 74
Brevicoryne brassicae, 74, 99
Bruceine B, 69, 86, 87

C

Caffeic acid, 26, 28, 29
Caffeine, 9
    chemical data for, 42, 76
    effect on insect growth and development, 41
    effect on insect mortality, 73
    insects affected by, 4
    source of, 4
    structure of, 2, 40, 72
Caffeylaldaric acid, 27—29
Calamus oil, 162
Calliphora
    erythrocephala, 75
    vicina, 153
Callosobruchus
    chinensis, 4, 75, 82, 157, 164
    maculatus, 73, 79, 82, 94, 101, 106, 157
cAMP, 10, 11
Camptotheca acuminata, 4
Camptothecin, 2, 4, 9, 12
L-Canaline, 23—25
Canavalia ensiformis, 4
Canavanine, 9, 12, 68
    chemical data for, 25, 80
    effect on insect growth and development, 24
    effect on insect mortality, 79
    insects affected by, 4
    source of, 4
    structure of, 2, 23, 79
Canavanine sulfate, 25
Canavainosuccinate, 23—25
Capric acid, 103
Caproic acid, 103
Caprylic acid, 103
S-(β-Carboxyethyl)-cysteine, 23—25
Cardenolides, 22
Δ-3-Carene, 88, 90, 91

Carpocapsa pomonella, 74, 75
Carvacrol, 51, 52, 54
Carvone
  chemical data for, 33, 91
  effect on insect growth and development, 32
  effect on insect mortality, 90
  structure of, 30, 88
Caryedes brasiliensis, 21
Caryophyllene, 31, 32, 34
Castor oil, 162
Catechin, 44, 46, 47
Catechol, 50, 52, 53
8-Cedren-13-ol, 89, 90, 92
Ceratitis capitata, 4, 145—146
Ceroplastes, 161
Cerotomia catalpae, 74
Cevadine, 73, 75, 77
Chavicol, 27—29
Chemosterilants, 1—13
  causing nutritional deficiencies, 9
  classification of, 1—5
  future prospects for, 12—13
  hormonal regulation by, 11—12
  mode of action of, 5—12
  mutagenic, 5—9
Chilo
  partellus, 145
  suppressalis, 52
Chinaberry tree, see also Azadirachtin
  flowers of, 123—124
  fruit of, 123—124, 165
  geographic distribution of, 122—123
  leaves of, 156—157, 165
  pests of, 161—162
  seed kernels of, 123—124
Chironomus, 98
  riparius, 11
Chitin synthesis inhibitors, 1
Chloramphenicol, 6, 8
Chlorogenic acid, 26, 28, 29
Chlortetracycline, 6, 8
Chorotoicetes terminfera, 43
Chrysanthemum cinerariaefolium, 84, 164
Chrysomya megacephala, 4
Cibarian, 105—107
Cimex lectularius, 4
Cimicidine, 73, 75, 78
Cinerin I, 84, 85
Cinerin II, 84, 85
Cinnamonium zeylanicum, 90
Cinnzeylanine, 89, 90, 93
Cinnzeylanol, 89, 90, 93
Citowett, 142, 164
Citral, 30, 32, 33
Citronella, 33
Citronellal, 30, 32
Cnaphalocrocis medinalis, 127, 139, 142—143, 160
Coccinella septempunctata, 160
Cocculolidine, 73, 75, 77
Cocculus tribolus, 75

Coccus hesperidum, 161
Cochliomyia hominivorax, 4, 9, 11
Colchicine, 1, 5
  chemical data for, 76
  effect on insect mortality, 73
  insects affected by, 4
  source of, 4
  structure of, 2, 72
Colchicum autumnale, 4
Coleopsis lanceolata, 28
Coleoptera, 132—139, 146
Conotrachelus nenuphar, 4, 11
Corn, 46
Coronarian, 105—107
Coronilla varia, 106
Costelytra zealandica, 106
m-Coumaric acid, 26, 28, 29
o-Coumaric acid, 26, 28, 29
p-Coumaric acid, 26, 28, 29
Coumarin(s)
  chemical data for, 29, 95
  effect on insect growth and development, 28
  effect on insect mortality, 70, 94
  insects affected by, 4
  source of, 4
  structure of, 2, 26, 94
Crocidolomia binotalis, 145
Crotalaria spectabilis, 4
Cruciferae, 56
Cruciferous crops, 104
Cubé, 70
Culex, 98—99
  pipiens, 82
  pipiens pallens, 82, 90
  pipiens quinquefasciatus, 9, 104
Custard apple seed extract, 164
β-Cyano-L-alanine, 79, 80
Cycloheximide, 6, 8
Cydia pominella, 106
Cyrtorhinus lividipennis, 160
Cysteine, 9—11
Cytotoxic natural products, 1
Cytovirin, 8

D

Dacrydium intermedium, 86
Dactinomycin, 6, 8
Dacus
  cucurbitae, 4
  dorsalis, 4
  oleae, 5, 9
Danaus chrysippus, 41
Danettia oil, 67—68
Daphnis magna, 84
Daucus carota, 106
Daunomycin, see Daunorubicin
Daunorubicin, 6, 8
Degulin, 98—100
Dehydrotrewiasine, 105—107

Demecolcine, 2, 4, 5
Demethyltrewiasine, 105—107
Dendroctonus
    brevicomis, 90
    frontalis, 90
Dermestes maculatus, 11
Derride, see Elliptone
Derris, 98
Derris root, 70
7-Desacetyl-7-benzoyl-azaradione, 120
7-Desacetyl-7-benzoyl-gedunin, 120
Development
    allelochemics affecting, 21—23
    postembryonic, effect of azadirachtin on, 127—146
Diallyl disulfide, 104
Diallyl trisulfide, 104
Diaraetiella rapae, 160
Dicoumarol, 94, 95
Diflubenzuron, 153
Digitalin, 35—37
Digitalis purpurea, 36, 86
Digitoxin, 35—37, 86, 87
Dihydropipercide, 81—83
Dihydroquercetin, 44, 46
2,5-Dihydroxybenzoic acid, 50, 54
2,6-Dihydroxybenzoic acid, 50, 52, 54
2,4-Dihydroxy-7-methoxy-1,4-benzoxazin-3-one (DIMBOA), 48, 49
Dill-apiol, 101, 102
DIMBOA, see 2,4-Dihydroxy-7-methoxy-1,4-benzoxazin-3-one
Dimethylsciadinoate, 31, 32, 34
Dimilin, see Diflubenzuron
Dioclea megacarpa, 68
Diparopsis castanea, 5, 8
Dipel, 164
Diptera, effect of azadirachtin on, 145—146
L-DOPA, 23—25
Doryphera decemlineata, 99
Drosophila, 73
    funebris, 43
    hamatofilia, 73
    melanogaster
        compounds affecting growth and development of, 28, 32
        compounds affecting mortality of, 90, 94, 101, 104
        compounds affecting sterility of, 4—5, 8, 11
    pachea, 73
    phalerata, 11
Dysdercus, 128
    cingulatus, 4—5, 8, 11, 128
    fasciatus, 36, 127—131, 146, 148, 150
    flavidus, 4
    koenigii, 4, 129
    similis, 4

E

Earias

    insulana, 139, 145
    vittella, 39
Ecdysone, 13
α-Ecdysone, 2, 4, 11
β-Ecdysone, 2, 4, 11
Ecdysteroid receptor, 153
Ecdysteroids, 11, 153—154, 156
Echinacein, see Neoherculin
Egg, effect of azadirachtin on, 126—127, 146—151, 153—156
Eidolon helvum helvum, 161
Elliptone, 98—100
Embryonic development, effect of azadirachtin on, 126—127
Emetine, 2, 4
Empoasca fabae, 74
Enzyme inhibitors, effect on insect growth and development, 49
L-Ephedrine, 72, 74, 76
Ephedrus cerasicola, 160
Ephestia kuehniella, 139—141, 157
Epilachna
    corrupta, 99
    varivestis, 4, 36, 75, 104
        azadirachtin in, 126, 132, 135—139, 146—147, 150—156, 162—163
Epiphyas postvittana, 90
Episyrphus balteatus, 160
Epitrix parvula, 75
cis-(β-Epoxy)azadiradione, 120
Eriodictyol, 44, 46, 47
Eriosoma americanum, 74
Esterine, see Physostigmine
Estigmene acrea, 38, 90
Ethionine, 23—25
Eugenol, 26, 28, 29
Eupatorium japonicum, 32, 94
Euponin, 31, 32, 34
Eurygaster intergriceps, 11

F

Fagara, 82
    macrophylla, 82
Fagaramide, 81—83
Farnesol, 30, 32, 33
Fatty acids, 70, 103
Fecundity, effect of azadirachtin on, 146—151, 153—156
Female sterility, 1, 5
Fervenulin, 6, 8
Filimarisin, see Filipin
Filipin, 6, 8, 9
Fish poison, 165
Flavomycin, see Bambermycin
Flavonoids
    chemical data for, 47, 100
    effect on insect growth and development, 46
    effect on insect mortality, 70, 98—99
    structure of, 44—45

4-Fluorophenylalanine, 23—25
Folic acid, 9—11
Follicle-stimulating hormone, 11, 12
Formica polyctena, 148
FSH, see Follicle-stimulating hormone
Fungi, 1

G

Galleria mellonella, 56, 74
Gallic acid, 50, 52, 53
Gambusia, 165
Gentistic acid, see 2,5-Dihydroxybenzoic acid
Geraniol, 30, 32, 33
Gibberella fijikuroi, 8
Gibberellic acid, 31, 34
Gibberellin, 8
Glaucarubinone, 69, 86, 87
Glaucolide A, 31, 32, 34
Glossina morsitans, 4, 11
D-Glucosamine, 10, 11
Glycine, 49, 56
Gossypium, 70, 90, 99
    hirsutum, 38—39, 43, 46, 90
Gossypol
    chemical data for, 39, 92
    effect on insect growth and development, 38—39
    effect on insect mortality, 90
    structure of, 38, 89
Gramine, 72, 73, 76
Griffonia simpliciplia, 24
Griseofulvin, 7, 8
Growth, allelochemics affecting, 21—23
γ-Guanidinobutyric acid, 105—107
Guineensine, 81—83

H

Haematobia irritans, 24, 79
Hallactone A
    chemical data for, 33, 91
    effect on insect growth and development, 32
    effect on insect mortality, 90
    structure of, 30, 88
Hallactone B, 88, 90, 91
Haplophytine, 73, 75, 78
Haplophyton cimicidum, 75
Helianthus, 32
    annus, 32
Heliocide H₁
    chemical data for, 39, 92
    effect on insect growth and development, 39
    effect on insect mortality, 90
    structure of, 38, 39
Heliocide H₂
    chemical data for, 39, 92
    effect on insect growth and development, 39
    effect on insect mortality, 90
    structure of, 38, 39

Heliocide H₃, 89, 90, 92
Heliopsis longipes, 82
Heliothis, 90
    armigera, 39, 98, 145
    punctegera, 39
    virescens, 5, 90, 99
        azadirachtin in, 127, 142, 145
        compounds affecting growth and development
            of, 32, 36—39, 43, 46
    zea, 8, 11
        azadirachtin in, 127, 142, 145
        compounds affecting growth and development
            of, 28, 36, 38, 41, 46, 56
        compounds affecting mortality of, 90, 99, 106
Heliothrips femoralis, 75
Heliotrine, 2, 4, 5
Heliotropium lasiocarpum, 4
Hemigossypolone
    chemical data for, 39, 92
    effect on insect growth and development, 39
    effect on insect mortality, 90
    structure of, 38, 89
Heteroptera, 128—131, 146
Higher plants, chemosterilants from, 4—5
Hippelates collusor, 8—9
Holotrichia consanguinea, 161
Homoeosoma electellum, 32
Homoptera, 131—132, 146
Honey bee, 160
Hormonal regulation, by chemosterilants, 11—12
Hyalophora cecropia, 41
Hydrocoumarin, 26, 28, 29
Hydroquinone, 50, 52, 54
17β-Hydroxyazadiradione, 120
3-Hydroxybenzaldehyde, 50, 52, 54
o-Hydroxybenzoic acid, 52
3-Hydroxybenzyl alcohol, 50, 52, 53
20-Hydroxyecdysone, 119
5-Hydroxytryptophan
    chemical data for, 25, 80
    effect on insect growth and development, 24
    effect on insect mortality, 79
    structure of, 23, 79
Hygromycin B, 7, 8
Hylotropus bajulous, 90
Hymenaea, 32
Hymenoptera, 132
Hyoscamine, see Atropine
Hypogaster exiguae, 41
Hypolactin-3′,4′-dimethyl ether, 46, 47
Hypolactin-3′,4′-methyl ether, 44

I

Imidazole, 55—57
Indian lilac, see Neem tree
Indoleacetonitrile, 55—57
Insects, chemosterilants from, 11
Invertebrates, chemosterilants from, 11
N-Isobutyl-2E,4E-octadienamide, 81—83

Isobutyramides, unsaturated, 69, 81—83
3-(4-Isobutyryloxy-3-methoxyphenyl)-2,3-epoxypro-
pan-1-ol, 27—29
3-(4-Isobutyryloxy-3-methoxyphenyl)-1-propen-3-ol,
27—29
Isoorientin, 44, 46, 47
Isoquercitrin, 98—100
Isoscutellarein, 44, 46, 47
Isothiocyanates, 104

J

Jasmolin II, 84, 85
JH, see Juvenile hormone
Juniperus recurva, 90
Justica procumbens, 97
Justicidin A, 96, 97
Justicidin B, 96, 97
Juvabione, 2, 4
Juvenile hormone (JH), 11, 153, 156
Juvenile hormone (JH) I, 10, 11
Juvenile hormone (JH) III, 10, 11

K

Karakin, 105—107
(−)-Kaur-16-en-19-oic acid, 30, 32, 33
Kobusin, 27—29, 96, 97

L

Lachnosterna serrata, 161
Larix, 21
Lasiocarpin, 3—5
Lasioderma serricorne, 4
Laspeyresia
funebrana, 5
pomonella, 8—9, 11, 86, 90
Lecithin, 131, 164
Lepidoptera
chemosterilants from, 11
effect of azadirachtin on, 139—146
Leptinotarsa decemlineata, 8, 98
azadirachtin in, 126—127, 136, 146—150, 163—
164
Leptocarisa oratorius, 131—132
Letharia vulpina, 28, 52
Leucenol, see L-Mimosine
LH, see Luteinizing hormone
Libocedrus biswillii, 97
Lignins, 43, 70, 96, 97
Likuden, see Griseofulvin
Limonene, 88, 90, 91
D-Limonene, 88, 90, 91
1-Limonene, 88, 90, 91
Liriomyza
sativae, 139
trifolii, 139, 145, 148

Lissapol, 142
Locusta
migratoria, 4, 75, 79
migratoria migratorioides, 43, 52, 86
azadirachtin in, 126—127, 150, 153, 156
Lonchocarpus, 98
Longevity, effect of azadirachtin on, 146—151
Lophocereine, 72, 73, 76
Lophocereus schotti, 73
Lotus pedunculatus, 106
Lower plants, chemosterilants from, 8—9
Lucilia sericata, 75
Lupinidine, see Sparteine
Luteinizing hormone, 11, 12
Luteolin, 44, 46, 47
Lycopersicon, 28, 41, 46
hirsutum, 106
Lycosa pseudoannulata, 160
Lygus kalmii, 75
Lymantria dispar, 144, 162

M

Macrosiphium pisi, 74
Macrosiphoniella sanborni, 74, 99
Macrosiphum euphorbiae, 28
Magnolia kobus, 28, 97
Makisterone A, 10, 11
Malacosoma americana, 99
Malaria, 165
Male sex pheromone, 145
Male sterility, 1, 5
Mamestra brassicae, 145
Mammea, 67, 70
americana, 94
longifola, 94
Mammea Xa, 94, 95
Mammea Xb, 94, 95
Mammein, 70
Manduca sexta, 4, 24, 41, 106
azadirachtin in, 126, 139, 144—145, 150, 153
Margosan A, 164
Margosa tree, see Neem tree
Mating propensity, 145
Maysin, 44, 46, 47
Maysin-3′-methyl ether, 44, 47
Maytansanoids, 71
Medicagenic acid, 31, 32, 34
Medicago, 32
Melanoplus femur-rubrum, 75, 98
Melia, 74
azedarach, see Chinaberry tree
cortex, 74
volkensii, 123
Meliatoxin, 165
Mentha pulegium, 90
Metabolism, effect of azadirachtin on, 156
Metamorphosis, effect of azadirachtin on, 151—153
Methionine, 10, 11
6-Methoxybenzolinone, 48, 49

8-Methylpsoralen, see Xanthotoxin
N-Methyltrenudone, 105—107
N-Methyltyrosine, 79, 80
Microvelia douglasi atrolineata, 160
Millettia, 98
L-Mimosine, 79, 80
Mitomycin C, 7—9
Monocrotaline, 3—5
Monogamy factor, 10, 11
Morin, 98—100
Mundulea, 98
Musca
  autumnalis, 4—5, 24, 79
  domestica
    chemosterilants from, 11
    compounds affecting growth and development
      of, 32, 41
    compounds affecting mortality of, 75, 82, 84,
      90, 94—101, 104
    compounds affecting sterility of, 4—5, 8—9,
      11
    domestica nebulo, 4, 9, 164
    domestica vicina, 5
Mustard beetle, 94
Mutagenesis, by chemosterilants, 5—9
Mutagens, 1
Mutations, 5
Myrcene, 88, 90, 91
Myricetin, 45—47
Myristica fragans, 28
Myristicin
  chemical data for, 29, 102
  effect on insect growth and development, 28
  effect on insect mortality, 101
  structure of, 26, 101
  Mythimna separata, 139, 141—142
Myzus
  persicae, 8—9, 74, 99, 160
  porosus, 74
  solanifolii, 74

N

Nagilactone C
  chemical data for, 33, 91
  effect on insect growth and development, 32
  effect on insect mortality, 90
  structure of, 30, 88
Nagilactone D, 88, 90, 91
Nagilactone E, 88, 90, 92
Naringenin, 44, 46, 47
Neem bark, 165
Neem ingredients, see Azadirachtin
Neem oil, 121, 137, 158
Neem products, practical uses of, 156—159
Neemrich I, 164—165
Neem seed cake, 158
Neem seed extract, 137
Neem tea, 165
Neem tree, see also Azadirachtin, 36

characteristics of, 119—120
  edible, 123, 165
  flowers of, 122
  fruit of, 120—122
  geographic distribution of, 119—120
  leaves of, 122—123, 156—157
  medicinal uses of products of, 165
  pests of, 161—162
  seed kernels of, 120, 122, 157—158
  seeds of, 121—122
Neem twig, 165
Neoherculin, 81—83
Neoquassin, 69
Nephotettix
  bipunctatus cinticeps, 75
  virescens, 132—133, 137, 149—150, 163
Nephrolepis exaltata, 5
Neriifolin, 86, 87
Neuroregulator, 13
Neurosecretory hormone, 13
Nicotiana, 74—75
Nicotine, 69
  chemical data for, 42, 76
  effect on insect growth and development, 41
  effect on insect mortality, 74—75
  structure of, 40, 72
Nicotinoids, 68
Nikkomycin, 7, 9
Nilaparvata lugens, 75, 126, 132, 137, 146, 149—
  150, 153
3-Nitropropionic acid, 105—107
Nocardia lurida, 9
trans-2-Nonenal, 105—107
Nornicotine, 72, 75, 77
Nostoc, 158
Nutritional deficiency, caused by chemosterilants, 9

O

Oncopeltus fasciatus, 4, 11, 22, 36, 74—75
  azadirachtin in, 131, 151—152
Oostatic hormone, 11, 12
Operophtera brumata, 43
Oporinia autumnata, 49
Orobol, 44, 46, 47
Orthezia insignis, 74
Orthoptera, 127—128
Oryza, 52
Oryzaephilus
  acuminatus, 161
  surinamensis, 99, 161
Ostrinia
  furnacalis, 145
  nubilalis, 48—49, 56, 68, 86, 106
Oxytetracycline, 7, 9

P

Pactamycin, 7, 9

Panonychus
  citri, 8
  ulmi, 8
Paper factor, 119
Papilio polyxenes asterius, 28, 56
Parthenium hysterophorus, 5
Pastinaca sativa, 101
Pectinophora gossypiella, 32, 36, 38—39, 82, 90,
    99, 145
Pellitorine, 81—83
Peltatin methyl ether A, 96, 97
Penicillium griseofulvum, 8
Pepper, 69
Periplaneta americana, 4, 74—75, 84, 98, 127
Permanent larvae, 135, 150
Permanent nymph, 156
Persea americana, 32, 56
Persian lilac, see Chinaberry tree
Persistence, of azadirachtin, 162—163, 166
Pests
  of chinaberry tree, 161—162
  of neem tree, 161—162
Phacelia tanacetifolia, 132, 160
Phaedon cochlearis, 84
α-Phellandrene, 88, 90, 91
Phellondendri cortex, 74
Phellondendron, 41
Phenococcus gossypii, 74
Phenolics, 50—54
2-Phenylethylisothiocyanate, 104
β-Phenylnitroethane, 67
Phenylpropane derivatives, 26, 28, 29
Phenylpropanoids, 70, 101, 102
Phloretin, 51, 52, 54
Phlorizin, 51, 52, 54
Phloroglucinol, 50, 52, 53
Phormia regina, 8, 11, 75
Physostigma venenosum, 75
Physostigmine, 68, 73, 75, 77
Pieris brassicae, 36, 126—127, 139, 142—143,
    163
Piesma quadratum, 131, 139
Pilocereine, 72, 73, 76
Pimpinella anisum, 101
β-Pinene, 88, 90, 91
D-α-Pinene, 88, 90, 91
L-α-Pinene, 88, 90, 91
Pinitol, 55, 56
Pinus, 90
  ponderosa, 90
  taeda, 90
Piper, 82
  nigrum, 82
Piperaceae, 69
Pipercide, 81—83
Piperine, 69, 81—83
Piperlonguminine, 81—83
Piperonyl butoxide, 163—164, 166
Piprotal, 163
Pisum sativum, 5
Plant-insect interactions, 21

Plants
  compounds affecting insect growth and develop-
    ment from, 21—23
  insect toxicants from, 67—71
Plutella xylostella, 36
  azadirachtin in, 126—127, 139, 143—144, 162—
    164
Podocarpus, 70
  gracilior, 5
  halii, 32, 90
  nerifolius, 90
  nivalis, 32
Podolactone A, 30, 32, 33
Podolactone C
  chemical data for, 33, 92
  effect on insect growth and development, 32
  effect on insect mortality, 90
  structure of, 30, 88
Podolactone E, 30, 32, 33, 88, 92
Podolide, 88, 90, 92
Popillia japonica, 74, 99, 139
Population management, 1
Porfiromycin, 7, 9
Precocene, 119
Precocene I, 3, 4, 12
Precocene II, 3, 4, 12
Precocene epoxide, 12
Pride of India, see Chinaberry tree
Pristiphora erichsonii, 21
Prodenia
  eridania, 24, 74—75, 79
  litura, 11
2-Propenylisothiocyanate, 104
Propylisothiocyanate, 104
Prostaglandin, 11
Prothoracicotropic hormone, 13, 153
Protocatechuic acid, 50, 52, 54
Protoparce quinquemaculata, 75
Pseudosarcophaga affinis, 24, 56, 103
Psila rosae, 106
Pteridophyta, 1, 4—5
Pulegone, 88, 90
Pulvinaria, 161
Putrescine, 10, 11
Pyrethrin I, 84, 85
Pyrethrin II, 84, 85
Pyrethrins, 1, 4
Pyrethroids, 69, 84, 85
Pyrethrum, 69
Pyrogallol, 50, 52, 53
Pyrrhocoris apterus, 4

Q

Quassia, 86
Quassin, 69
Quassinoids, 69
Quebracho, 43
Queen substance, 10, 11
Quercetin

chemical data for, 47, 100
effect on insect growth and development, 46
effect on insect mortality, 99
structure of, 44, 98
Quercitrin, 98, 100
Quercus, 43

R

Rantankin, see, *N*-Methyltyrosine
Rauwolfia serpentina, 5
Rayanodine, 68
Repellent, 67
Reproduction, regulation of, see Chemosterilants
Reserpine, 9
chemical data for, 76
effect on insect mortality, 73
insects affected by, 5
source of, 5
structure of, 3, 72
Resistance, to azadirachtin, 161
Resorcinol, 50, 52, 53
Reticulotermas flavipes, 75
Rhagoletis pomonella, 4
Rhodnius prolixus, 41
Rhyzopertha dominica, 161, 164
Ricinus communis, 140
Ristocetin, 9
Robinetin, 45—47
Rotenoids, 98—100
Rotenone, 69—70, 98—100
Rutaceae, 94
Rutin
chemical data for, 47, 100
effect on insect growth and development, 46
effect on insect mortality, 99
structure of, 44, 98
Ryania, 68, 75
Ryanodine, 72, 75, 77

S

Sabadilla, 68
Salicylic acid, 50, 52, 53
α-Sanshool, see Neoherculin
Santalum album, 5
Schistocerca gregaria, 4, 43, 139
Schizaphis graminum, 28, 43, 46, 52
Schoenocaulon, 75
Sciadin, 31, 32, 34
Sciadinone, 31, 32
Scopoletin, 26, 28, 29
Scutellarein, 44, 46, 47
α-Selinene, 31, 32, 34
β-Selinene, 31, 32, 34
Sellowin A
chemical data for, 33, 92
effect on insect growth and development, 32
effect on insect mortality, 90

structure of, 30, 88
Sendanin, 35—37
Sesame oil, 131, 164
Sesamin, 27—29, 96, 97
Sinapic acid, 26, 28, 29
Sitophilus
oryzae, 4, 82, 157
zeamais, 157
β-Sitosterol, 35—37
Sogatella furcifera, 132—133, 137, 146, 149—150, 153
Solenopsis, 161
invicta, 8, 12
Sparsamycin, see Tubericidin
Sparteine, 72, 74, 77
Spermatophyta, chemosterilants from, 1, 4—5
Spiders, 159—160
Spilanthes
acmella, 56, 82
oleraceae, 82
Spilanthol
chemical data for, 57, 83
effect on insect growth and development, 56
effect on insect mortality, 82
structure of, 55, 81
Spodoptera
eridania, 24, 32, 56, 94
exigua, 32, 38
frugiperda, 32, 36, 90
azadirachtin in, 126—127, 142, 145, 162, 164
littoralis, 5, 11, 79
azadirachtin in, 126, 139—141, 146—150, 162
compounds affecting growth and development, 32, 36, 38
litura, 140, 160—161
ornithogalli, 28, 32, 52
Startle activity, 145
Stemofoline, 73, 75, 78
Stemona japonica, 75
Stemonine, 73, 75, 78
Stemospironine, 73, 75, 78
Sterculia foetida, 5
Sterilant, see Chemosterilants
Steroids, 69, 86, 87
Stomoxys calcitrans, 4, 79
Store pests, 156—157
Streptomyces
alanosinicus, 8
ardus, 9
aureofaciens, 8
avermitilis, 8
bambergiensis, 8
caespitosus, 8
fervens, 8
filipenensis, 8
fradiae, 9
griseus, 8—9
hygroscopicus, 8
orientalis, 9
pactum, 9
peucetius, 8

refuineus, 8
rimosus, 9
tendae, 9
tubericidus, 9
venezuelae, 8
viridifaciens, 9
Streptomycin, 7, 9
Streptovitacin A, 7, 9
Strychnine, 72, 73. 76
Sulfur compounds, 70—71, 104
Sumatrol, 98—100
Supernumerary sixth larval instar, 144—145
Surangin B, 70, 94, 95
Surinamine, see *N*-Methyltyrosine
Synergisms, insecticide, 163—164
Syringic acid, 51, 52, 54

T

Taeniothrips simplex, 74—75
Tannic acid, 43
Tannins, 43
Telenomus remus, 160—161
Tenebrio
    castaneum, 5, 84
    molitor, 74, 82
Tephrosia, 98
Terpenes, 88—93
Terpenoids, 30—34, 69—70
Terramycin, see Oxytetracycline
Tetracycline, 7, 9
5,7,2′,3′-Tetrahydroxyflavone, 44, 46, 47
Tetranychus
    atlanticus, 9, 104
    bimaculates, 75
    cinnabarinus, 9
    pacificus, 8
    telarius, 8, 99
    urticae, 8—9
Thai neem tree, 122—123
Thallophyta, 1, 8—9
Theobromine, 72, 73, 76
Theophylline, 40—42
Thermobia domestica, 4, 75
Thevetia thevetioides, 86
Thiocyanates, 70—71, 104
Thrips, 99
    nigropilosus, 75
    tabaci, 74, 99
β-Thujaplicin, 89, 90, 93
Thuja plicta, 90
Thujopsene, 89, 90, 92
Thomatine, 40—42
Toxicants, 67—71
Toxicarol, 98—100
Toxicology, of azadirachtin, 164—165
Trachyloban-19-oic acid, 30, 32, 33
Treflorine, 105—107
Trenudine, 105—107
Trewia nudiflora, 106

Trewiasine, 105—107
Trialeurodes
    packardi, 74
    vaporariorum, 74, 99
Triallyl trisulfide, 104
Tribolium
    castaneum, 24, 29, 32, 49, 84
    confusum, 4—5, 8, 103
Tricetin, 45—47
Trichilia roka, 36
Trichoplusia ni, 38, 106
2-Tridecanone, 105—107
Triterpenes, 69, 86, 87
Triterpenoids, steroidal, 35—37
Trogoderma granarium, 4
Tropital, see Piprotal
Trypsin inhibitor, 49
Tubericidin, 7, 9
Tylosin, 7, 9
Typhlociba comes, 99
Tyramine, 72, 74, 77

U

Umbelliferae, 28, 94
Uragoga ipecacuanha, 4
O-Ureidohomoserine, 23—25
UV light, 162—163

V

Vancomycin, 7, 9
Vanessa cardui, 98
Vanillic acid, 50, 52, 54
Veratridine, 73, 75, 77
Vernonia, 32
Viburnum japonicum, 28
Vicia faba, 131, 149
Vinblastine, 3, 5, 40—42
Vinca rosea, 5
Vincristine, 3, 5
Vitamin E, 9—10
Vitellogenin, 156
Vulpinic acid, 27—29

W

Wing disk lesion, 151

X

Xanthotoxin, 94, 95
Xiphinema basiri, 161
Xyleborus ferrugineus, 4, 11
m-Xylohydroquinone, 3, 5

Z

Zanthoxylum, 82
Zea mays, 48
Zonocercus variegatus, 43

Milton Keynes UK
Ingram Content Group UK Ltd.
UKHW020821141024
449569UK00008B/501

9 781138 596962